P9-AEZ-713

BIOPROCESS PRODUCTION OF FLAVOR, FRAGRANCE, AND COLOR INGREDIENTS

BIOPROCESS PRODUCTION OF FLAVOR, FRAGRANCE, AND COLOR INGREDIENTS

Edited by

Alan Gabelman
Tastemaker
Cincinnati, Ohio

A WILEY-INTERSCIENCE PUBLICATION

JOHN WILEY & SONS, INC.

New York • Chichester • Brisbane • Toronto • Singapore

CHEM

This text is printed on acid-free paper.

Copyright © 1994 by John Wiley & Sons, Inc.

All rights reserved. Published simultaneously in Canada.

Reproduction or translation of any part of this work beyond
that permitted by Section 107 or 108 of the 1976 United
States Copyright Act without the permission of the copyright
owner is unlawful. Requests for permission or further
information should be addressed to the Permissions Department,
John Wiley & Sons, Inc., 605 Third Avenue, New York, NY
10158-0012.

Library of Congress Cataloging in Publication Data:
Bioprocess production of flavor, fragrance, and color ingredients /
Edited by Alan Gabelman.
     p.   cm.
    "A Wiley-Interscience publication."
    Includes bibliographical references (p.   –   ) and index.
    ISBN 0-471-03821-0 (acid-free)
    1. Flavoring essences—Biotechnology. I. Gabelman, Alan.
TP418.B565 1994
664'.5—dc20   93-46797
                                        CIP

Printed in the United States of America
10 9 8 7 6 5 4 3 2 1

TP418
B565
1994
CHEM

■■■■■■■ CONTRIBUTORS

David W. Armstrong, Institute for Biological Sciences, National Research Council, Ottawa, Canada

Leslie A. Brown, Institute for Biological Sciences, National Research Council, Ottawa, Canada

David C. Eaton, Universal Flavors, Indianapolis, Indiana

Alan Gabelman, Tastemaker, Cincinnati, Ohio

Gunnard Jacobson, Universal Foods Corporation, Milwaukee, Wisconsin

Charles H. Manley, Takasago International Corporation (USA), Teterboro, New Jersey

Tilak W. Nagodawithana, Red Star Specialty Products, Universal Foods Corporation, Milwaukee, Wisconsin

Om Sahai, ESCAgenetics Corporation, San Carlos, California

Eugene W. Seitz, Chr. Hansen's Laboratory, Inc., Milwaukee, Wisconsin

John Wasileski, Universal Foods Corporation, Milwaukee, Wisconsin

Frank W. Welsh, Bureau of Food Regulatory, International and Interagency Affairs, Food Directorate, Health Canada, Ottawa, Canada

■■■■■ CONTENTS

Although fermentation has been used since prehistoric times to prepare cheese, bread, wine, and beer, in recent years the food, flavor, and fragrance industries have shown a renewed interest in fermentation and other bioprocesses. One reason for this surge is the increasing demand for healthy foods. Today's consumer expects to find food products that are low in sugar, salt, cholesterol, and saturated fat, high in fiber, and all natural. This preference for natural foods opens the door for bioprocessing because flavor ingredients prepared using fermentation or enzymatic transformation are considered *natural*.

Other factors encouraging development of biological rather than chemical routes to ingredients include increased concern over depletion of fossil fuels and the corresponding emphasis on use of renewable resources. Still another factor is today's highly competitive business climate, which has encouraged consideration of a wide range of technologies to obtain a competitive edge. Often biological processes, which in the past may not have even been considered, are found to be cost effective.

Concurrently, impressive advances in biotechnology have also occurred. For example, today's fermenters are equipped with highly sophisticated instrumentation capable of on-line analysis of a multitude of substances. Analytical data then are fed to a computer, which calculates controller set points based on a sophisticated mathematical model. The organism being cultured might be a recombinant bacterium or yeast, engineered to allow inexpensive, large-scale production of a material that would otherwise be scarce and expensive. Advances such as these have made biological production of many ingredients feasible, which in turn has further encouraged development of biological solutions to problems.

However, advances like these also are raising important regulatory questions. For example, the definition of *natural* found in the U.S. Code of Federal Regulations is subject to interpretation, so that across the flavor industry one finds differing opinions regarding what is and is not natural. Another pressing regulatory issue is the use of ingredients derived from genetically engineered organisms, or genetically engineered organisms themselves, in foods.

In light of the surging interest in fermentation and other bioprocesses, a book that presents the latest information on the subject is timely and appropriate. The chapters that follow offer thorough coverage of a wide range of technical and regulatory topics that should be useful and interesting to anyone associated with the food, flavor, or fragrance industry.

ALAN GABELMAN

Cincinnati, Ohio

Overview of Bioprocess Flavor and Fragrance Production

FRANK W. WELSH

Bureau of Food Regulatory, International and Interagency Affairs, Food Directorate, Health Canada, Ottawa, Canada

Many chemicals interact with human olfactory receptors to induce flavor and fragrance stimuli. These substances range from simple organic acids, aldehydes, and esters to complex di- and triterpenoid structures, multiple-ring structures, and the like. Many of these substances can be chemically synthesized; however, they often do not produce the complexity found in natural flavors and fragrances. On the other hand, all flavor and fragrance substances are produced by biological processes in nature. The challenge is to identify and develop these biological processes into cost-effective, industrial-scale biosyntheses for the production of natural flavorants. The purpose of this book is to provide a comprehensive, timely discussion of the biotechnological, engineering, and business considerations for such developments. This chapter provides an overview of this complex and challenging endeavor; more in-depth discussions of specific topics are found in the chapters that follow.

1.1 INTRODUCTION

Plants and certain microorganisms have produced flavors and fragrances since the beginning of time, but the first published report of the flavor- and fragrance-producing capabilities of selected bacteria and fungi did not appear until 1922 [1]. However, this research did not identify the chemical constituents of the fragrance, and the relationship between microbial physiology and the production of odorous metabolites was not identified until the 1950s [2]. Since that time, a significant body of literature has addressed the production of flavor and fragrance chemicals by microorganisms and is reviewed elsewhere [3–7]. This

Bioprocess Production of Flavor, Fragrance, and Color Ingredients, Edited by Alan Gabelman, ISBN 0-471-03821-0 © 1994 John Wiley & Sons, Inc.

research is primarily directed at the identification of specific flavor or fragrance chemicals and optimization of their biosynthesis using microbial physiology and genetic manipulation techniques.

Most natural flavors and fragrances are the result of mixtures of chemicals, which are found at low concentration; they have more rounded, complex flavor attributes than those obtained with a single chemical. The substances found in these mixtures include terpenes, aldehydes, esters, lactones, higher alcohols, and other complex molecules that result from the secondary metabolism of plants or may be obtained from certain animal sources. Certain fungi and yeasts also possess the potential for secondary metabolism and can produce flavors or fragrances. Examples are presented in Table 1.1.

TABLE 1.1 Examples of the Flavors and Fragrances Produced by Selected Microorganisms and Their Chemical Constituents

Organism	Sensory Descriptor	Volatiles Produced
Bacteria		
Lactic acid bacteria *Streptococcus* *Lactobacillus* *Leucanostoc*	Sharp, buttery, fresh	Acetaldehyde, diacetyl, acetoin, lactic acid
Propionibacterium	Sour, sharp	Acetoin, dienals, aldehydes
Pseudomonads	Malty, milky	Acetoin
Bacillus	Granary	3-Methyl-1-butanal
Corynebacterium		2-Methyl-2-hydroxy- 3-keto-butanal, pyrazines
Actinomycetes		
Streptomyces	Damp forest, soil odor	(2-Me)-3-isopropyl pyrazine, geosmin
Yeasts		
Saccharomyces	Aroma associated with bread and alcohol fermentations	Higher alcohols, lactones, thio-compounds
Kluyveromyces	Fruity, rose	Phenylethanol and esters, terpene alcohols, short- chain alcohols, and esters
Geotrichum	Fruity, melon	Ethyl esters, higher alcohol esters
Hansenula	Floral, soil odor	Ethyl esters, higher alcohol esters
Dipodascus	Apple, pineapple	Higher alcohol esters
Sporobolomyces	Peach	Lactones

TABLE 1.1 (*Continued*)

Organism	Sensory Descriptor	Volatiles Produced
	Molds	
Aspergillus	Fungal, musty, mushroom	Unsaturated alcohols
Penicillium	Mushroom, blue cheese, rose	1-Octene-3-ol, Methyl ketones, 2-phenyl-ethanol, thujopses, nerolidine
Ceratocystis	Banana, pear, peach, plum	Alcohols, esters, monoterpene alcohols, lactones
Trichoderma	Coconut, anise, cinnamon	6-Pentyl-α-pyrone, sesquiterpenes, cinnamate derivatives
Phellinus	Fruity, rose, wintergreen	Methyl benzoates and salicylates, benzyl alcohol
Septoria	Anise or cinnamon	Cinnamate derivatives
Lentinus	Aromatic, fruity	Higher alcohols, sesquiterpenes

Certain fragrance substances also can be synthesized chemically, but substances produced by this approach cannot be considered natural and do not have the same economic value as chemicals from natural sources. In addition, the complex structure of certain flavor and fragrance chemicals makes them difficult to synthesize chemically. As a result, the majority of natural fragrances are still produced by more traditional extraction processes from plant and animal sources. Such sources, however, suffer from a diminishing supply of raw material, expense of isolation, and variability in the amount and quality of final product from different geographic regions. In addition, many important flavors and fragrances originate in developing countries where political and socioeconomic factors may lead to supply uncertainties. As a result, several companies have established active research and development programs to examine the microbial production of flavor and fragrances [8,9]. These programs address two main areas: (1) microbiologically mediated syntheses and (2) plant production methods. Recent trends in these areas are reviewed here, followed by a discussion of selected scientific, technical, and nontechnical considerations.

1.2 MICROBIOLOGICALLY MEDIATED SYNTHESES

Schindler and Schmid [10] have provided an extensive review of microorganisms that biosynthesize fragrances and aromas. The fragrance-producing species are primarily fungi, perhaps because unlike bacteria, they are able to carry out secondary metabolism. Culture conditions (media composition, seed culture generation, pH, fermentation time, temperature, and the like) were identified as factors in determining the amount and type of fragrance or flavor substance produced. For example, *Ceratocystis moniliformis* can produce banana-, citrus-, or peachlike aromas. The aroma produced was dependent on the carbon and nitrogen source (Table 1.2) and was produced only after nitrogen was depleted in the medium. After 5 days of fermentation 50 μg of monoterpenes were produced per milliliter of culture. The production of complex chemicals following nutrient depletion is typical of secondary metabolism. Similar production levels have been reported for citronellol, linalool, and geraniol (citrus aromas) biosynthesis by *Kluyveromyces lactis*. For this process, chemical yields were increased by increasing the temperature and by using higher concentrations of asparagine as the nitrogen source [11].

On the other hand, the yeast *Ambroisiozyma* sp. may accumulate 60 to 100 mg/L of mixed monoterpenes using the mevalonate pathway [12], which synthesizes mevalonic acid from acetate. In animals, mevalonic acid is the first compound in the cholesterol biosynthesis pathway, whereas in plants this five-carbon hydroxy acid may be converted to a variety of terpenoid compounds. Large-scale production of monoterpenes by yeast may be preferable to fungal biosynthesis, because the homogenous growth of unicellular yeast does not present the rheological problems encountered with filamentous fungi. The data presented in the previous references [11,12] also show that many microbiologically mediated fragrance and aroma syntheses produce low amounts of the secondary metabolite, as these metabolites inhibit cell or pathway activity (end-product inhibition). The development of novel fermentation procedures, such as continuous extraction methodology or the use of nonaqueous reaction media, may overcome this problem. Such approaches are discussed later in this chapter.

The low yield of microbiological terpene biosyntheses, coupled with the ready availability of natural source terpenes, makes the development of such processes unnecessary at this time. The real value of these processes may be

TABLE 1.2   Effect of Carbon and Nitrogen Source on Aroma Production by *Ceratocystis moniliformis*

Carbon Source	Nitrogen Source	Aroma	Chemicals Identified
Dextrose	Urea	Fruity, banana	Acetate esters, ethanol
Dextrose	Leucine	Fruity, over-ripe, banana	Isoamyl acetate, ethanol
Galactose	Urea	Citrus, grapefruit, lemon	Monoterpenes, ethanol
Cornstarch	Urea	Cantaloupe, tropical flower, banana	
Dextrose	Glycine	Pineapple, lemon, sweet	
Dextrose	Methionine	Weak potato	
Glycerol	Urea	Canned pear, peach	Decalactones, ethanol

the biotransformation of less expensive natural source terpenes to higher value flavors and fragrances. (See Chapter 4 for an in-depth discussion of this topic.) This goal may be achieved by the stereospecific oxidation or reduction of complex mixtures of terpenes to increase the concentration of the active isomers. One example is the stereospecific biotransformation of L-menthone to L-menthol [13]. During the flowering stage of the mint plant (*Mentha piperita*), the essential oil is composed primarily of L-menthone. Approximately 40 percent of the L-menthone is converted to L-menthol between floral initiation and full bloom. On the other hand, *Pseudomonas putida* YK-2 has also been reported to biotransform L-menthone to L-menthol. It has been proposed that menthone-rich essential oil be extracted from the plant during floral initiation and be converted to L-menthol by the more efficient and rapid microbial process. The potential usefulness of this process would be dependent on the cost of developing and commercializing this process and the acceptability of using a potentially harmful microorganism for a food-related process.

Biosynthetic processes consisting of esterifications and stereospecific hydrolyses also have been used to purify racemic mixtures of aroma terpenoids. By first esterifying a mixture of terpenoid alcohols and then stereoselectively hydrolyzing that mixture, mixtures of fragrance and flavor compounds may be produced that are easily separable. Such a process (Figure 1.1) has been developed for the large-scale stereospecific hydrolysis of DL-menthyl succinate to L-menthol by *Rhodotorula minuta* var. *texensis* [14].

Similarly, a variety of microorganisms can carry out *de novo* synthesis of lactones, but yields are low (see Chapters 3 and 6). Research has been conducted to develop improved biosynthetic processes for such conversions and to determine the effect of media composition on lactone synthesis. For example, *Trichoderma viride*, when grown in potato dextrose media, produced 165 mg/L of 6-pentyl-α-pyrone, a lactone characterized by a strong coconut fragrance. However, when *T. viride* was grown in a modified Czapek Dox medium, 10 to 30 mg/L of 6-(pent-1-enyl)-α-pyrone was formed. This difference was attributed to the difference in growth medium [15]. Several similar biosyntheses have been identified in the recent literature, but further development of these transformations has been hindered by the low yields.

One exception exists. Various *Candida* species have been shown to convert ricinoleic acid (the major fatty acid of castor oil) to 4-decanolide [16]. These microorganisms are capable of degrading castor oil triglycerides, thereby releasing ricinoleic acid, which is then degraded by β-oxidation to 4-hydroxydecanoic acid. Heat and acid treatment are then used to convert the hydroxy acid to 4-decanolide. The lactone was purified by solvent extraction; yields were claimed to be 5 g/L [17].

Ketones represent an additional class of fragrance and aroma chemicals that can be produced by microbiological processes (see Chapters 3 and 6). The aroma note of these compounds is most often found in cheese; a number of saturated and unsaturated aliphatic, aromatic, and cyclic ketones have been isolated from cheese, although the odd-numbered C_5 to C_{11} 2-alkanones pre-

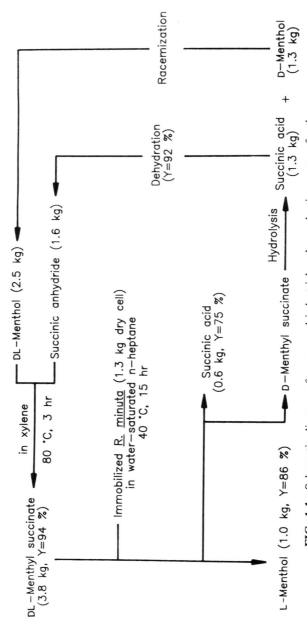

FIG. 1.1 Schematic diagram of a proposed industrial-scale production system for the conversion of menthyl succinate to L-menthol. *Source* Reprinted by permission from Omata, T., Iwamoto, N., Kimura, T., Tanaka, A., and Fukui, S., *Appl. Microbiol. Biotechnol.* **11**, 199 (1985).

dominate. These compounds give *Penicillium*-ripened cheeses their distinctive flavors.

The biosynthesis of 2-alkanones is accomplished by β oxidation of fatty acids followed by the addition of an acyl group and decarboxylation to form 2-alkanones one carbon shorter than the original fatty acid. This approach has been applied to the production of blue and Camembert cheese flavors and blue cheese food material. Here, submerged fermentations of *Penicillium* sp. were developed using lipolyzed fat or vegetable oil as the substrate. For blue cheese flavor concentrate production, whey is supplemented with fats, oils, or specific fatty acids and fermented with *Penicillium roqueforti* to produce a fully balanced cheese flavor that may contain up to ten-fold the amount of ketones found in blue cheese [18]. Similar technology can be applied to the synthesis of other mixtures of ketones if the required combination of fatty acids is available to produce the desired structures.

The production of diketones is another area of investigation that may be applied to fragrance and aroma production. For example, diacetyl (2,3-butanedione) has a buttery, nutlike flavor and is a high-value flavor compound. Natural diacetyl may be obtained by distillation of lactic acid starter cultures. However, significant effort has been expended to develop alternate microbial processes to synthesize this flavor, because of its high economic value and wide use. Microorganisms such as *Streptococcus lactis* and *S. cremoris* have been used in media containing a humectant to raise diacetyl production yield; the highest diacetyl concentration reached was 10.5 mg/L [19]. A high-efficiency method has also been developed for converting pyruvate to acetoin using these microorganisms. The acetoin can then be oxidized to diacetyl by chemical processes [20], although the end product may not be considered natural in this case. A concentration of 6.87 g/L acetoin was achieved. High yields (35 g/L) of acetoin and diacetyl also also been claimed for a fermentation carried out by *Enterobacter cloacae* in medium containing 50 g/L sucrose, 5 percent (w/v) peptone, and yeast extract. The acetoin was chemically oxidized to diacetyl at an overall recovery rate of 60 percent based on the sugar consumed (production of 30 g/L) [21]. Additional discussion of dairy flavors and fragrances can be found in Chapter 6.

A number of organic acids, particularly low-molecular-weight acids, contribute to aroma and fragrance complexity and may be produced by microbiological processes. Processes for the biosynthesis of acetic acid and butyric acid are well known and have been described elsewhere [7]. Recently, microbiological processes were developed for the production of complex Swiss and Emmental cheese flavors (primarily low-molecular-weight fatty acids) and propionic acid. In the production of Swiss cheese flavor, *Streptococcus thermophillus*, *L. helviticus*, or *L. lactis* is used as a starter culture to produce lactic acid. The lactic acid and residual sugar are then fermented by *Propionibacterium* species to carbon dioxide, propionic acid, and acetic acid. The ratio of propionic acid to acetic acid is typically in the range of 3:1 to 5:1 (w/w). These acids, plus background levels of other acids, result in the nutty cheese flavor of Swiss and Emmental cheese. Continuous production of these flavors has been achieved by immobilizing *Propionibacterium* sp. in calcium alginate with sodium lactate as the carbon source. This system produced 6.3 g/L propionic acid and 2.8 g/L acetic acid in a 5-day batch fermentation. In continuous mode, the process ran for two months, and higher productivity than was obtained in batch mode was maintained throughout the fermentation period [22]. Production levels of propionic acid were increased 30 percent by media manipulation.

Alcohols are also important components of certain flavors and fragrances. For example, mushroom fragrances are composed primarily of C_8 saturated and unsaturated alcohols. The ability to produce these alcohols is widespread among fungi. *Trichothecium roseum* has been reported to produce octan-3-ol, oct-1-en-3-ol, oct-1,5-dien-3-ol, octan-1-ol, and oct-2-en-1-ol. Similarly twelve other fragrance alcohols have been identified from species of *Aspergillus, Penicillium,* and *Fungi imperfecti. Penicillium caseicolum* and *P. camemberti* can also produce oct-1-en-3-ol, oct-1,5-dien-3-ol, and octan-3-ol. However, all of these alcohols are believed to form the corresponding methyl ketones by the reductive action of one or more dehydrogenases [7]. To produce these alcohols by biological processes, biosynthetic pathways would have to be manipulated to inhibit dehydrogenase activity. Additional discussion of microbiological production of acids and alcohols is presented in Chapter 3.

1.3 PLANT PRODUCTION METHODS FOR FRAGRANCE SYNTHESIS

Plants used for the production of fragrances can be manipulated either by breeding the plant to produce greater amounts of the fragrance or aroma of interest or by identifying, isolating, and culturing selected cell lines from the plant that produce the desired substance (plant cell culture).

The first approach is of interest, because well-known plant breeding techniques, used in conventional agriculture, can be applied directly to this area. Table 1.3 presents some of the traditional and novel plant breeding techniques in practice today. Traditional breeding methods benefit from being known, effective, and of little concern from a regulatory standpoint. However, the length of time leading to the development of new plant strains (estimated at 5 to 7 years) and the cost (approximately $5 million to $7 million) may be excessive given the potential annual sales of some target fragrances and the volatility of market demand. On the other hand, novel plant breeding methods, using genetic manipulation, provide the opportunity for shorter development periods, greater precision, and the potential for interspecific gene transfers. However, developments using these methods may be subject to more stringent regulatory control (as described later in this chapter).

The development of plant strains with enhanced fragrance and flavor production has been the subject of recent research, as demonstrated by efforts to

TABLE 1.3   Some Commonly Used Approaches to Plant Breeding

Conventional Plant Breeding Methods

Intraspecific crossing
Hybridization
Nonspecific mutagenesis
  Chemical mutagenesis
  Ultraviolet irradiation
  Irradiation
Wide hybridization

Novel Plant Breeding Methods

Tissue culture techniques
  Clonal propagation
  Somoclonal variation
  Gamitoclonal variation
Protoplast fusion techniques
Recombinant DNA techniques

---

enhance menthol production by mint (*Mentha piperita*). This project has involved increasing the expression of genes coding for the enzyme responsible for conversion of L-menthone to L-menthol. Attempts have also been made to modify the genetic structure of the plant so that the enzymes needed for this conversion could be found in cells other than callus cells. Both of these approaches should result in enhanced menthol production by the plant. Similar efforts have been directed toward other herbs and spices; however, the biochemical processes and underlying genetic constructions that lead to the development of these fragrances often are understood poorly. As a result, significant basic research may often be required prior to developing such new plant species [23].

Plant cell culture techniques, however, may represent a cost- and time-effective approach to developing consistent supplies of fragrance and aroma substances. These techniques involve the identification and isolation of differentiated cells responsible for fragrance synthesis from a plant. The cells are then maintained and grown on specialized media that will support the production of the desired metabolite by the cell. Genetic manipulation of these cells may be used to enhance fragrance production or to transfer the genetic material responsible for fragrance production to more stable or faster growing cell lines. By isolating and growing the selected cells in a fermentationlike process, the cell's metabolism may be directed away from primary metabolism (growth and replication) to secondary metabolism, thereby increasing the potential for fragrance or aroma production. These cultures often require specific nutrient compositions, environmental conditions, strain maintenance procedures, and fermenter designs to ensure consistent production. Metabolite composition and concentration even may vary as a result of production scale-up. In addition, many of the desired secondary metabolites may limit the biosynthesis either by inhibiting steps in the metabolic pathway leading to the metabolite or by disrupting the cell. Thus, care must be taken to limit direct contact between the cell and the metabolite. Finally, the metabolites often are produced in low concentration, such that large fermenter volumes may be required to obtain commercial-scale quantities of the desired substance.

One possible commercial application of plant tissue culture is the production of capsaicin, the active principal in hot chili flavor [24]. An immobilized, continuous-flow bioreactor system has been developed for the biosynthesis of capsaicin using *Capsicum frutescens*. A reticulated polyurethane foam matrix to immobilize the cells has been used in the production of 0.1 mg capsaicin/g dry weight per day when the reaction was coupled with simultaneous oil extraction. Selected fragrances, such as jasmine, strawberry, and grape, have also been targeted for production by plant cell cultures.

From an industrial standpoint, the most advanced application of plant cell culture techniques is the production of shikonin, a secondary metabolite from the plant *Lithospermum erythrorhizon*. Shikonin can be used as a dye and pharmaceutical and is worth approximately $4500/kg [25]. In developing this process, the pigment concentration was increased from 1 to 2 percent to 12 to 15 percent on a dry weight basis.

An in-depth discussion of plant tissue culture is presented in Chapter 8.

1.4   TECHNICAL CONSIDERATIONS

Two aspects that must be considered before using microbiological or plant cell culture methods are the slow growth rate and low final concentration of metabolites. Both issues can be addressed by microbial physiology and genetic manipulation approaches.

1.4.1   Microbial Physiology Approaches

Manipulation of cell physiology may be the easier and quicker approach to increasing flavor and aroma productivity. As has happened for antibiotics, the initial step may be the large-scale screening of microorganisms for production of the desired aroma or fragrance. Then the quality of the fragrance from the microbial source must be compared to that of the aroma or fragrance derived from the traditional source. Should this comparison be favorable, the physiological conditions have to be established for optimizing production of the desired substances. In addition, the biochemical pathway for production should be identified. In this way the production mechanism will be understood and the potential rate- and concentration-limiting steps can be identified [26]. Further enhancements may be obtained through changes in reactor design, cell immobilization, nonaqueous systems, and extractive processes coupled with the fermentation. Cell immobilization and reactor design implications have been described elsewhere [27,28].

Nonaqueous and continuous extraction processes may represent a simple approach for enhancing fragrance and flavor production. Many fragrances and flavors are composed of relatively high-molecular-weight compounds that are poorly soluble or insoluble in water. Because fermentation media are composed primarily of water, the potential exists for poor diffusion of these substances away from the cell. This effect could limit production due to feedback inhibition or metabolite damage to the cell. The use of solvents as components of the reaction system may help alleviate this problem [7,29,30]. For example, in an aqueous system the yeast *Pichia pastoris* could produce only small amounts of benzaldehyde from benzyl alcohol; a shortened reaction time and ninefold increase in benzaldehyde yield were obtained when hexane was used as the reaction medium [31].

For nonaqueous fluids to be effective, a monolayer of water must be maintained about the cell to maintain cell integrity while permitting the rapid diffusion of substrate and metabolites to and from the cell. For this reason, water-miscible solvents are less desirable that water-immiscible solvents; that is, the latter allow the formation of the necessary monolayer of water, whereas water-miscible solvents may disrupt the monolayer and damage the cell. However, in certain situations, the solubility of the product in a water-immiscible solvent may be lower than its solubility in aqueous media. Such a product can concentrate in the aqueous phase surrounding the cell and limit the biosynthesis [32]. In this case water-miscible solvents may be more useful if they increase the solubility of the metabolite in water without damaging the cell. Similar considerations also exist for the use of continuous extraction processes in concert with batch or continuous fermentation processes. The use of nonaqueous systems is discussed further in Chapters 3, 6, and 9.

In addition to nonaqueous systems, other modifications have been made to fermentation processes to enhance production rates and product concentrations. These processes, such as the use of humectants to alter water activity of the reaction system or resins to adsorb selected metabolites are described elsewhere [27].

1.4.2 Genetic Manipulation Approaches

Other methods for enhancing flavor and fragrance biosynthesis include genetic manipulation techniques such as nonspecific mutation, enhanced gene replication, and the cloning of genetic material into more rapidly growing or stable cell lines. Mutations introduced by genetic manipulations can induce changes in pathways that cause accumulation of desirable metabolic intermediates or direct carbon flow down an alternate pathway to form products of greater value. Similarly, a mutation may decrease the susceptibility of a microbe to end-product inhibition or catabolite repression, thereby resulting in increased product formation. Nonspecific mutagenesis can be used to accomplish these tasks; although it is a random process, it does increase the rate of mutation in the cell. Site-specific mutagenesis is more efficient and can also be used for gene amplification, whereby increased copies of a gene can increase the output of specific enzymes and result in increased product formation. The scientific basis of directed mutations have been described elsewhere [33,34].

Alternatively, genetic material responsible for coding specific microbial traits can be removed from a donor organism, transferred via an appropriate vector, and inserted into a host organism that is more hardy or has more desirable physiological or fermentation properties. In so doing, production of the desired metabolite can be enhanced. However, the genetic structure of the donor and host organisms must be known before using these procedures. Because the genetic structures of many flavor- and fragrance-producing plants or microorganisms are elucidated poorly, the first steps are often identification of the pathways for production of the desired metabolites and genetic material that codes for those pathways. With this information, appropriate genetic constructs can be developed to produce the desired fragrances and aromas.

1.5   NONTECHNICAL CONSIDERATIONS

The large-scale biosynthesis of flavor and fragrance compounds using either microbiological or plant cell technologies is technically feasible. However, nontechnical issues such as regulations for the introduction of products of biotechnology into food or the environment, the definition of a natural fragrance, and cost considerations may limit the acceptance of such technologies. These issues will be discussed separately.

1.5.1   Regulation of Biotechnology

A more thorough discussion of this topic is presented in Chapter 2; however, some salient points are discussed here. It is evident that certain segments of society perceive a risk associated with introducing biotechnology products as food or into the environment. Scientifically this risk may or may not exist. However, until the potential for pleiotropic effects, issues associated with the use of specific marker genes, and the like are addressed, such perceptions will continue to exist. These potential risks must be evaluated and the results must be communicated to the public.

To address these and similar issues, several governments and international organizations have prepared regulatory or administrative proposals that will control the introduction of products of biotechnology into the food supply. The frontrunner in this endeavor was the United Kingdom, which has established guidelines for the voluntary premarket notification for novel foods and food processes [35]. A decision-tree approach was used to identify the information requirements for such assessments. The information requirements vary depending on the degree of change from similar traditional products and processes and the approach used to modify the organism. As a result, substances produced by conventional microbial or plant breeding methods will require less information on safety concerns than substances produced by a genetically modified microorganism. For example, a fragrance substance produced by conventional plant cell culture that will be used in food may require the information shown in Table 1.4. On the other hand, more information would be required if the same fragrance was produced by a genetically modified microorganism (Table 1.4). In addition, the United Kingdom has proposed that substances containing products of biotechnology be labeled as such.

The United States has also published policy related to the introduction of novel plant varieties as food which uses existing regulations to address the issues concerning foods derived by genetic engineering [36]. In addition, the Organization of Economic Cooperation and Development (OECD) has published guidelines for the safety assessment of foods derived from new plant varieties [37], while the Council of Europe is preparing regulations to require premarket notification prior to the sale of novel foods [38]. Canada is also preparing regulations that would require premarket notification for novel foods, and guidelines for the safety assessment of genetically modified plants and microorganisms. There is also existing legislation (the Canadian Environmental Protection Act) which requires that the Canadian Government be informed of,

TABLE 1.4 Suggested Information Requirements for Approval of a Fragrance for Use in Food According to the United Kingdom Novel Foods Regulations[a]

Notification Requirements for Plant Cell Culture

Intake and extent of use
Technical detail of processing and product specifications
Nutritional studies
Toxicological assessment

Notification Requirements for Genetically Modified Cells

Instructions for use
Evidence of previous human exposure
Intake and extent of use
Technical details of processing and product specifications
Nutritional studies
History of organism
Characteristics of derived strain
Toxicological assessment
Assessment of genetic modification procedure
Genetic stability of the modified organism

[a] These recommendations are provided only as a guideline and are not a rigid check list.

and safety assessments conducted on all substances that are new to Canada, including products of biotechnology.

1.5.2 Definition of the Term *Natural*

A second major concern is the use of the term natural to describe fragrances and flavors. For flavors that are mixtures of simple low-molecular-weight organic compounds, a synthetically produced alternative is often less expensive and more readily available then the natural counterpart. To gain a competitive edge, the natural product must have a benefit that enhances its financial viability. This advantage is often associated with being able to call the substance "natural"; consumers often perceive that a natural substance is preferable to a synthetic product. Thus, the definition of natural may play a significant role in determining the viability of certain microbiologically produced fragrances.

In the United States, two classes of flavor chemicals exist: natural and artificial. The Code of Federal Regulations (CFR Chapter 21, Part 101.22.a.3, 1990) defines a *natural flavor* as:

> . . .the essential oil, oleoresin, essence or extractive, protein hydrolysate, distillate of any product of roasting, heating or enzymolysis, which contains the flavoring constituents derived from a spice, fruit juice, vegetable or vegetable juice, edible yeast, herb, bud, bark, root, leaf or similar plant material, meat, seafood, poultry, eggs, dairy products, or fermentation products thereof whose significant function in food is imparting flavoring rather than nutrition.

Thus, compounds produced by biosynthetic processes may be considered natural as long as the starting materials are considered natural.

In the United Kingdom, no distinction is made between natural and artificial flavorings by the 1984 Food Labelling Regulations. However, recent concern over consumers being misinformed by use of the term natural has led to a report by the Local Authorities Coordinating Body on Trade Standards (LACOTS) to the Food Advisory Council, which suggested that:

> . . .the term natural should be used without qualification to describe single foods of traditional nature which have not been processed; food ingredients from such sources and permitted food additives and flavorings from recognized food sources.

Such a definition would limit the use of the term natural in describing flavors from biosynthetic processes.

A third classification, which exists in most other developed countries, is *nature identical*. These compounds are produced by chemical processes but are identical in all aspects to substances identified in nature. Confusion does arise as a flavor identified as natural in Italy may be artificial in France and nature identical in Germany. The European Economic Community (EEC) is attempting to develop a consensus on definitions of natural, nature identical, and artificial. Until such consensus is reached, each country will maintain its own unique set of definitions. This subject is discussed in-depth in Chapter 2.

1.5.3 Cost Considerations

A third consideration is the cost–benefit analysis for microbiologically produced flavors and fragrances. One estimate has indicated that microbial routes of production would become economical only for compounds with a market value of $200 to $500 per kilogram [26]. Fragrances, such as jasmine, have prices in the $5000/kg range and may be appropriate for development, whereas fragrances such as spearmint (priced at approximately $30/kg) would probably not be suitable.

A break-even analysis was developed for the microbiological production of a hypothetical flavor or fragrance with potential sales of either 1000 kg/year, 10,000 kg/year, or 100,000 kg/year (Table 1.5). This example is not meant to be a detailed cost analysis; instead it represents an example of the price and supply considerations necessary to ensure cost effectiveness, and, as a result, some liberties have been taken in the analysis. A research and development expense of $5,000,000 was charged to the process and amortized over 5 years with no discount rate. A production level of 1 g/L in the fermenter and a fermenter variable operating cost of $0.10/L were assumed. The variable operating cost of the extraction was assumed to be $50/kg product. Fixed costs such as labor, maintenance, and the like were assumed to be: $50,000 for the 1000-kg fermentation, $100,000 for the 10,000-kg fermentation, and $200,000 for the 100,000-kg fermentation. Annual depreciation was estimated using a 5-year amortization period and no discount rate. It was also assumed that the facility would be dedicated to the fermentation.

TABLE 1.5   Estimated Price Needed to Recover the Cost of Developing and Producing a Flavor or Fragrance, Using Three Different Sales Potentials

	Potential Sales (kg/yr)		
Expense	1,000	10,000	100,000
Fixed costs ($)			
Research and Development	1,000,000	1,000,000	1,000,000
Depreciation	40,000	400,000	4,000,000
Labor, Maintenance, etc.	50,000	100,000	200,000
Subtotal	1,090,000	1,500,000	5,200,000
Variable costs ($)			
Fermentation	100,000	1,000,000	10,000,000
Extraction	50,000	500,000	5,000,000
Subtotal	150,000	1,500,000	15,000,000
Total costs ($)	1,240,000	3,000,000	20,200,000
Breakeven price ($/kg)	1,240	300	202

The results show that a product with a potential market of 1000 kg/year would require a selling price of $1240/kg; a 10,000-kg/year market would require a selling price of $300/kg; and a 100,000-kg/year market would require a selling price of $202/kg.

The estimates in Table 1.5 may be questioned because the assumptions fail to include any potential economies of scale that could be achieved in the fermentation process. They also do not address any potential differences in the cost of extraction for the different compounds; there are no costs for the regulatory process; and the research and development costs are held constant. However, they do indicate that a flavor or fragrance substance must have a significant value and market potential for production by a microbiological process to be cost effective.

The perceived value of the product further complicates the pricing strategy. When a flavor or fragrance is available in small quantities, the consumer may perceive that product as being something rare and thus of higher value. Should the product become more available, the perceived exotic nature might dissipate; that is, the perception would be that the product is more common and thus should have lower value.

1.6 SUMMARY

The technological potential exists for the production of aroma and fragrance materials by microbiological processes. However, in microbiological or plant cell cultures the concentrations of the fragrance chemicals are often low and production rates may be slow. To effectively produce these flavors and fragrances, methods must be developed to enhance production rate, yields, and recovery efficiency. Should these developments be achieved, regulatory considerations may affect the potential implementation of these technologies, particularly if these substances result from cultures produced by genetic manipulation. The final concern may be the value of the desired flavor or fragrance and the elasticity of its price, given the enhanced availability of the substance and potential changes in consumer perception of the product.

REFERENCES

1. Omelianski, V.L., *J. Bacteriol.* **8**, 393 (1922).
2. Gordon, M.A., *Mycologia* **142**, 167 (1952).
3. Kempler, G.M., *Adv. Appl. Microbiol.* **29**, 29 (1983).
4. Sima Sariaslani, F., and Rosazza, J.P.N., *Enzyme Microb. Technol.* **6**, 242 (1984).
5. Sharpell, F.H., in *Comprehensive Biotechnology: The Principles, Applications, and Regulations of Biotechnology in Agriculture and Medicine* (M. Moo-Young, ed.), Vol. 3. Springer-Verlag, Berlin, 1985.
6. Gatfield, I.L., *Food Technol.* **42**, 110 (1988).

7. Welsh, F.W., Murray, W.D., and Williams, R.E., *Crit. Rev. Biotechnol.* **9**, 105 (1989).

8. Dziezak, J.D., *Food Technol.* **40**, 108 (1986).

9. Lugay, J.C., *ACS Symp. Ser.* **317**, (1986).

10. Schindler, J., and Schmid, R.D., *Process Biochem.* **4**, 2 (1982).

11. Drawert, F., and Barton, H., *J. Agric. Food Chem.* **26**, 765 (1978).

12. Klingenberg, A., and Sprecher, E., *Planta Med.* **3**, 264 (1985).

13. Nakajima, O., Iriye, R., and Hayashi, T., *Nippon Nogei Kagaku Kaishi* **52**, 67 (1978).

14. Omata, T., Iwamoto, N., Kimura, T., Tanaka, A., and Fukui, S., *Appl. Microbiol. Biotechnol.* **24**, 199 (1985).

15. Wong, H.P., Ellis, R., and LaCroix, D.E., *J. Dairy Sci.* **58**, 1437 (1975).

16. Okui, S., Uchiyama, M., and Mizugaki, M., *J. Biochem. (Tokyo)* **54**, 536 (1963).

17. Farbood, H., and Willis, B., European Patent PCT 1072 (1983).

18. Watt, J.C., and Nelson, J.H., U.S. Patent 3,072,488 (1963).

19. Troller, A., U.S. Patent 4,304,862 (1981).

20. Montville, T.J., Hsu, A. H.-M., and Meyers, M.E., *Appl. Environ. Microbiol.* **53**, 1789 (1987).

21. Gupta, K.G., Yadav, N.K., and Dhawan, S., *Biotechnol. Bioeng.* **20**, 1895 (1978).

22. Boyaval, P., and Corre, C., *Biotechnol. Lett.* **9**, 801 (1987).

23. Klausner, A., *Bio/Technology* **3**, 534 (1985).

24. Mavituna, F., Wilkinson, A.K., Williams, P.D., and Park, J.M., *Nat. Conf. Bioreact. Biotransforms.*, Gleneagles, Scotland, *1987*.

25. Curtain, M.E., *Bio/Technology* **1**, 649 (1983).

26. Sprecher, E., and Hanssen, H.-P., *Top. Flavour Res., Proc. Int. Conf.*, *1985* (1985).

27. Roffler, S.R., Blanch, H.W., and Wilke, C.R., *Trends Biotechnol.* **23**, 129 (1984).

28. Knorr, D., Miazga, S.M., and Teutonica, R.A., *Food Technol.* **39**, 135 (1985).

29. Butler, L.G., *Enzyme Microb. Technol.* **1**, 253 (1977).

30. Klibanov, A.M., CHEMTECH **16**, 354 (1989).

31. Duff, S.J.B., and Murray, W.D., *Biotechnol. Bioeng.* **34**, 153 (1989).

32. Welsh, F.W., Williams, R.E., and Dawson, K.H., *J. Food Sci.* **55**, 129 (1990).

33. Wasserman, B.P., Montville, T.J., and Korwek, E.L., *Food Technol.* **42**, 133 (1988).

34. International Food Biotechnology Council, *Regul. Toxicol. Pharmacol.* **12**, Part 2. (1990).

35. Ministry of Agriculture, Fisheries and Food, *Guideline No. 38.* Minist. Agric. Fish. Food, London, 1990.

36. Food and Drug Administration, *Federal Register*, **57**, 22984 (1992).

37. OECD, *Safety Evaluation of Foods Derived by Modern Biotechnology, Concepts and Principles.* OECD, Paris (1993) 79 p.

38. Commission of European Communities, *OH. J. Europ. Commun.* C **190,** 3 (1992).

The Development and Regulation of Flavor, Fragrance, and Color Ingredients Produced by Biotechnology

CHARLES H. MANLEY

Takasago International Corporation (USA), Teterboro, New Jersey

During man's evolution there has been a steady increase in the understanding, use, and diversification of natural fermentation products. There are many examples of the traditional use of fermentation in foods and beverages, from cheese, bread, and meat to wine, beer, and bourbon. It is only in recent times that man has started to appreciate the sciences of biochemistry, microbiology, and genetics and to apply this knowledge to the development of flavor, fragrance, and color ingredients.

This new biotechnology is being used to create such ingredients either as "whole substances" based on existing foods, food components, spices or herbs, or as purified or semipurified isolates that have high aroma or color impact. The driving force is partly the need to develop specific components that are difficult to produce by synthetic chemical methods or to produce them under what are considered more natural, milder methods of processing. Biotechnology methods also allow for the manufacturing of complex yet uniform materials, at higher yields from readily available substrates without the sometimes adverse conditions existing in agricultural production.

A major scientific issue related to these products, in general, is their consideration as safe materials for human consumption. The basic philosophy of judging the safety of complex mixtures such as foods in the raw state, prepared foods, and flavors has not been well established. Currently a number of trade, consumer, governmental groups are working on such a philosophy; this will

Bioprocess Production of Flavor, Fragrance, and Color Ingredients, Edited by Alan Gabelman, ISBN 0-471-03821-0 © 1994 John Wiley & Sons, Inc.

lead to regulating the use of biotechnology for products that are consumed by humans as well as for labeling such products. These are worldwide rather than just domestic issues, and a rational policy, harmonized for the world's industry and consumers, will assure the future success of this new technology.

2.1 HISTORY

2.1.1 The Ancient Beginning

Two ancient pursuits of mankind are brought together in this text. Each pursuit has contributed to the shaping of the history of human civilization. The first pursuit is the art of fermentation developed by empirical reasoning of intelligent humans. It was an invention that allowed people to travel long distances far from their basic food sources. The fermentation of foods stabilized them against spoilage and thereby allowed their high-valued nutrients to be carried long distances. This spared the great armies and explorers the burden of hunting and gathering food as they traveled. Cheeses from milk were carried by the nomadic tribes in their wanderings around the world, alcoholic beverages gave rise to social problems, and, with "Niro fiddling while Rome burned," led to the downfall of great civilizations from an overindulgence in the fermented drinks of the gods!

Much of the early food industry of the world was based on the curing, pickling, or fermentation of foods and juices. Wine, beer, bread, pickles, pickled meats, cheeses, vanilla, chocolate, tea, and coffee are now preeminent materials ("foods") in the human diet.

2.1.2 The Romance of the Industry

At the same time that people were learning to control the fermentation of foods and juices, they discovered that the process led to the development of more desirable flavors and aromas. This led to the second great pursuit of humankind—the development of flavors and fragrances! The desire for materials bearing flavors and aromas opened up the land trade routes to the Asian continent, the great camel caravans of the East carrying the gold, frankincense, and myrrh of biblical times. And traders to and explorers of Asia, such as Marco Polo, brought new spices and herbs for flavor, color, and fragrance to a European civilization emerging from the Dark Ages. The spice and fragrance trading companies of the early colonial period gave rise to the famous city of Grasse in southern France and the start of the fine perfume industry of Europe.

The world was changing by these events and with the coming of ocean travel, and the search for over-water routes to the spice islands of the East, the world would be permanently altered. Perhaps no force on earth at that time, or even in modern times, had so much influence on the course of human history. That was the romance of flavor, fragrance, and color, when these materials were regarded as gold. Even today some of these materials have that glitter.

2.1.3 The Advent of Science

During the middle of the 19th century, other events caused a decline in the romance of the industry. Scientific theory and method came of age and with it the understanding of the secrets of flavors, fragrances, colors, and fermentation. First to deal a blow to the romance of the industry was the development of chemistry, organic chemistry to be precise, with Friedrich Wohler's synthesis of urea in 1828. It brought our understanding of the things around us to a much higher level. Breakthroughs in the understanding of fermentation came in 1864 when Louis Pasteur, after stormy scientific arguments questioning the theory of spontaneous generation, invented the process of pasteurization. He proved that microorganisms were the cause of the phenomenon of fermentation. A year later, the famous German chemist F.A. Kekulé explained the structure of aromatic compounds through his benzene ring theory.

Science was assembling its power, and the romance of the industry was starting to decline. The science of chemistry was used first to develop synthetic organic chemicals for use in fragrances, flavors, and colors. The color tar dye research of the German chemical industry led to the first colors used in foods, first without a concern for human safety, and then, with the food laws of the United States and other countries, a clear establishment of safety (toxicology) considerations and regulation of these food additives and their applications was put in place.

The large chemical industries of Switzerland, Germany, and Holland generated the know-how for the creation of synthetic chemicals for use in making flavors and fragrances. The use of aromatic chemicals for flavors hit its peak in the 1960s when over 1500 synthetic materials were used and considered generally recognized as safe (GRAS) for use in foods and beverages in the United States.

Chemistry also generated methods for the extraction of natural flavors and fragrances from herbs and spices. One of the methods, distillation, was used in ancient times. The great historian Herodotus (484–425 B.C.) reported on the methods for creation of the oil of turpentine and camphor. The first systematic study of essential oils, those oils extracted or distilled from herb and spices, was made by the French chemist J. B. Dumas (1800–1884). He published his first treatise devoted to essential oils in Liebig's *Annales der Pharmaci* in 1833. One of the last great treatises in essential oils was published in 1948 by Dr. Ernest Guenther, the head of research for the then Fritzsche Brothers Company [1]. These natural extracts and distillates are still the bases of some of the finest flavors and fragrances created by the industry.

2.1.4 Recent Consumer Trends

In the last 20 years, there has been a great consumer preference for natural materials. The 1958 Food Additive Legislation, administered by the U.S. Food and Drug Administration (FDA), allows for the use of fermentation and enzymolysis as natural methods for the creation of flavors [2]. Because of the increasing interest in natural products, more attention has been focused on the production of natural flavoring materials from raw materials, such as the use of the essential oils of herbs and spices.

However, such raw materials contain only small amounts the aromatic components useful in making flavors and fragrances. Table 2.1 shows the percent yield of aromatic material from various spices and herbs [3]. As one can see, the yields are not impressive. This obviously means that tremendous amounts of raw material must be grown to produce useful amounts of aromatic material. These raw materials often are in limited supply because of seasonal variation, climatic factors, and political problems. Their prices may be high, and their availability and quality variable. The disadvantages of this situation and the desire for naturally derived materials make biotechnology an attractive alternative to the traditional natural raw materials.

More specifically, one may look to biotechnology to produce materials:

1. With natural status
2. With higher yields from common readily available substrates
3. That are complex yet uniform
4. That are produced under mild conditions.
5. Without adverse external conditions, which profoundly influence the quality and quantity of the products of agriculture.

TABLE 2.1   Amount of Volatile Oil in Various Botanicals Used for Flavor

Item	Percent of Volatile Oil
Anise	1.5–3.5
Basil	0.4
Caraway	2.5–7.5
Cassia	0.5–4.0
Cinnamon	0.5–1.0
Clove	15–20
Coriander	0.4–1.0
Garlic	0.1–0.25
Grapefruit	0.06–0.08
Leek	0.005–0.02
Lemon, pressed	0.2–0.3
Lemon, distilled	0.6
Lime	0.18–0.32
Mace	12–15
Nutmeg	6.5–15
Onion	0.02–0.03
Orange	0.07–0.14
Pepper, black	2.0–4.5
Peppermint	0.3–0.4
Rosemary	0.5–2.0

Source   Pearson [3].

2.2 THE BIOTECHNOLOGY AGE

2.2.1 The Meaning of Biotechnology

The term *biotechnology* has come into the literature in the last 10 years or so. Basically, it is a high-tech term for the age old processes of fermentation. Table 2.2 reviews the elements of the field identified as biotechnology.

In the flavor, fragrance, and color industry, one finds that most of the research effort and, recently, most of the manufacturing efforts are directed at producing (1) aromatic chemicals that would be considered naturally derived; (2) specific chemicals with unique properties (e.g., chiral centers) that cannot be produced by synthetic means; (3) mixtures of chemicals that can be used as components (e.g., a methyl ketone mixture for blue cheese flavors); and (4) total flavor/fragrance profiles generated by fermentation or cell expression of secondary metabolites using plant tissue culture.

2.2.2 Aromatic Chemicals

Of the over 1500 synthetic chemicals that are used by the U.S. flavor industry, only 68 compounds have been reported to have annual consumption rates of more than 3000 kg/year [4]. Furthermore, only 20 or so have been produced commercially by fermentation routes [5].

The groups of chemicals listed below are of major interest to the flavor industry. These chemical groups are reviewed in Table 2.3, which gives examples from each group and the organisms used in their production [6–10]:

1. Acids, used mostly in cheese flavors
2. Alcohols, used mostly in fruit flavors, alcoholic beverages, and mushroom flavor
3. Lactones, used in dairy flavors
4. Esters, used in fruit and dairy flavors
5. Aldehydes, used in citrus and fruit aromas
6. Ketones, used in cheese and butter flavors
7. Pyrazines, for their nutty and roasting aromas
8. Terpenoids, for their citrus and mint aromas

TABLE 2.2 Elements of Biotechnology

Traditional Biotechnology	High-Tech Biotechnology
Fermentation	Genetic engineering
Enzymolysis	Recombinant DNA methods
	Cell culture
	Immobilized enzymes
	Unique uses of enzymes
	Cell fusion
	Synthetic enzymes

TABLE 2.3 Major Flavor Substances Developed by Fermentation Methods

Substance Category	Flavor Character	Flavor Substances	Microorganisms
Acids	Dairy, cheese, butter	Butyric acid	Clostridium butyricum
		Propionic acid	Streptococcus thermophilus
		Lactic acid	Lactobacillus lactis
Alcohols	Fruity, mushroom	Propanol	Clostridium acetobutylicum
		3-Octanol	Trichothecium roseum
		Vanillyl alcohol	Saccharomyces cerevisiae
Lactones	Dairy, peach, butter	6-Pentylpyrone	Trichoderme viride
		4-Decanolide	Sporobolomyces odorus
		4-Butanolide	Polyporus durus
Esters	Fruity, berrylike	Ethyl acetate	Saccharomyces cerevisiae
		C_2 to C_4 allyl esters	Geotrichum candidum
		Long-chain fatty acids	Pseudomonas genus
		Methyl salicylate	Phellinus genus
Aldehydes	Fruity, cherry	Aliphatic aldehydes	Candida utilis
		Benzaldehydes	Acinetobacter calcoaceticus
Ketones	Dairy, blue cheese, cheese	2-Alkanones	Penicillium roqueforti
		Diketones	Pseudomonas genus
		Diacetyl	Streptococcus lactis
Pyrazines	Nutty, roasted, green	Various pyrazines	Corynebacterium glutamicum
		2-Methoxy-3-isopropyl pyrazine	Pseudomonas perolors
Terpenoids	Citrus, mint	Various (including α-terpineol)	Ceratocystis moniliformis
		Linalool oxides	Botrytis cinerea

24

Pure aromatic chemicals have an important role to play in the creation of flavors with strong aromas and with the ability to survive the rigors of modern-day food preparation processes, storage conditions, and home preparation methods.

2.2.3 Specific Chemicals

Biotechnology can be used to convert inexpensive, readily available aroma chemicals to higher value materials that have specific uses in the flavor and fragrance industry. One example is the conversion of linalool to a series of other terpenoids as well as to the furanoid and pyranoid linalool oxides by the use of *Botrytis cinerea* [11]. Other biotransformations of terpenes have been used in *in vitro* plant cell production of specific chemicals from lower cost organic substrates. These are good examples of the use of suspension cell cultures that are able to perform specific transformations. Further discussion and examples of the use of plant tissue culture may be found in Chapter 8.

2.2.4 Simple and Complex Mixtures

Currently the creation of so-called pure aromatic chemicals is one of the major applications of biotechnology to flavor and fragrance material production. However, simple or complex mixtures of aroma components are also useful. Table 2.4 presents examples of such mixtures produced by the action of microorganisms or enzymes on substrate material(s) and commercialized as flavors.

One of the major uses of enzymes and microorganisms has been the development of strong, aged cheese flavors. Cheese flavors are very complex mixtures of both volatile and nonvolatile organic chemicals. Most of these materials have, as their substrates, the fatty acids that are found in the milk used to produce the cheeses. Such components as free fatty acids, aldehydes, methyl ketones, and diacetyl are major contributors to cheese and dairy flavors. Specific differentiation of cheese varieties is established by the amount of each component produced during the fermentation or added in the final formulation. Table 2.5 reviews the various chemicals produced, their production route, and the particular character they contribute to the cheese or dairy flavor [12]. A detailed discussion of the use of fermentation to produce dairy flavors may be

TABLE 2.4 Commercial Development of Flavors by Biotechnology

Cheese flavors
Dairy and cheese flavors
Fruit flavors
Mushroom flavors
Fish flavors
Flavor enhancers

TABLE 2.5 Cheese and Dairy Flavors

Chemical Type	Mechanism of Production	Flavor Character Developed
Fatty acids	Lipases and esterases	C_2 to C_4, butterlike
		C_6 to C_{10}, cheddarlike
		$>C_{10}$, aged cheddar type
Methyl ketones	*Penicillium roqueforti*	Subthreshold, milk/cream
		Higher level, blue cheese
Diacetyl	*Lactobacillus*	Butter top-notes
Lactones	Lipases	Creamy notes
Peptides	Proteases	Acid peptides, sour notes
		Hydrophobic, bitter
		Glutamic and aspartic,
		umami/flavor enhancing
		(All give aged character)

found in Chapter 6. Clearly aromatic chemicals are not the only success story for biotechnology in the food and beverage area.

2.2.5 Flavor and Fragrance Profiles

The production of secondary metabolites with plant cell cultures represents a very complex issue [13]. A great deal more must be learned about the biochemical and genetic regulations of plant secondary metabolites before large-scale production becomes a commercial reality. Someday in the future, we may be able to switch on the appropriate enzyme systems to have the fruit, herb, or spice cell produce the full profile of aromatic materials for which the particular botanical is well known. Unfortunately our current abilities are more limited. See Chapter 8 for a more complete discussion of this topic.

2.2.6 Colors

For at least a century, and perhaps much longer, red rice has been prepared in Southeastern Asia. The color from the rice has been used as a food colorant and for rice wine. In 1895, Went determined that the pigments were produced by a species of the fungal genus *Monascus* [14]. In the 1950s, chemists determined the chemical identity of the pigments from the *Monascus* fungus, and during the 1970s the manufacture of color from the fungus was commercialized in Japan [15]. The production of colors by *Monascus* and other organisms is discussed in detail in Chapter 7.

The FDA has not permitted the use of these materials in foods in the United States, yet the uncontrolled use of red rice for coloring foods and beverages has been practiced in Southeastern Asia for more than a century and the use of the purified pigment for more than 15 years in Japan. There has been no observation of negative health effects in either use.

These examples are but a few of the many applications of biotechnology in the production of complex flavor, fragrance, and color materials; others are discussed in later chapters of this book.

2.3 LEGAL AND REGULATORY ISSUES

The question of how these traditional materials of fermentation are or should be regulated and judged safe is a major question that must be faced by the industry and the government regulators. As the industry turns from these more traditional uses of microorganisms and enzymes to the new biotechnology based on genetic alteration of microorganisms and plant cells, the question of the safety of these materials becomes an even more important issue.

In the United States, the safety of most foods, food ingredients, and fragrance materials is regulated by the FDA under various provisions of the Federal Food, Drug and Cosmetic Act (FDC Act) [16]. The existing food safety laws provide the FDA with a comprehensive, flexible set of tools for regulating the safety of every component of the food supply and cosmetics. It is the policy of the U.S. government to use the existing laws to regulate the food and food ingredient products of biotechnology [17]. The current law recognizes that natural materials, including foods, may contain many substances that, if consumed at high levels, are toxic, but are not harmful when consumed as inherent constituents of foods. The FDA is empowered to act against such substances, but only it if finds that they render the natural material(s) ''ordinarily injurious'' to health [18].

This concept is an important consideration in judging the safety of food ingredients or cosmetic materials produced by biotechnology. The U.S. Congress does not wish to require premarket approval of these materials when such approval is not necessary to assure safety. Therefore, materials that are generally recognized as safe (GRAS) need not be reevaluated and approved by the FDA unless there are reasons to believe that the material derived from a biotechnology process has a hazard associated with it.

Foods and food ingredients may be divided into four major categories: whole food, simple chemically defined mixtures, biological ingredients and processing aids, and chemical additives (single chemicals). Biotechnology, in creating flavors, fragrances, and colors can produce products that fit into each of these categories.

2.3.1 Whole Foods

In more specific biotechnology terms, this category represents the total material from cell culture or enzyme conversion of substrates, with the residual substrate and organism included. These whole substances are subject to regulation as foods by the FDA under section 402(A)(1) of the FDC Act. The Act establishes two standards: one for inherent natural substances of the food and another for substances that are added. Substances occuring naturally in a food only render a food misbranded or adulterated if the substances make the food "ordinarily injurious" to health. The added substances render the food adulterated if they "may" make the food injurious to health, that is, if there is a reasonable possibility that any consumer will be injured by consuming the food.

Over the years, new foods (i.e., new strains and varieties of existing foods) have been developed by the "conventional" methods of cross-breeding techniques. These foods are accepted as safe to eat unless a new toxic substance or an elevated harmful level of an existing toxin indicates a reasonable possibility that the food would be harmful. As mentioned earlier, new foods developed through biotechnology will be subject to the same regulation.

2.3.2 Simple Mixtures of Chemical Substances

Simple mixtures of aromatic substances are well represented in the creation of flavors, fragrances, and colors. Usually, in developing flavors, fragrances, or colors, the fermentation, enzymolysis, or cell culture expression creates mixtures of substances that are useful only when isolated in a purer form. For centuries it has been the practice of the industry to isolate flavors, fragrances, and colors from foods, spices, herbs, flowers, and so on. This has been, as indicated earlier, the historical basis of the industry. Biotechnology has only given the industry the advantages of creating these extracts, isolates, or distillates in a more controlled manner.

The 1958 Food Additives Amendment to the FDC Act focuses on additives used in the food supply. The act established requirements for these additives prior to their introduction to the market. The need for premarket clearance based on the scientific establishment of safety is part of the act. That is, the additive must be shown to be harmless under its intended use. However, because many substances (such as flavor extracts) have a long history of safe use in foods, the act excludes such materials from the safety review required for food additives and establishes a group of substances that are GRAS [19]. The GRAS status of a material is achieved by recognition among a panel of qualified experts that the substance is safe (i.e., there is a reasonable certainty of no harm under intended conditions of use) and is only used for materials that are used at very low levels in the food system. This category includes the substances used to formulate flavors.

For materials introduced after the 1958 law was enacted, the GRAS status must be based on qualified scientific procedures. On the other hand, for substances used prior to the 1958 law (grandfathered materials), GRAS status may be given based on either scientific evidence or historical experience based on common use in foods [20,21]. Therefore, a mixture that was derived from a substrate via fermentation or enzyme conversion prior to the 1958 law would not need premarket approval. Reviews to determine the GRAS status of such materials should consider the FDA published list of criteria for GRAS status [22]. It would also be prudent if the industry would consider a GRAS review of biotechnologically derived material similar to the review of data given by an expert panel such as the Flavor and Extract Manufacturers Association's (FEMA) Expert Panel or the Research Institute for Fragrance Materials (RIFM) Panel of Experts. Further comments on this approach are made in a later section.

2.3.3   Biological Ingredients and Processing Aids

The traditional use of enzymes and microorganisms for the production of flavors and fragrances is well known and have been in use for many centuries. However, biotechnology has led to new uses of these traditional processes. As we saw in the earlier discussions, certain enzymes and microorganisms that have not been used in foods or were used on nonconventional substrates now are used typically to create simple mixtures, single aromatic chemicals, or colors.

Many of these materials have been used traditionally and therefore have attained GRAS status. New (nontraditional) enzymes or microorganisms or new uses for well-known microorganisms would have to be reviewed and be the subject of a GRAS affirmation petition. The basis of the scientific review will be discussed in a later section.

2.3.4   Chemical Additives

Perhaps the easiest materials to qualify as safe products of biotechnology are pure chemical substances that currently have GRAS Status. Those materials may be food flavor materials that appear on the FDA or FEMA GRAS list or those fragrance substances appearing on the Research Institute for Fragrance Materials list [23,24]. The Food Additive Amendment establishes the regulation of a substance (additive) and not the process by which it is produced. However, in some cases, the FDA identifies manufacturing processes that it regards as assuring safety of the additive. On the other hand, if the manufacturing method is not defined, the industry is allowed to use any process as long as the resulting material meets all of the identity and purity requirements of the approved GRAS material and does not introduce new substances that represent a hazard.

2.4   LABELING ISSUES

As noted above, one of the major advantages of biotechnologically-produced products is attainment of natural status and the ability to make such a claim on the product label and the ingredient listing. The nature of the substrate(s) used in biotechnology processes also has an impact on labeling. If the substrate is of animal origin, the U.S. Department of Agriculture (USDA) plays a role in regulating the product. Materials of animal origin also have an impact on the religious approval and use of the product. The kosher law of the Jewish religion plays an important role in food product labeling. Needless to say, the use and labeling of materials from biotechnology in the United States will be influenced by regulations being developed in other parts of the world. The success of marketing these products will depend on our understanding of the basic issues of safety and product labeling.

2.4.1 Natural Claims

The success of many products of biotechnology may hinge on the use of the term natural. The processes of biotechnology are considered by many regulatory groups, in both the United States and around the world, as natural processes, although there is some dispute as to whether a so-called pure chemical component produced from fermentation may be called natural. Before we discuss the issues of what natural is and how it is regulated, let us see why the issue of natural is so important.

The concept of natural has an important place in the eyes of the consumer. Although this place is not kept for all things that the consumer comes into contact with, it is, without a doubt, an important issue when it comes to diet. In the case of foods, the consumer usually believes that natural materials, including flavors, are more healthy and safer than their synthetic counterparts. Although scientific evidence does not support this view, the consumers' beliefs are very strong. The proof is in the sales of natural flavors, which have steadily increased over the last 15 to 20 years, to the point where today the majority of flavors sold are natural.

But what exactly is a natural flavor? In particular, how is natural related to those materials produced by biotechnology? As mentioned earlier, the major regulatory group involved in regulating flavors is the FDA. In 1973 the FDA defined *natural flavor* as:

> . . . the essential oil, oleoresin, essence or extractive, protein hydrolysate, distillate, or any product of roasting, heating, or enzymolysis, which contains the flavoring constituents derived from a spice, fruit or fruit juice, edible yeast, hulls, bark, bud, root, leaf or similar plant material, meat, seafood, poultry, eggs, dairy products, or fermentation products thereof, whose significant function in food is imparting flavoring rather than nutrition [25].

It is interesting to note that this is the only place in the food regulations where a food or food ingredient has been so defined. The definition was established well before biotechnology was put to use in making specific substances for the flavor and fragrance industry. Therefore, although there is an acceptance of fermentation and enzymolysis as natural processes, the regulations did not foresee the development of pure natural chemicals or specific mixtures for flavor or fragrance use. Many of these materials may be substituted for the synthetic ones used to make traditional artificial flavors.

The classical approach of isolating flavors from foods, spices, or roasted, fermented, or enzymatically converted food(s) or spices are clearly covered in the 1973 regulation as natural. But what about benzaldehyde made from the fermentation of phenylalanine, which in turn was isolated from an acid-hydro-lyzed soy protein? This concept is an issue now being considered by the regulatory groups and the flavor industry.

Although the fragrance industry is not inundated by requests for natural fragrances, the trend toward natural fragrances continues to grow.

2.4.2 *Natural Flavor* Definition

Flavors made from meat and poultry or any protein-based product fall under the jurisdiction of the USDA. In 1990, the USDA issued a regulation titled "Ingredients that may be identified as *flavors* or *natural flavors* when used in meat and poultry products" [26]. The purpose of the regulation was to restrict the use of meat, seafood, or any material containing protein (e.g., yeast or yeast extract) that is used to make a flavor without proper identification of the substrate used in making the flavor. The regulation was motivated by the observation that some proteinaceous materials cause certain health-related problems such as allergic reactions or are related to cultural or religious preferences.

During 1990, the flavor industry worked with the USDA to develop a protocol for labeling flavors that contained these materials. It was decided that the FDA was the lead regulatory agency for defining what constitutes a *natural* flavor. Therefore, the labeling on flavors would come under FDA review and be subjected to the FDA's existing natural certification procedure. That procedure requires that a manufacturer legally certify (have on file in his office) a flavor as natural (as defined in the Code of Federal Regulations) and that the flavor does not, to the best of his knowledge, contain any artificial materials.

The USDA, however, still retains control of food products made with meat or poultry and therefore can review flavor labels to see if they properly identify the proteinaceous substrates used to prepare the flavor. Assistance in getting a label reviewed and accepted for use in a USDA-inspected plant is gained by submitting a "sketch" of the formulation to the USDA's Proprietary Mix Committee (PMC) in Washington, D.C. During such a review, the flavor manufacturer needs to reveal the nature of the proteinaceous substrate used in preparing the flavor. This would be typical for the case where enzymolysis or fermentation is used to create a flavor from meat, dairy, seafood, or other proteinaceous materials.

2.4.3 Kosher Status

One of the issues developed by the USDA in their regulation was the identification of materials that are of concern because of cultural or religious practice. The kosher dietary laws, *kashruth*, of the Jewish faith is the most important one to consider for the food industry in the United States. There is a strong trend within the American food industry to label products *kosher*, and consequently to require that ingredients used to manufacture these products also be kosher. The kosher food market is worth $1.5 billion, with 40 percent of supermarket shelves in the Northeast now containing certified kosher food. A 1977 study showed 1000 kosher products on the market; a 1989 study showed that the number of such products had grown to 18,000 [27]. This is somewhat amazing when one considers that the country's Jewish population is approximately six million, only 2 to 3 percent of the total, but it is understandable when one considers that many people believe the kosher review process adds quality and nutrition to the food product.

Kashruth are not "health laws" as such, although many of them make sense from a health point of view [28]. The four major rules in the kosher laws are as follows:

1. Plants—all are considered kosher;
2. Flesh and blood—proper procedures for killing the animal must be followed to ensure a kosher product;
3. Separation of dairy and meat during processing and consumption is required;
4. Acceptable species or animals:
    Meat—mammals that chew their food and are split-hoofed
    Foul—no birds of prey
    Fish—with fins and removable scales only (no shellfish)

Many of materials that may be used to create flavors from biotechnology may involve nonkosher substrates or the use of enzymes from animal origin. For example, the use of pancreatic lipases on dairy substrate to produce a butter or cheese flavor would not be considered a kosher process whereas the use of a lipase of microbiological origin might be considered kosher. The best way to assure that a process or ingredient will meet kosher requirements is to contact a rabbinical group that specializes in kosher food product review.

2.4.4   International Regulations

The three major world markets for flavor, fragrance, and color ingredients are the United States, Europe and Japan. We have already reviewed the position of the United States regarding natural materials and the use of biotechnology in making these materials.

Japan has recently decided that there will be no natural claims made on food products. This action removes the need to develop new processes for creating natural flavors. However, the science of biotechnology and fermentation has been practiced for years in Japan with the manufacture of soy sauce and many other foods and flavors. Therefore the development of new ingredients via biotechnology continues at a high rate even though labeling is not the cause of the pressure to create natural ingredients for use in flavors. Instead the pressure to use biotechnology comes from the desire to create new materials or produce existing materials that are hard or impossible to create by purely synthetic routes.

The European Community (EC) represents what used to be a patchwork of differing views on the subject of flavor and, in particular, natural flavor. A great diversity of regulations existed across Europe with regard to the definition of natural flavor, but, with the advent of the new EC regulations, that diversity will yield to a single regulation. For example, the Germans had the strictest view in allowing as natural only material isolated from substrates originating from natural (traditional) food processes. Some German experts stated that fermentation involves chemical changes and therefore would not be considered natural.

The EC view will allow materials (mixtures and single substances) to be called natural if they are obtained exclusively by physical, microbiological, or enzymatic processes from materials of vegetable or animal origin, either in the raw state or after processing for human consumption by traditional food preparation processes (including drying, roasting, and fermentation). The International Organization of the Flavor Industry (IOFI), which is the worldwide organization representing the flavor industry, has defined the biochemical processes for natural flavor concentrates and substances by the organisms or enzymes, substrates, and growth factors that may be used to produce such materials [29]. Their definition includes the use of any bacterium, yeast, fungus, or animal or plant cell, in whole or in part, and enzymes derived therefrom. Enzymes and microorganisms not traditionally accepted as constituents of food would need to have their safety in use adequately established. Flavorings containing incompletely inactivated enzymes and/or microorganisms would also need to have their safety evaluated.

The substrates must be natural, but materials added to the substrate necessary for the growth and function of the organism(s) or enzymes such as co-factors, minerals, nutrients, vitamins, hormones, and acid or base for pH adjustment are not restricted in origin. However, levels of these materials may not exceed the levels required for proper function of the organism(s) or their parts. A carrier system is allowed to be aqueous or nonaqueous, but if a nonaqueous system is used it cannot be the substrate, or remain in the final mixture unless the nonaqueous compound is natural or permitted as a carrier solvent for natural flavoring.

The definition of natural is becoming more uniform around the world, and it would seem that food manufacturers will one day be faced with only a few major regulatory rules rather than the current dissonance of regulations.

In addition to natural and artificial, a third category of flavor ingredient exists in Europe, that is *nature identical*. A synthetic compound is considered nature identical if it is identical to the same compound found in nature. The term artificial is reserved for those synthetic components that are not found in nature. However, the growth of natural claims for food flavor is increasing currently in Europe. The nature-identical concept is being challenged, and with it, the simple label definition of "flavoring," used in many European countries to indicate that the flavor is compounded from nature-identical yet artificial components.

2.4.5 Future Views of Natural Products

In a recent poll of the food industry top 50 processors indicated that flavor adulteration was a very to extremely important issue (83 percent) with the issues of natural labeling (69 percent), kosher labeling (45 percent) and biotechnology (39 percent) as more secondary issues [30]. In the same poll 54 percent of the respondents in the flavor industry indicated that natural certification was an extremely important issue to the food industry. The kosher issue was believed to be extremely important by only 2% of the respondents.

Although the FDA, USDA and other world government groups may attempt to redefine natural, it is believed that the definition will continue to cover materials made from foods and other materials traditionally accepted in the food system by routes that will continue to include fermentation and other methods of biotechnology. It would appear that the consumer, the food industry and the flavor industry consider the term *natural* an important one, and one that will continue to use the scientific results of biotechnology.

2.5 NEW BIOTECHNOLOGY

Do the prior-established regulations for additives impact on the methods of regulating materials produced by biotechnology for use in FDA-regulated products? There are indications that the lawmakers are considering that the current regulations are, indeed, sufficient to insure the safety of the U.S. public. The FDA's Biotechnology Coordinating Committee has submitted to the agency's policy board a proposed definition of the term *new biotechnology* with a view to adopting terminology throughout the agency [31]. The term will be used to differentiate between the traditional practice of biotechnology (classical fermentation and enzymolysis) and the new biotechnology based on recombinant DNA technology, cell fusion technology, novel applications of cells, tissues, or their components. This would separate the traditionally accepted industrial practice from the new area which has been and will continue to be poorly understood and feared by the public.

If biotechnology is to continue to grow, the new biotechnology must be accepted and understood by the regulators and the public. For that to happen, a scientifically sound approach must be taken to understand and control the exposure of materials created by the new biotechnology. Furthermore, the public must be reassured in the popular media that this new technology is being used in a safe and useful way to advance mankind.

Dr. David A. Kessler, the Commissioner of the FDA, stated in his testimony to the Senate that he believes that ". . . there should be written guidance provided to industry on the policies and procedures FDA uses for the regulation of biotechnology products." He further stated, ". . . that these need not necessarily be regulations.'' Then-Senator Albert L. Gore (D-TN), at that same hearing, said, "The FDA has not claimed exactly how the Food, Drug and Cosmetic Act will be applied to genetically engineered foods," noting that ''. . . an industry council has even taken the step of issuing a report on regulation of genetically engineered foods, in an apparent attempt to prompt the FDA to establish a clear regulatory framework'' [32].

The FDA has proposed a policy on food biotechnology consistent with Dr. Kessler's statements [33]. The FDA cited a ''history'' of safety in the development of new plant varieties and proposed that genetically engineered foods be regulated primarily under their postmarket authority, conducting premarket safety reviews only when the agency considers it necessary. Premarket evaluation of safety and nutritional concerns is left primarily to industry, and the FDA proposes to establish guidelines as suggested by the Commissioner. The FDA plans to hold a workshop for industry, the scientific community, and the public to discuss specific scientific issues relating to the agency's proposed policy.

2.5.1 How to Regulate the New Biotechnology

The report noted by Gore, "Biotechnology and Food: Assuring the Safety of Foods Produced by Genetic Modification," was issued by the International Food Biotechnology Council (IFBC). A summary document, "Biotechnologies and Food: A Summary of Major Issues Regarding Safety Assurance," has been published in a peer-reviewed journal [34].

The specific objective of the report and the IFBC was to provide a comprehensive, scientifically based report, with extensive literature references, and a glossary of terms for safety criteria of foods and food ingredients derived from genetically modified plants and microorganisms. Drafts of the report have been reviewed by over 150 representatives of government agencies in 13 countries, industrial scientific organizations, professional societies, congressional-legislative staffs, public interest and consumer groups, and academicians.

Through a basic understanding of the methods of genetic modification, an intelligent approach may be made to the management of the safety and regulatory issues associated with food ingredients derived from the new biotechnology. Within the regulatory framework that we have discussed above, the IFBC suggests that a flexible, tiered-approach system of safety evaluation, guided by a decision tree, can be made. Table 2.6 indicates the basic levels of information and confidence that must be developed for the system to work. It would rarely be necessary to pursue all three levels exhaustively. Threshold values would need to be established below which one would need not pursue the generation of further safety-oriented information. Prior use of traditional biotechnology has been based upon this approach and has led to the development of a great variety of foods and ingredients that are consumed without problems or fear by the public. However, the extension of the traditional methods to the new biotechnology based on scientifically directed genetic modification at the gene level gives rise to a basis for considering that the process is

TABLE 2.6 Safety of Food Ingredients

Information and confidence levels needed to assure the safety of food ingredients:
1. Scientific knowledge of and confidence in the genetic background and the procedures of genetic modification.
2. Relevant information regarding the composition of the food or food ingredient, including its potential toxicants, nutrients and/or functional properties.
3. All relevant toxicological data.

dramatically different from the traditional method rather than an extension. With many consumer groups actively challenging the new biotechnology, it is important that a scientifically sound approach to safety evaluation be accepted.

2.5.2 Concerns in Understanding of Biotechnology

Issues about biotechnologically developed ingredients such as bovine somatotropin hormone (BST) are currently hot topics for consumer groups. Although BST is a product of the new biotechnology, the major source of criticism is the residual level of BST in cow's milk and its downstream exposure to humans, even though experts indicate that BST always has been in milk with no observed harm to humans [35]. We should not allow issues such as this to be misconstrued by the public as biotechnology issues. Some consumer groups will use issues such as BST convince the public that biotechnology is the culprit, when in fact it is only a method for efficient production of a protein–hormone that has been in existence for as long as cows have been in existence.

Recall that the process of irradiation of food for preservation of foods has been a subject of specific review for the last 35 years without any firm acceptance by the FDA due to lack of public confidence. Dr. Fred R. Shenk, Director of the FDAs Center for Food Safety and Applied Nutrition, in an address at the 1990 National Meeting of the Institute of Food Technologists, noted that public acceptance is the "biggest hurdle" for the use of food irradiation for preservation. He noted that there is a "positive acceptance" of food irradiation in Europe, but that the level of understanding in the United States is "low." After over 35 years of scientific evaluation, there is still a low level of understanding and acceptance of this process [36]. It may be said that if thermal processing of food (used since Nicholas Appert's development of canning of foods in 1810) was not considered traditional, it would be as poorly understood as the use of irradiation of foods.

Needless to say, there is a lesson in the irradiation story for the current deliberations on the new biotechnology.

2.6 THE DECISION TREE APPROACH TO SAFETY EVALUATION

The answer may be in the use of a decision tree to establish the degree of "newness" or potential of safety concerns regarding a new food, ingredient

TABLE 2.7 Decision Tree Approach to the Assessment of Safety of Whole Food or Complex Mixtures

Questions that should be asked about the mixture or food produced by the new biotechnology:

Is the genetic material from traditional foods?
Are the significant constituents found only in the parent or related species?
Do the constituents occur in their usual range?
Are new constituents produced?
Do these new constituents present a risk?
Are intake and eating patterns altered?
Are the significant nutrients at their expected levels?
Is there knowledge regarding the genetic material?

With answers to these questions the safety and risk associated with the material or mixture produced by the new biotechnology can be judged.

mixture, or pure substance produced by biotechnology. Table 2.7 reviews the IFBCs proposed decision tree [37]. The approach is designed to answer relevant safety issues including five major questions:

1. Does the microorganism used to create the material end up in the final system (food or ingredient) as consumed?
2. Is the microorganism free of transmissible antibiotic resistance markers?
3. Are the vectors that are used characterized and free of attributes that would render them unsafe for use in food?
4. Does the DNA insert code for a substance that is safe for use in food?
5. Is the microorganism free of an intermediate host DNA which could code for a toxic product?

As we mentioned before, the review of single components or simple, well-defined mixtures represents an easy review with no unique safety consideration needed. The safety evaluation of complex materials and foods from the "new" biotechnology will need close comparison of the material with traditional counterparts [38], other desired expression products and toxic constituents [39,40]. This will be a complicated process, but one within the potential of contemporary science.

2.7 CONCLUSIONS

As is seen in the chapters to follow, the use of biotechnology in the flavor, fragrance, and color industry is no longer just an academic pursuit. From the production of pure chemicals and complex flavor systems, such as dairy flavors, the industry is moving forward, both in production and research, to continue to create new processes and/or components useful to the industry. The procedures maybe as ancient as mankind or as "high tech" as the most recent research paper on genetic engineering. For the industry to continue to produce more unique, economical, stable, cheap, uniform, and plentiful materials from this technology, safety guidelines are necessary and their soundness should be reviewed in view of further scientific developments. The industry, in concert with the biotechnology industry and other users of biotechnology, must continue to educate the public regarding the nature of biotechnology and to avoid the problems of developing areas of major misunderstanding. It will also be necessary to continue to work with government on all levels so that the regulatory system will react in a thoughtful manner regarding real issues related to the products of biotechnology.

This is a major effort that all professionals involved in the use of biotechnology must continue. As ancient as our pursuit of aroma and fermentation is, we must look to all professions in the industry to continue to place it on a safe and scientific basis, and yet allow mankind to share its romance.

REFERENCES

1. Guenther, E., *The Essential Oils*, Vols. I–VI. Krieger Publ. Co., Huntington, NY, 1948.
2. Code of Federal Regulations, 21, Section 101.22.
3. Pearson, D., *The Chemical Analysis of Foods*, 7th ed. Churchill-Livingstone, London, 1976.
4. Stofberg, J., and Grundschober, F., *Perfum. Flavor.* **12,** 27 (1987).
5. Personal discussions with professionals in the industry.
6. Lanza, E., and Plamer, J.K., *Phytochemistry,* **16,** 1555 (1977).
7. Janssens, L., DePoorter, H.L., Demey, L., Vandammme, E.J., and Schamp, N.M., *Meded. Fac. Landbouwwet. Rijks univ. Gent* **53,** 2071 (1988).
8. Latrasse, A., Degorce-Dumas, J.R., and Leveau, J.Y., *Sci. Aliments* **5,** 1 (1985).
9. Welsh, F., Murray, W.D., and Williams, R.E., *Crit. Rev. Biotechnol.* **9,** 105 (1989).
10. Leete, E., Bjorklund, T.A., Reinneccius, G.A., and Cheng, T.B., *Spec Publ. R. Soc. Chem.* **95** (BIOFORM FLAVOURS), 75–95 (1992).
11. Schreier, P., and Winterhalter, P., in *Biogeneration of Aromas*, (T. Parliment and R. Croteau, eds.) Am. Chem. Soc. Washington, DC, 1986, 85–98.
12. Manley, C.H., and Kanisawa, T., The development of dairy flavors by biotechnology. Presented at the Annual Meeting of the American Institute of Chemical Engineers, Chicago, 1990.
13. Whitake, R.J., Hobhib, G.C., and Steward, L.A., in *Biogeneration of Aromas* (T. Parliment and R. Croteau, eds.). Am. Chem. Soc., Washington, DC, 1986, 347–362.
14. Ito, H., Japanese Patent 44,830 (1973).
15. Nakagawa, N., Japanese Patent 91,937 (1976).
16. 21 United States Congress (U.S.C.) Section 321 et seq.
17. Federal Register, *Fed. Regist.* **56,** 23,333 (1986).

18. 21 United States Congress (U.S.C.) Section 342 (a) (1).

19. 21 United States Congress (U.S.C.) Section 321 (s).

20. Code of Federal Regulations, 21, Section 172, subpart F.

21. Code of Federal Regulations, 21, Section 170, 30 (a) to (c).

22. U.S. Food and Drug Administration, *Toxicological Principles for the Safety Assessment of Direct Food Additives and Color Additives used in Foods.* FDA, Washington, DC, 1982.

23. *Flavor and Fragrance Materials—1989.* Allured Publ. Co., Wheaton, IL. 1989.

24. *List of Fragrance Material.* Research Institute for Fragrance Materials, Englewood Cliffs, NJ, 1989.

25. Code of Federal Regulations, 21, Section 101.22 (a) (3).

26. U.S. Department of Agriculture, Food Safety and Inspection Service, *FSIS Notice, Labelling of Flavors and Control of Proprietary Flavoring Mixtures*, 6–90. USDA, Washington, DC, 1990.

27. Eidus, I., *Adweek,* **30,** 1 (1989).

28. Regenstein, J.M., and Regenstein, C.E., *Food Technol.,* **33,** 93–98 (1979).

29. *Code of Practice.* International Organization of Flavor Industries (IOFI), Geneva, Switzerland, 1990.

30. Best, D., *Prepared Foods*, pp. 76–78 (1990). (Based on a survey of the food and beverage processors and Flavor and Extract Manufacturers Association members.)

31. *Food Chemical News*, **32**, 19 (1990).

32. *Food Chemical News*, **32,** 49 (1990).

33. *Federal Register*, **57**, 22,984 (1992).

34. Lindemann, J., *Regul. Toxicol. Pharmacol.* **12,** 96–104 (1990).

35. *Food Chemical News*, **32,** 26 (1990).

36. *Food Chemical News*, **32,** 10, (1990).

37. *Biotechnology and Food*: *Assuring the Safety of Foods Produced by Genetic Modification*. International Food Biotechnology Council, Washington, DC, 1989.

38. Sovci, S. W., Fachmann, W., and Kraut, H., *Food Composition and Nutrition Tables*. Wiss. Verlagsges., Stuttgart, 1981.

39. Cheeke, P.R., *Toxicants of Plant Origin*, Vols. I–IV. CRC Press, Boca Raton, FL 1989.

40. Liener, I.E., *Toxic Constituents of Plant Foodstuffs*, Academic Press, New York, 1980.

Aliphatic, Aromatic, and Lactone Compounds

DAVID W. ARMSTRONG and LESLIE A. BROWN

Institute for Biological Sciences, National Research Council, Ottawa, Canada

Aliphatic, aromatic, and lactone compounds play a pivotal role in flavor formulations on their own or as precursors of other important compounds. This chapter highlights the commercially important members of these groups. For the biological production of any of these substances, it is important to remember that success depends on more than just technical breakthrough. Other factors, including market influences, regulatory considerations, economics, and the growing need for environmentally friendly processing routes, also are integral parts of biological flavors production. A number of technical bottlenecks exist for biological processing, including end-product inhibition, substrate supply problems, and product recovery. Possible solutions to these and other impediments to commercialization are highlighted with examples from the chemical groups discussed in this chapter.

3.1 INTRODUCTION

It has long been known that microorganisms can produce odors reminiscent of the perfume of flowers or of the taste of fruit. Initially this was merely a novelty of no practical use except perhaps for classification. Omelianski [1] was one of the first microbiologists to study how the presence of microorganisms affects the taste and aroma of foods and how certain aromas are associated with specific organisms.

Aliphatic, aromatic, and lactone flavor substances play an important role in the overall aroma presentation of many of our foods and beverages. This chap-

Bioprocess Production of Flavor, Fragrance, and Color Ingredients, Edited by Alan Gabelman, ISBN 0-471-03821-0 © 1994 John Wiley & Sons, Inc.

ter examines high potential biological routes for production of such compounds and suggests means of optimizing their production. For successful commercialization, it is important to understand something of the forces that have had and will have a growing influence on the biotechnological production of these compounds. These forces not only include technical aspects but also markets, regulatory considerations (see Chapter 2), economics, and more recently, heightened environmental responsibility for chemical processing.

Food-related industries for the most part operate large-scale, commodity transformation operations characterized by low profit margin [2]. In other words, inexpensive raw materials are converted by inexpensive processing into inexpensive products. Because of the low profit margin, there are limits to the acceptable developmental costs for food-related biotechnologies, so that research programs are usually small. However, somewhat more sophisticated approaches are likely to be acceptable for production of value-added materials used in smaller volumes, such as flavors and aromas [2].

In the last ten years, the demand for natural flavor ingredients has risen from 5 to 10 percent of food company requests to 75 to 80 percent [3]. This trend can be attributed to increasing health- and nutrition-conscious lifestyles, which have encouraged the development of natural food products [4]. Another factor is the growing consumer interest in companies and their products that meet certain environmentally friendly standards, that is, those products that are obtained through processes perceived as not being harmful to the environment.

Food processors have responded to the trend toward naturalness by an increased use of natural food flavorants found in plants [5]. However, plants cannot continue to be the major source of these materials because numerous problems are becoming increasingly evident. Often the desired flavor component is in low concentration in plants, leading to expensive extraction and recovery processes. Seasonal variation affecting yield and quantities of the flavor component is another factor that even the primary producer has little control over. The supply is also dependent upon the sociopolitical stability of the producing regions (primarily developing nations).

For the derivation of most commercially significant flavor esters, aldehydes, and ketones, the major routes have typically been by chemical synthesis. Most of the compounds are structurally simple (e.g., ethyl acetate, acetaldehyde) and very amenable to simple chemical synthesis. There are, however, several drawbacks to chemical synthesis. Biological routes to many flavor substances are simpler than synthetic routes because the latter can involve multistep processes with associated increments in cost per step. For example, the lactone 6-pentyl-α-pyrone, which imparts a coconutlike flavor impression, can be produced *de novo* by the fungus *Trichoderma viride*, whereas chemical synthesis requires seven steps [6]. Many fermentation-derived flavors produced from simple inexpensive sugars can rival synthetic routes, especially with fluctuating fossil fuel prices.

Another drawback is that chemical synthetic processes result in racemic mixtures where the need is for specific, optically active stereoisomers (see

Section 3.4.1). The problem with a racemic mixture is that the final functional property of the flavor or fragrance usually is not suitable. Isolation of the appropriate stereoisomer can be very expensive or even technically impossible [7]. On the other hand, a number of compounds (e.g., γ- and δ-lactones) can be produced in an optically active and pure form by relatively simple microbiological reduction [8].

The biggest obstacle to chemical synthesis is that although compounds made this way are identical structurally to the same compounds found in nature, the chemically produced ones do not qualify as being natural by legal definition. A natural flavor, according to the U.S. Food and Drug Administration (FDA) guidelines set down in 1958, must be produced from natural starting materials and the end-product must be identical to something already known to exist in nature. Furthermore, products of biological but not chemical transformation of natural substances can be legally labeled as natural. Therefore, to achieve natural status, even structurally simple flavor substances such as ethyl acetate must be produced by a means complying with the above legal definition. Clearly the biological production of flavors is driven not only by market demands but also by regulatory influences.

Even though legal definitions of natural have been established, a number of industries have created their own guidelines. For example, McCormick & Strange made the decision a few years ago to change their label on almond extract (primarily benzaldehyde) from natural to imitation. This step was taken because the extract was derived from apricot pits and not from almonds. Even though the source of the ingredient was legally natural, corporate policy superseded regulatory guidelines where it appeared that some confusion could have arisen in the marketplace [9].

An interesting situation exemplifying the role of regulations on development of biological processing routes is seen for benzaldehyde. A private-sector claim was received by the FDA that benzaldehyde marketed as natural by a number of firms was in fact not natural as outlined in the FDA regulations cited above. The concern was related to the fact that the natural benzaldehyde was being derived from cinnamon bark or leaf. The major constituent, cinnamic aldehyde, was split into benzaldehyde and a coproduct (presumably acetaldehyde), often using a mineral acid or base catalyst. It is the use of the acid/base catalyst that caused the concern over the natural status of the final benzaldehyde. The self-regulating body for flavor manufacturers, Flavor Extract Manufacturers Association (FEMA), brought in their own interpretation and recommended that a moratorium be placed on this type of processing route. Flavor companies subsequently put more effort into developing other routes including biological ones. This situation has resulted in natural bitter almond oil selling for upward of \$400/kg versus \$100/kg for the disputed technological route described above. Either route greatly overshadows the price of \$2/kg for totally synthetic benzaldehyde derived from oxidation of toluene. It should be recognized that a number of flavor esters are also produced from the corresponding natural alcohols and fatty acids using mineral acid catalysts. These products are also in question as a result of the benzaldehyde controversy, which is driving the development of other processing routes including biological ones.

The demand for natural fragrance derivatives is not as obvious as it is for flavors. In many cases, synthetic compounds are acceptable; in fact, in some cases, they are more desirable than natural substances. For example, bergamot oil (obtained from *Citrus bergamia*, with bergamol or linalyl acetate as the most valuable constituent) is used in perfumery for imparting a floral fragrance. Recently it was selling for up to $150/kg and was used in a wide range of perfumes, including Chanel no. 5. Some perfumers have now switched from the natural oil to the synthetic bergamol because of the presence of bergapten in the former, which causes skin irritation when exposed to sunlight [10]. The use of the synthetic bergamol also resulted in a significant cost savings because the price is only $10/kg. No loss of consumer appeal occurred because natural perfumes are preferred only if they give a unique sensory impression. This example highlights the marked difference between the flavor and fragrance markets and how the challenge of providing natural substances has not been as important for fragrance products.

One interesting and profitable segment of the market for natural foods is the kosher market. Kosher foods find customers not only among Orthodox Jews but also among the general populace, because the kosher preparatory procedures often ensure a higher quality of food product. Kosher requirements can have an influence on the manufacture of certain specialty chemicals. For example, alcohol is used in the production of a variety of specialty chemicals such as acetaldehyde, ethyl acetate, and acetic acid (vinegar). According to kosher law, the source of the alcohol is very important. Alcohol from grapes requires strict rabbinical supervision at every step of its preparation. The shipping container, the equipment used to make the product, and the media ingredients used in the fermentation must all be kosher [11]. There are further restrictions during Passover in that alcohol derived from grains cannot be used—only those alcohols derived from sugar, molasses, or petrochemical substrates are permissible [12]. Ingredients from nonanimal sources are not a problem, but calcium sources cannot include nonkosher bones or shells [11].

To summarize, even though compounds such as certain alcohols, esters, and aromatics may be relatively easy to produce synthetically, there is a combined market and regulatory "push" for biological routes. The adoption of such processes also could satisfy simultaneously growing requirements for environmentally friendly "green" products and processing along with large kosher markets.

3.2 REQUIREMENT FOR PHYSIOLOGICAL AND GENETIC STUDIES

To manipulate effectively microorganisms to produce aliphatic, aromatic, and lactone flavor substances, information about microbial primary and secondary metabolism as well as genetic characteristics must be obtained. Unfortunately, very little can be gained from the literature, because most of the information available deals with the mechanisms of antibiotic production.

Generally, it is assumed that many volatile metabolites useful in flavors and fragrances are the result of secondary metabolism and are found at their maximum levels after peak cell growth. There are, however, numerous examples of the contrary. For example, Jourdain et al. [13] found that *Sporobolomyces odorus* produced lactones while cells were still growing (primary metabolism) as well as after growth had stopped. In the latter case, nonproliferating cells produced secondary metabolites under limiting nutrient conditions.

Temperature and pH effects must be considered during the environmental manipulation process used to determine whether the organism selected can indeed produce the compound of interest. Temperature and pH are usually set at values that will give optimum growth, although it may be discovered later that the chosen conditions are not optimal for the enzyme involved in the bioconversion. For example, *Mycoacia uda* was found to produce an oil possessing a sweet almondlike odor attributable to *p*-tolualdehyde [14]. Although the optimum temperature for growth and oil yield was 30°C, the oil produced at this temperature did not possess the almondlike aroma. Subsequently, it was determined that a temperature of 25°C was the optimum for aroma production.

Carbon and nitrogen sources also influence production of aroma substances. Lanza et al. [15] did an extensive study on the effects of different carbon and nitrogen sources on aroma production by *Ceratocystis moniliformis*. They found that aroma intensity was, for the most part, correlated with growth. They also found four carbon/nitrogen source combinations that yielded strong, fruity aromas due to the production of short-chain esters such as isoamyl acetate, which gives a strong banana odor. Hanssen et al. [16] compared the production of short-chain esters, alcohols and phenylethyl derivatives by different strains of *Kluyveromyces lactis* using different nitrogen sources (Table 3.1). All three

TABLE 3.1. Maximum Cell Mass and Accumulation of Volatile Metabolites in Cultures of Three *Kluyveromyces lactis* Strains Grown on Glucose (5%), Yeast Extract (0.25%), and Various Nitrogen Sources.

	Nitrogen Source[a]					
Strain	Phenylalanine (0.05%)	Tyrosine (0.07%)	Asparagine (0.1%)	Leucine (0.1%)	Peptone (0.5%)	Calcium Nitrate (0.05%)
CBS 5670	31.07	20.82	21.47	46.24	6.75	13.90
CBS 4372	93.32	73.25	46.66	157.58	9.80	25.17
CBS 2359	28.90	9.31	31.62	102.82	24.66	15.02

[a]Short-chain alcohols and esters (mg/500 mL culture broth)

Source Adapted by permission of Verlag der Zeitschrift der Naturforschung from Hanssen et al.

strains examined showed the highest production of short-chain alcohols and esters when grown in media supplemented with leucine. One of the short-chain esters produced was 2-phenylethyl acetate, which has a rose scent. In a paper by Berger et al. [17] it was found that *Ischnoderma benzoinum*, when supplied with tyrosine, tended to form anise aldehyde. *p*-Anise aldehyde is frequently used in sweet blossom compositions (e.g., lilac type) as well as in flavor compositions for confectioneries and beverages. Sufficient studies have been completed such that amino acids can be separated into three groups with respect to the type of volatile compounds formed from them by certain microorganisms (Table 3.2) [18–21].

The effects of oxygen tension on the production of cellular metabolites are well known. As an illustration, aeration had a significant effect on ethanol utilization and ethyl acetate accumulation by *Candida utilis* [22]. Poor aeration (as indicated in Figure 3.1 by a low headspace-to-culture volume ratio of 0.46) resulted in slow ethanol utilization with very little accumulation of the ester. Greater ethanol utilization and ester production was obtained when the aeration

TABLE 3.2 Amino Acid Precursors of Volatile Compounds

Amino Acid	Compounds Produced
Group 1	
L-Glycine	Esters containing ethyl alcohol and
L-Alanine	*n*-propionic acids
L-Glutamic acid	
L-Glutamine	
L-Aspartic acid	
L-Asparagine	
L-Arginine	
L-γ-Aminobutyric acid	
L-Proline	
Group 2	
L-Valine	Esters containing isobutyl alcohol and isobutyric acid
L-Isoleucine	Esters containing isoamyl alcohol, isobutyric acid, and 2-methylbutyric acid
L-Leucine	Isoamyl alcohol
L-Serine	Esters of isobutyl alcohol
L-Threonine	Esters of isobutyl alcohol, isoamyl alcohol, isobutyric acid, and isovaleric acid
Group 3	
L-Phenylalanine	Veratryl alcohol, cinnamic acid, benzoic acid, and phenethyl alcohol

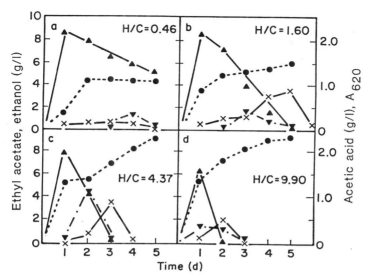

FIG. 3.1 Effect of aeration in flasks on accumulation of ethanol, acetic acid, ethyl acetate, and cell mass by *Candida utilis*. H/C represents headspace/culture volume ratio, so that higher H/C indicates greater aeration. The symbols refer to (▲) ethanol, (▼) acetic acid, (X) ethyl acetate, and (•) A_{620}. *Source* Reprinted by permission of John Wiley & Sons, Inc., © Copyright 1984, from Armstrong, D. W., Martin, S. M., Yamazaki, H., *Biotechnol. Bioeng.* **26,** 1038 (1984).

was increased (H/C = 1.6 and 4.37). However, at high levels of aeration (H/C = 9.90) the ester yield was decreased. The intermediate levels of aeration were thought to allow for higher intracellular metabolite accumulation (e.g., acetyl-coenzyme A, -CoA) concomitant with relatively efficient use of ethanol as a precursor. Fine tuning aeration rates to optimize flavor fermentations is a very important task for the biotechnologist. Many other aerobic metabolic flavor routes rely on the correct balance of oxygen to meet specific pathway demands while not allowing for full metabolic function. The latter condition imposed could allow for flavor compound accumulation and limited further metabolization and loss.

Phosphates and trace elements such as iron, zinc, and manganese are involved in many areas of metabolism and consequently can play a major role in aroma production. Not only their presence, but also their absence, can be important to the production of flavors. For example, in the case of the production of ethyl acetate by *C. utilis*, the presence of ferric chloride led to inhibition even at levels as low as 10 μM [23]. Under iron-limited conditions the yeast formed acetyl-CoA from ethanol, with a subsequent reaction occurring with ethanol to form ethyl acetate. However, when iron was present, the acetyl-CoA was oxidized through the tricarboxylic acid (TCA) cycle for growth and did not accumulate, leading to diminished ethyl acetate production. (See Section 3.3.2 for a further discussion of underlying mechanisms.)

Strain specificity can be very important to odor production, and there can be a great deal of variation between strains of a single organism. *Phlebia radiata*, for example, has at least 25 strains but only two are known to generate aroma compounds [24]. Strain A produced γ-decalactone (peach) and 2-aminobenzaldehyde (pungent, malodorous), whereas strain B did not produce these compounds at all. Furthermore, both strains produced the ketone 3-hydroxy-2-butanone (creamy, fatty) but strain A did so at levels four times greater than those of strain B. Consequently, when screening such organisms, it is important to examine as many strains as possible and not to eliminate a particular organism as unproductive when only limited strain evaluation has been conducted.

One way of enhancing production of desirable compounds is genetic modification. This can involve classic mutational approaches or the use of recombinant deoxyribonucleic acid (rDNA) genetic engineering. Benefits can include not only yield improvement but also resistance to toxic by-products and improved genetic stability. For genetic engineering to be productive, however, many hurdles must be overcome and a thorough understanding of the relevant metabolic pathways is necessary. The situation can be complex because metabolic pathways for the production of a particular flavor substance can involve multiple genes. Additionally, enzyme and/or product synthesis is under tight metabolic control in many species, including carbon and nitrogen catabolite repression. The development of a regulatory mutant could circumvent these latter problems [25]. Greater insight into the use of genetic manipulations should broaden the possibilities for gene transfers between different species and genera of microorganisms.

In the following sections, the biological production of aliphatic, aromatic, and lactone flavors is discussed. Most of the biotechnologies for producing these compounds are whole cell fermentations. As such, these classes of flavors are discussed under whole cell bioconversions. This section is followed by a discussion of enzyme-based technologies (Section 3.4). In the latter section, key representative biotransformations are used to illustrate the potentials and pitfalls of this technological approach.

3.3 WHOLE CELL BIOCONVERSIONS AND BIOTRANSFORMATIONS

Generally, isolated enzymes are well suited for single- or limited-step processes, whereas whole cell technology can be used for very complex metabolic syntheses. The ability of whole cells to convert simple sugars *de novo* to relatively complex flavor molecules is of particular interest to the natural flavor industry. A number of examples related to whole cell production of aliphatic, aromatic, and lactone flavor substances are presented in this section.

3.3.1 Flavor Aldehyde Production by Whole Cells

Considerable interest has been shown in the whole cell production of acetaldehyde, which is used to impart "freshness" to food and beverage products

[26] and also in formulations for nutty and roasted flavor notes. It has been shown that the food-grade yeast *C. utilis* (one of three yeasts allowed for use in food for human consumption) is able to oxidize ethanol to acetaldehyde in appreciable yields (ca. 70 percent of theoretical) [27]. The highest yield occurs when the medium is iron-limited and the level of feed alcohol is approximately 65 g/L. The imposition of iron-limiting conditions results in the TCA cycle being severely rate limited (Fig. 3.2). That is, certain iron-requiring enzymes in the TCA cycle, including succinate dehydrogenase and aconitase, become activity limited. Accumulation of acetyl-CoA results, which in turn leads to increased ethyl acetate accumulation via an alcohol transferase mechanism when the feed ethanol concentration is 35 g/L or less. Through the same metabolic scheme, when the level of ethanol is elevated above 35 g/L, acetaldehyde begins to accumulate. That is, the rate of oxidation of ethanol to acetaldehyde exceeds that of the oxidation of acetaldehyde to acetic acid, so the aldehyde accumulation is favored. This intracellular accumulation of acetaldehyde inhibits acetyl-CoA synthetase, which results in a reduction in ethyl acetate formation and a change in product distribution to acetaldehyde. Postprocessing, the spent yeast can also be used in meaty and savory formulations, adding significantly to the economics of the total process.

A strain of *Zymomonas mobilis* having altered alcohol dehydrogenase activity has also been used to produce acetaldehyde from glucose via pyruvate [28]. The possibility of producing acetaldehyde directly from fermentable sugars is very attractive because it would bypass the requirement for a separate ethanol fermentation and an alcohol separation/recovery step. The strain was screened by a relatively simple approach using allyl alcohol (2-propenol) (Fig. 3.3). Normally alcohol dehydrogenase will convert allyl alcohol to acrylaldehyde

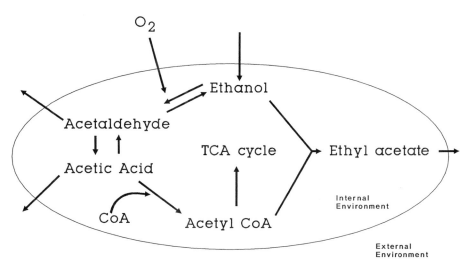

FIG. 3.2 Proposed pathway of of ethanol metabolism in *Candida utilis. Source* Reprinted by permission of Elsevier Science Publishers from Murray et al. [27].

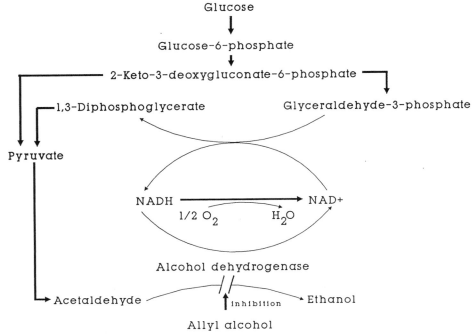

Glucose

↓

Glucose-6-phosphate

↓

2-Keto-3-deoxygluconate-6-phosphate

1,3-Diphosphoglycerate   Glyceraldehyde-3-phosphate

Pyruvate

NADH ⟶ NAD+

1/2 O_2 H_2O

Alcohol dehydrogenase

Acetaldehyde   Ethanol

Inhibition

Allyl alcohol

FIG. 3.3   The Entner-Duodoroff pathway for glucose metabolism in *Zymomonas mobilis*. *Source*   From Wekker and Zall [28].

(acrolein), a substance highly toxic to cells. However, if the organism has a reduced level of alcohol dehydrogenase activity, it will survive treatment with allyl alcohol. Similarly, such an organism can accumulate acetaldehyde, because no conversion to ethanol can occur. The particular strain selected used 21 g/L glucose to produce approximately 4 g/L acetaldehyde (about 40 percent theoretical yield) in 24 hours.

Another whole cell approach to acetaldehyde production is the use of *Pichia pastoris*, a methylotropic yeast, to oxidize ethanol to acetaldehyde (Fig. 3.4) [29]. When this yeast or other methylotropic organisms are grown on methanol as the sole carbon and energy source, peroxisomes form, which contain alcohol oxidase, catalase, and the cofactor flavin adenine dinucleotide (FAD) [30]. The first enzyme in the pathway, alcohol oxidase, catalyzes the oxidation of methanol to formaldehyde, which is then further metabolized using other enzymes. However, alcohol oxidase has a broad specificity and can oxidize several low-molecular-weight alcohols, including ethanol, to the corresponding aldehydes. When ethanol is added to a culture of *P. pastoris* grown on methanol, acetaldehyde accumulates, because the second enzyme in the pathway, formaldehyde dehydrogenase, is specific for formaldehyde.

Various manipulations, including the use of higher cell densities, hyperbaric

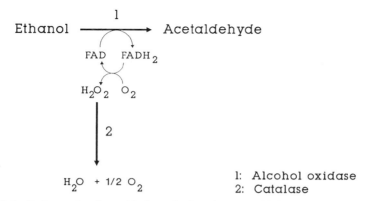

FIG. 3.4 Pathway for the oxidation of ethanol to acetaldehyde by *Pichia pastoris*.

oxygen, fed-batch feeding of ethanol, and continuous *in situ* trapping of the acetaldehyde result in significant improvement over a simple batch approach. The addition of TRIS buffer to the production medium allows for formation of an addition product between the amine and aldehyde groups. Free aldehyde is then obtained by an alkaline shift of the TRIS medium. However, with recent changes in regulatory guidelines, the use of an inorganic chemical pH shift to release the addition product may not be considered a natural process, so that other *in situ* means to continuously remove or sequester the acetaldehyde may have to be developed. Furthermore, despite promising initial specific productivities of up to 1.38 g of acetaldehyde/g of cells per hour, there is the problem of short enzyme half-life. The k_i for the inhibition of the whole cell process has been determined to be 2.6 g of acetaldehyde/L. Increasing the TRIS buffer level to sequester more product does not lead to an increase in acetaldehyde levels after a certain point [27].

Another problem with the alcohol oxidase approach is that serine proteases can be activated by the use of other substrates besides methanol [31,32]. These proteases can degrade the peroxisomes containing the alcohol oxidase and catalase. However, selective inhibition of the protease activity would potentially bring marked improvements to the stability of the alcohol oxidase process. Production of other flavor-active aldehydes using this method is discussed later in this section.

More examples of the use of nonconventional flavors bioprocessing are beginning to become evident. A recent development for the production of flavor aldehydes and possibly other volatile compounds is the use of a whole cell gas-phase reaction in supercritical fluids [33]. Supercritical fluids exist in a nebulous state between a liquid and gas and are formed by bringing a gas such as carbon dioxide to its critical temperature (31°C) and pressure (72 bar). Small changes in temperature and/or pressure result in large changes in functional properties of the supercritical fluid (solvency, solubility, etc.). In this type of system, ethanol was converted via a yeast alcohol oxidase system to acetaldehyde, although no report of process details or yields was given [33]. It is anticipated that hydrophobic precursors could be contacted with the cell/enzyme system more efficiently using this approach, because supercritical fluids are particularly good at increasing mass transfer of apolar species.

There is also the possibility of conducting *de novo* reactions involving many biosynthetic steps under other nonconventional processing conditions. Schneider et al. have demonstrated the production of ethanol from solid glucose using *Saccharomyces cerevisiae* in a controlled humidity gas-phase reaction system [34]. In this system dry preparations of the yeast were mixed with solid glucose under conditions of low relative humidity. This mixture was placed in a vial with small holes to allow for vapor permeation of ethanol into an external container. The container on the outside contained a saturated salt solution to control relative humidity. The ethanol produced in the reaction system in the inner vial was trapped in the aqueous phase of the external saturated salt solution. Analysis of the ethanol produced showed conversion efficiencies greater than 80 percent of theoretical yield. This work underlines the possibility of using unique environments for complex multistep bioconversions, including cofactor regeneration, and conversion of adenosine diphosphate (ADP) to adenosine triphosphate (ATP). The limited availability of water in this particular study ensured a greater concentration of product (i.e., no dilution). Efforts are ongoing to establish the mechanisms involved and possible applications: considerable potential for the production of aldehydes and other flavor components exists.

Apart from acetaldehyde, there are a number of other natural aldehydes that have significant commercial potential. One example is vanillin (vanillic aldehyde or 4-hydroxy-3-methoxy benzaldehyde), which imparts a vanilla-type flavor to many food products. Most of the vanillin available today is derived synthetically from eugenol, guaiacol, or lignin (a waste by-product from the wood pulp industry). Natural vanillin is derived from vanilla (vanilla beans), and sells for over \$600/kg. Vanillin can also be obtained from plant cell culture, but there are a number of technical processing problems (expensive complex media, cell clumping, etc.) with this approach, and yields are low [35]. Despite these problems, however, a number of companies worldwide are pursuing this route for vanillin production. For example, ESCAgenetics Corporation (San Carlos, CA) has developed a plant cell culture process that achieves a selling price for vanillin of approximately \$12/kg (compared to synthetic vanillin at \$3/kg). This cost is based on data obtained from 72-L bioreactors; development of larger scale operations for commercialization of natural vanillin is still underway. (An in-depth discussion of plant tissue culture is presented in Chapter 8.)

With the growing demand for natural vanillin, other biotechnological routes are also being examined. For example, it is known that many microorganisms can synthesize and/or metabolize aromatic compounds. A biological process to produce vanillin either *de novo* or via a biotransformation of an inexpensive precursor (e.g., wood-derived lignin) is likely to be developed in future. The myriad of microbially based transformations of aromatic compounds should allow for this possibility.

Another aldehyde in great demand is benzaldehyde, which is used as an oil-of-bitter-almond substitute for almond or cherry-type flavorings. Before the strong consumer demand for natural products developed, the bulk of the cherry-type flavors used in foods and beverages was synthetic (derived from the oxidation of toluene). Attention was drawn to benzaldehyde when a major source of the natural aldehyde from France was discovered to be derived from toluene. This incident, along with stronger consumer demand for natural products, encouraged flavor houses to seek alternate process routes to natural benzaldehyde.

Traditionally, the industry has produced natural benzaldehyde from cherries, peaches, apricots, and other related fruit kernels, where the benzaldehyde is bound through a glycosidic linkage in amygdalin. The breakage of this bond yields benzaldehyde and highly toxic hydrogen cyanide. On a commercial scale it is not trivial to deal with the latter by-product. More recently flavor houses have used cinnamon oil (derived from cinnamon bark or leaf), which contains upward of 72 percent cinnamic aldehyde, a precursor of benzaldehyde. A patent has been issued for conversion of cinnamic aldehyde to benzaldehyde using a hydrolytic reaction catalyzed by high pressure (600 to 800 psi) and/or base and/or elevated temperature [36]. Until the mid-to-late 1980s neither of these processes was practiced to any great extent, which underlines the recent intense industrial drive for natural benzaldehyde technologies. Another approach, the splitting of cinnamic aldehyde into benzaldehyde and acetaldehyde by acid–base catalysis, came under scrutiny by the FDA. As it stands now, both FEMA and FDA seem to agree that there is some reason for concern over the naturalness of these types of processes.

Clearly the ideal route to benzaldehyde would be the *de novo* production from simple, readily available sugars (e.g., glucose, sucrose, etc.) or at least the biotransformation of a suitable natural precursor to the aldehyde. Considering the number of aromatic metabolic pathways in microorganisms, there is a distinct possibility that the metabolic capacity exists to carry out one of these production schemes. However, although it is known that various microbial systems can produce benzaldehyde, little advancement on the control and optimization of the basic underlying pathways has occurred. The *de novo* production and transformation processes known for benzaldehyde, along with co-produced flavor compounds, are discussed below.

It is known that the bacterium *Acinetobacter calcoaceticus* oxidizes benzyl alcohol to benzaldehyde [37]. Although some bacteria such as *Acinetobacter* can transform aromatic alcohols, the occurrence of aromatic aldehydes is more common among higher fungi [38–40]. Birkinshaw [38] was the first to recognize anise aldehyde (4-methoxy benzaldehyde) as the principle odor constituent of the surface cultured Basidiomycete *Ischnoderma*. However, this discovery was not pursued because the research was focused on antibiotics production. A more recent study [17] demonstrated the potential application of *Ischnoderma benzoinum* (*Polyporus resinosus*). Identification of the odorous products of this culture revealed the presence of a wide range of metabolites, including aliphatics, lactones, terpenes, and aromatics. Of the aromatics present, anise aldehyde and benzaldehyde were produced in significant quantities on a defined medium (Fig. 3.5). Anise aldehyde was produced at a maximum yield of about 350 mg/L, whereas benzaldehyde was present at levels of around 130 mg/L. Unfortunately, this Basidiomycete is known to be a very slow growing organism, so that the time required for production was quite extended (>20 days).

Other flavor aldehydes, including benzaldehyde, can be produced from whole cell conversions. For example, the *Pichia* system, discussed above for acetaldehyde, can produce a series of other aldehydes from their respective alcohols (Table 3.3) [41]. With these types of transformations, one system could produce a series of useful products depending upon market requirements. Unfortunately, when oxidation of benzyl alcohol was attempted, it was found that both benzyl alcohol and the oxidized product benzaldehyde inhibited the reaction such that only 5 g/L maximum benzaldehyde accumulated [42]. This type of end-product inhibition is common for aldehydes because of their reac-

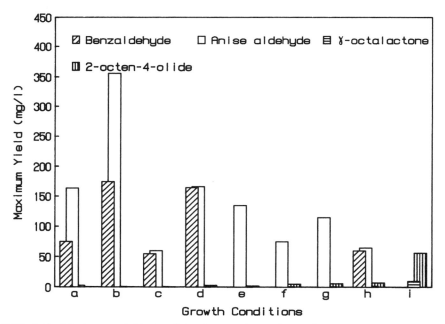

FIG. 3.5 Maximum yields (mg/L) of aromatics under various growth conditions of *Ischnoderma benzoinum*: (a) synthetic medium, surface culture; (b) synthetic medium, submerged culture; (c) as in (b) but phenylalanine substituted for asparagine (d) as in (b), plus 10 mmole/L phenylalanine; (e) as in (b), plus 10 mmole/L tyrosine; (f) as in (b), plus 1% (v/v) triglyceride; (g) yeast-malt medium, surface culture; (h) yeast-malt medium, submerged culture; (i) as in (h) plus 1% (v/v) triglyceride. *Source* Reprinted by permission of John Wiley & Sons, Inc., Copyright 1987, from Bereer, R. G., Neuhaeuser, K., Drawert, F., *Biotechnol. Bioeng.* **30**, 987 (1987).

TABLE 3.3   Bioconversion of Alcohols to Flavor Aldehydes Using Alcohol Oxidase

Substrate	Product	Initial Substrate Concentration (g/L)	Final Product Concentration (g/L)[a]
Ethanol	Acetaldehyde	50	28.4
Propanol	Propionaldehyde	50	25.4
Butanol	Butyraldehyde	50	25.1
Isobutanol	Isobutyraldehyde	50	16.5
Pentanol[b]	Valeraldehyde	20	18.7
Isoamyl alcohol[b]	Isovaleraldehyde	20	6.2
2-Methyl butanol[b]	2-Methyl butyraldehyde	20	6.7
Hexanol[b]	Hexanal	5	5.0

[a]Twelve-hour batch conversion, 30°C, 0.5 M TRIS HCl, pH 8.0, methanol-grown *P. pastoris* cells, 5 g/L.
[b]Substrate was at or near its saturation level.

Source   Reprinted by permission of Annals of the New York Academy of Sciences from Williams et al.

tivity with proteins (e.g., enzymes). Another major limitation of this type of bioconversion is the restricted availability of natural alcohol precursors other than ethanol (see Section 3.3.3).

3.3.2   Flavor Ester Production by Whole Cells

In most foods, especially fruits, straight- and branched-chain esters contribute greatly to the overall flavor. Traditionally these esters are extracted directly from plant material. An alternative is the use of plant tissue culture; however, this is not cost-effective at this time because of high processing costs from expensive media, dilute product levels, and so on.

For the formation of simple esters, the use of heat refluxing, which would qualify under FEMA and FDA guidelines as natural processing, could be used. Of course, the acid and alcohol components would have to be natural. For ethyl acetate, as an example, one would require fermentation-derived ethanol and acetic acid for a direct esterification.

The use of whole microbial cells to produce aromatic esters such as methyl salicylate (wintergreen) or ethyl or methyl benzoate (fruity notes for flavor and/ or fragrance) is also possible, although in many cases the immediate commercial prospects are limited. For example, it is known that certain members of the fungal genus *Phellinus* can produce the latter compounds but at low levels [43]. However, despite short-term limitations for commercialization of certain microbial-derived esters, several applications of microbial systems have commercial potential. They are described below.

The biological production of flavor esters by whole cells can proceed via two routes: (1) the alcoholysis of acyl-CoA compounds [44], or (2) esterification of an organic acid with an alcohol [45]. In the former case the oxidative decarboxylation of pyruvate and the ATP activation of acetic acid leads to acetyl-CoA. Alcohol acetyl transferase, a membrane-associated enzyme responsible for the transfer of alcohols to acetyl-CoA, then mediates the formation of esters [46]. Normally the levels of acetyl-CoA are regulated tightly, because it is a major metabolite consumed by the TCA cycle, but under some conditions the intracellular pools can be increased significantly. Physiological manipulations leading to the latter situation are discussed below.

The practical potential for use of whole cell technology for production of flavor esters has been demonstrated with acetyl-CoA as an intermediate. For example, *C. utilis* was found to be capable of ethyl acetate production from fermentable sugars or ethanol under iron-limiting conditions via acetyl-CoA as an intermediate [22]. The organism demonstrated aerobic fermentation (also known as the Crabtree effect) in that it converted most of the added sugar to ethanol under aerobic conditions. The fermentation kinetics showed clearly the sequential conversion of sugar to ethanol, followed by accumulation of ethyl acetate at the expense of the alcohol. It was proposed that once the sugar had been converted to ethanol or when ethanol was added directly under iron-limited conditions, ethyl acetate would be formed via an alcohol acetyl transferase mechanism. This could be a protective mechanism for the yeast, intended to control the intracellular pools of acetyl-CoA. That is, under the iron-limited conditions used, it would be expected that acetyl-CoA pools could grow to significant levels as a result of the bottleneck created in the TCA cycle (induced by the iron limitation). In the presence of ethanol, the yeast can produce a less toxic product (i.e., ethyl acetate) that is quite volatile. The optimal level of ethanol for yields of ester was approximately 10 g/L. Over 10 to 15 g/L ethyl acetate accumulated in fed-batch systems. As previously discussed, acetaldehyde was formed as the principal product if the alcohol level was shifted above approximately 35 g/L. This process exhibited excellent stability; that is, fed-batch operations of several months have been conducted. An additional attractive process feature is the potential of the residual biomass as a source of a high-volume, value-added by-product for use in yeast autolysates.

It has also been reported that the yeast *Hansenula anomala*, which is similar to *C. utilis*, is capable of conversion of sugars to ethyl acetate [47]. The yields were comparable to those obtained with *C. utilis*, although the rate of production was significantly less with *H. anomala*. In another study [48] it has been shown (Table 3.4) that *C. utilis* can attain maximum ester yields after only 72 hours, whereas *H. anomala* requires almost 2 weeks on a complex medium or 50 days on a minimal-salts medium. In practical terms, one must always examine the rate of production as this will translate into major additional processing costs if the time frame is unduly extended. Additionally, *H. anomala*, unlike *C. utilis*, cannot give a by-product credit for yeast autolysates/hydrolysates for use in imparting meaty/savory flavors to food because this yeast is not approved for human consumption.

An alternative route for the production of other natural flavor alkyl esters is

Table 3.4 Production of Ethyl Acetate by Various Yeasts

Yeast	Glucose Concentration (%)	Growth Medium	Yield[a]	Time to Max. Ester (d)
Hansenula anomala	5	YE-salts	24.2	12
H. Anomala	5	NH_4^+-salts	23.7	50
Candida utilis	2	NH_4^+-salts	51.3	3
C. utilis	10	NH_4^+-salts	28.9	3

[a]Percent of theoretical yield.

detailed in a recent patent and incorporates the deamination of certain amino acids to their corresponding carboxylic acids (C_4 to C_5), from which acetyl-CoA derivatives are formed [49]. Esterification of the acids with added alcohols (C_2 to C_5) can occur then. Although the patent lists a number of organisms with this capability, the yeast *Geotrichum fragrans* is highlighted as the model system.

The second means of producing natural flavor esters is direct esterification of the appropriate alcohol and organic acid by whole cells. This esterification can lead to a number of economically important flavor esters such as ethyl butyrate (pineapple note). Ethyl butyrate produced synthetically sells for under $1/kg, whereas that produced from natural sources such as from orange peel can sell for up to $100/kg. Total U.S. markets for ethyl butyrate have been reported to be approximately 137,000 kg/year [50].

This process of direct esterification can occur naturally, for example, with lactic acid bacteria and *Pseudomonas* in dairy products [51]. However, certain esters (primarily ethyl butyrate and ethyl hexanoate) cause flavor defects in dairy products, and, as a result, early investigations into the production of esters were concerned more with elimination of these flavor esters than with developing them for use in flavors. Ester production in dairy products is mediated by esterase activity using presumably lipase-derived fatty acids from hydrolyzed triglycerides in milk (see Chapter 6). For flavor ester production, potential augmentation of this natural process could occur by the addition of alcohol and fatty acid precursors directly for subsequent esterification, rather than relying upon the endogenous fermentation to provide these precursors.

Another type of whole cell direct esterification involves organisms having high levels of the enzyme lipase. Normally, lipase functions in a hydrolytic manner (Fig. 3.6) [52], yielding the alcohol and acid moieties of the hydrolyzed ester. However, it has been demonstrated that in organic solvents, due to low water activity, the equilibrium can be shifted so that a condensation-type reaction occurs (see Section 3.4.1). Organic solvents also allow for greater solubility of hydrophobic species (i.e., long-chain fatty acids and their derived esters). The increase in solubility of longer chain fatty acids could lead to better enzyme–substrate interaction and hence better reaction efficiencies. Bell et al. [53] demonstrated the utility of using dried fungal mycelium from *Rhizopus*

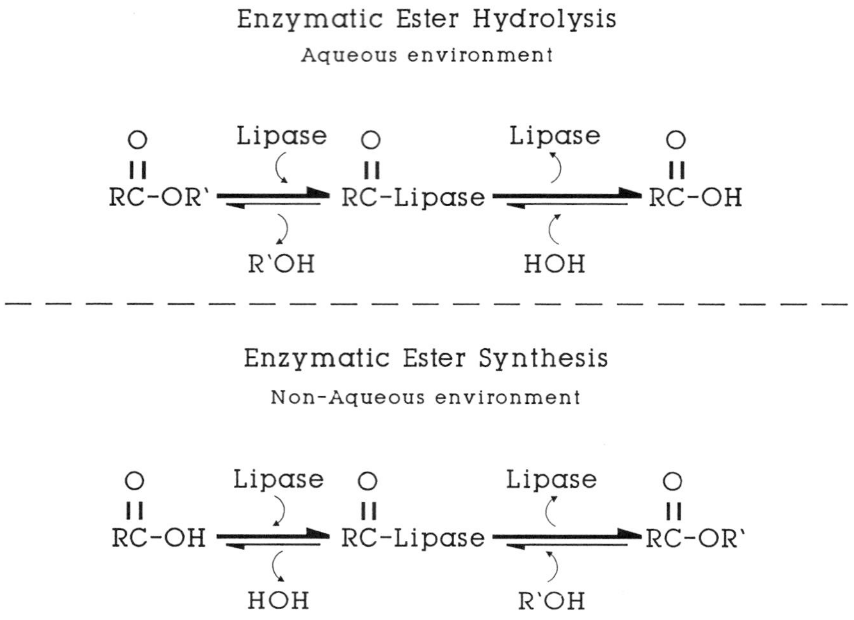

Enzymatic Ester Hydrolysis
Aqueous environment

$$\underset{RC-OR'}{\overset{O}{\parallel}} \underset{\text{Lipase}}{\rightleftharpoons} \underset{R'OH}{} \underset{RC-\text{Lipase}}{\overset{O}{\parallel}} \underset{\text{Lipase}}{\rightleftharpoons} \underset{HOH}{} \underset{RC-OH}{\overset{O}{\parallel}}$$

Enzymatic Ester Synthesis
Non-Aqueous environment

$$\underset{RC-OH}{\overset{O}{\parallel}} \underset{\text{Lipase}}{\rightleftharpoons} \underset{HOH}{} \underset{RC-\text{Lipase}}{\overset{O}{\parallel}} \underset{\text{Lipase}}{\rightleftharpoons} \underset{R'OH}{} \underset{RC-OR'}{\overset{O}{\parallel}}$$

FIG. 3.6 Lipase-mediated ester hydrolysis/synthesis. *Source* Reprinted by permission of the American Oil Chemists' Society from Miller et al. [52].

arrhizus in solvents such as octanol for production of esters. Their model system demonstrated a 90 percent conversion of palmitic acid to octyl palmitate. Excess water was removed using a molecular sieve, thereby enhancing esterification. The approach is interesting, because it could eliminate the need for several preparatory steps, including enzyme solubilization, purification, and immobilization, which are sometimes required for enzyme-based processes. Additionally, the association of lipase with the cellular biomass could stabilize the endogenous enzyme system, as most enzymes are known to be more stable when associated with membranes and other biological structures.

Recently Novo Laboratories developed a natural flavor and fragrance process using acetone-dried mycelium of *R. arrhizus* [54]. Significant production rates (up to 30 g/L per gram of mycelium) of isopentyl hexanoate were obtained in column reactors. Other esters, including benzyl butyrate and terpene esters, also were produced with various yields of between 30 and 80 percent of theoretical depending on the nature of the ester produced. It is anticipated that industrial use of bioesterification will become a commercial reality in the very near future.

3.3.3 Alcohol Production by Whole Cells

Alcohols in general play a minor role in flavors, although there are some important uses. For example, octan-3-ol and oct-1,5-dien-3-ol from various fungi (including *Aspergillus* and *Penicillium*) are used in mushroom flavors [55]. Other alcohols are important as precursors for further bioconversions (e.g., to esters and aldehydes).

In traditional alcohol fermentations, apart from ethyl alcohol production, a whole series of volatile compounds are produced including longer chain alcohols. These fusel alcohols include propanol, isobutanol, amyl and isoamyl alcohol. A general scheme for the production of fusel alcohols in yeast is shown in Figure 3.7 [56]. It is known that fusel alcohols originate when there is a reduction of α-keto acids resulting from the biosynthesis or degradation of amino acids. In general, amino acids undergo transamination followed by decarboxylation of the resulting keto acid into the corresponding aldehyde. Alcohol dehydrogenase brings about a reduction generating an alcohol with one

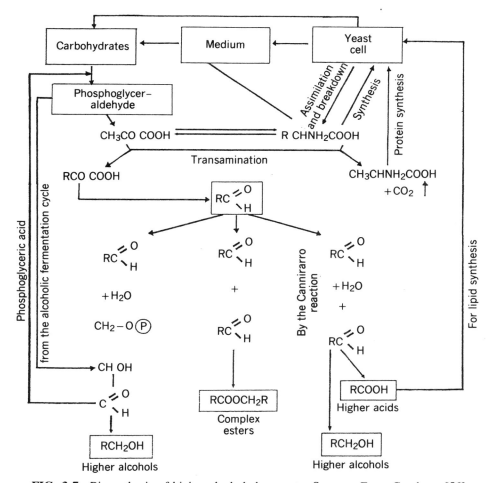

FIG. 3.7 Biosynthesis of higher alcohols by yeasts. *Source* From Gracheva [56].

carbon less than the corresponding amino acid. An exception to this is propanol, which cannot be formed from L-threonine. The various control mechanisms related to fusel alcohol production have been given in a comprehensive overview by Berry and Watson [57].

The fusel alcohols mentioned above can be obtained as a waste product from beverage alcohol distilleries and in turn used in whole cell lipase-mediated ester generation (see Section 3.3.2). It is not unreasonable to consider the possibility that these basic alcohol fermentations could be adapted and manipulated to produce excessive amounts of certain higher alcohols. For example, magnesium limitation of cell growth is known to result in excessive longer chain alcohol production [58]. Based upon the potential for value-added products, distilleries could incorporate this into their conventional beverage alcohol fermentation operations.

The current interest in specific aromatic flavor aldehydes such as benzaldehyde could be satisfied if enough natural benzyl alcohol could be obtained at an acceptable price. Alcohol oxidase or alcohol dehydrogenase cellular biotransformation of the alcohol to the aldehyde could be used then. Benzyl alcohol is produced by many fungi, including various species of *Phellinus* [59]. Unfortunately, the levels produced are extremely low (only several parts per million) and the growth rates of the producing microbes are very slow. It is possible, however, that these deficiencies could be resolved through strain improvement since the metabolic machinery is in place for *de novo* synthesis of benzyl alcohol.

3.3.4   Organic Acid Production by Whole Cells

Organic acids, including fatty acids (acetic, butyric, propionic, etc.) can play a significant role in flavors, not only on their own, such as in dairy flavors, but also as substrates for other flavor biosyntheses. Specifically, flavor esters production using lipase-based whole cell conversions (discussed in Section 3.3.2) relies heavily upon an adequate natural source of inexpensive organic acids. Three major organic acids (acetic, propionic, and butyric) used directly or indirectly for other flavors are discussed below.

Probably one of the oldest fermentation-derived organic acids is acetic acid. Various acetogenic bacteria are used to produce 13 percent acetic acid in fermentation vessels with volumes ranging up to 50 m^3 [60]. Organisms used include *Acetobacter aceti* and *Gluconobacter oxydans*. The process of submerged fermentation (Frings process), commercialized in 1949, is used extensively, although a surface process is also in commercial use. The latter trickling filter is filled with beechwood shavings and can have a total volume of up to 60 m^3. Ethanol is sprayed over the surface and trickles through the wood shavings containing bacteria. The substrate undergoes a partial conversion to acetic acid after which it is captured in a basin and recycled for further conversion through the microbial bed. Almost 90 percent of the alcohol added is converted to acetic acid in 3 days.

Acetic acid derived in this manner is an important precursor of natural ethyl acetate and numerous other acetate flavor esters. As an illustration, to produce ethyl acetate one simply would reflux fermentation-derived ethanol and acetic acid, under acidic pH, to generate the ester.

Propionic acid is another organic acid that can be derived in relatively high amounts from fermentation processes. A major area of potential application of this particular fatty acid is in the formation of various flavor esters including ethyl propionate (fruity/rumlike flavor), benzyl propionate (fruity flavors, floral fragrances), citronellyl propionate (rounds out citrus flavors and other fruit flavors, fresh fruity, roselike fragrances), and geranyl propionate (rose fragrances). Citronellyl propionate and geranyl propionate probably represent the major volume of products derived from propionic acid [61].

Propionic acid production via fermentation was recognized to have commercial potential in the early part of this century. Typically, the genus *Propionibacterium* was used to produce levels of propionic acid up to 20 g/L in a 14-day fermentation. More recent use of two-step fermentations, consisting of a lactic acid stage and a propionic acid stage, improved yields to approximately 41 g/L. Oddly, numerous proposals for commercial propionic acid (in pure form) have been developed, but none has seen commercialization. Clausen and Gaddy [62] proposed the use of renewable biomass (orchard grass) for the fermentative production of propionic acid. Preliminary designs were developed for a 200 ton/day plant incorporating fermentation and solvent extraction; based on a fermenter yield of 20 g/L of propionic acid, the projected product cost was approximately $0.30/kg. At the time of their proposal, market prices for propionic acid were somewhat higher at $0.50/kg. This process was not implemented at the time, but current interest in producing value-added propionate flavor esters could attract renewed attention.

Butyric acid is an especially important acid known for its applications in imparting buttery notes and for use in cheese flavors (see Chapter 6). Esters of butyric acid find wide use as fruit flavors especially in the soft drink and chewing gum industries. One can extract butyric acid directly from butter, because it contains between 2 and 4 percent butyric acid; however, a fermentative approach makes more economic sense. A number of organisms are known to produce butyric acid from simple sugars, including *Clostridium, Butyrivibrio, Eubacterium,* and *Fusarium* [63]. In the fermentation, butyric acid concentrations as high as 20 to 30 g/L can be achieved [61]. In the production of butyric acid by *Clostridium butyricum*, the pH must be held above 5, or the bacterium produces the solvent mix acetone/butanol [63]. If the pH is held above 5 the major product distribution shifts to butyric and acetic acids. After isolation of the acid, as discussed below, butyric acid can be used directly or for flavor ester production. Important esters that can be formed (see the discussion of ester production by lipase in Section 3.4.1) include amyl butyrate (strong ethereal, fruity odor reminiscent of apricots, bananas, and pineapples), isobutyl butyrate (fruity odor suggestive of pears, pineapples, and bananas), and ethyl butyrate (pineapple note).

Butyric and propionic acids, as with most fermentation-derived acids especially in their undissociated form, are toxic to the producing organism. Therefore the fatty acid(s) must be continuously sequestered from the production medium for maximum yield. Presumably, different resins could accomplish the continuous *in situ* trapping for subsequent isolation and concentration. For example, butyric or propionic acids could be adsorbed and concentrated onto silicalite (a zeolite analog) or other suitable material for subsequent removal and final purification. Other product recovery and purification protocols for low-chain fatty acids are discussed in detail elsewhere [61].

Because of their low volatility, many organic acids (especially polar ones) are very difficult to recover from complex fermentation media. One way to facilitate the recovery process is by *in situ* formation of more hydrophobic volatile esters. As an illustration, it might be possible to have an immobilized whole cell (lipase) preparation of *Rhizopus* mycelia through which the fatty acid-containing media is passed along with an appropriate alcohol. For example, if ethanol were added to such a butyrate-containing system, the product would be ethyl butyrate, which is recoverable by conventional distillation. The acid could then be recovered by a relatively simple hydrolysis, with the ethanol being available for reuse. Of course if the alcohol were natural, the end product, ethyl butyrate, would be highly desirable without any further processing. Presumably, a similar approach, giving enhanced recovery and end products directly, could be applied to other lower volatility fatty acids such as propionate (e.g., ethyl propionate).

Much opportunity exists in the area of fermentative production of various organic acids. Consequently, more studies should be done to capitalize fully on the many biological processes that have added to this potential, especially esterification technologies.

3.3.5   Ketone Production by Whole Cells

Ketones and diketones, like aldehydes, are characterized by the presence of a carbonyl group and are classified as either aliphatic, aromatic, or phenol derived [64]. Aliphatic ketones, such as acetoin and diacetyl, are important in cheese flavors, especially mold-ripened cheeses. Some aliphatic monoketones are used in perfumery for accentuation; 3-octanone, for example, is used in lavender notes. Aromatic ketones are of particular importance to the fragrance industry. Compounds like acetophenone, a sweet, orangelike liquid, and *p*-methylacetophenone, possessing a flowery, sweet odor, are used in soap perfumes and detergents. Phenol-derived ketones include only a few compounds that are of interest as flavor and fragrance ingredients. Two significant phenol ketones are acetanisole, used in soap perfumes, and raspberry ketone or 4-(4-hydroxyphenyl)-2-butanone, used in raspberry compositions. Table 3.5 outlines important naturally occurring ketones and their uses. Synonyms of these ketones are also listed to facilitate cross referencing among different studies [65]. As with lactones, multiple names for these compounds abound in the literature.

It is predominantly the aliphatic ketones that are produced in biological

TABLE 3.5 Important Naturally Occurring Flavor and Fragrance Ketones and Their Uses

Ketone	Description	Synonyms
Aliphatic		
2-Heptanone	Found in oil of cloves and cinnamon-bark oil; responsible for peppery odor in Roquefort type cheeses; used in perfumery as constituent of artificial carnation oils	Amyl methyl ketone Ketone C-7 Methyl amyl ketone
2-Nonanone	Present in attar of roses; clove oil and passion flowers	Methyl heptyl ketone Nonan-2-one
1-Octen-3-one	Fresh mushroom; metallic; fungal	Amyl vinyl ketone Vinyl amyl ketone
3-Octanone	Lavender note.	Amyl ethyl ketone Ethyl amyl ketone
2-Undecanone	Dark herbal odor reminiscent of hawthorn	2-Hendecanone 2-Oxoundecane Methyl nonyl ketone Rue ketone
Aromatic		
Acetophenone	Found in a large number of foods and essential oils; penetrating sweet odor reminiscent of orange blossom; used to perfume detergents and other industrial products	Acetylbenzene Acetylbenzol Methyl phenyl ketone Phenyl methyl ketone
Benzophenone	Flavor component of grapes; rosy, slightly geranium odor; used in flower compositions and as a fixative	Benzoylbenzene Diphenyl benzene Diphenylmethanone α-Oxodiphenylmethane Phenyl ketone
p-Methylaceto-phenone	Found in Brazilian rosewood and in pepper; flowery-sweet odor; milder than acetophenone; used for blossom in mimosa and hawthorn type soap perfumes.	1-Acetyl-4-methyl-benzene p-Acetyltoluene 1-Methyl-4-acetyl benzene Methyl p-tolyl ketone p-Tolyl methyl ketone
Methyl β-naphthyl ketone	Found in some essential oils; used in eaux de cologne, soap perfumes, and detergents; good fixative.	2'-Acetonaphthone β-acetylnaphthalene Cetone d β-Naphthyl methyl ketone Oranger crystals

TABLE 3.5 (*Continued*)

Ketone	Description	Synonyms
Aromatic		
4-Phenyl-3-buten-2-one	Volatile component of cocoa; sweet-flowery smelling; used in soap perfumes.	Benzilideneacetone Benzylacetone Benzylidene acetone Methyl styryl ketone
Phenol-derived		
Acetanisole	Occurs in anise oil; sweet odor reminiscent of hawthorn; used in soap perfumes	p-Acetylanisole p-Methoxyacetophenone Methyl 4-methoxy-phenyl ketone
4-(p-Hydroxy-phenyl)-2-butanone	Derived from fruit; used in fruit flavors, particularly raspberry compositions	p-Hydroxybenzyl-acetone 1-p-Hydroxphenyl-3-butanone Frambinon (Dragoco) Oxyphenylon (IFF) Raspberry ketone

systems. Certain fungi synthesize these ketones in response to the presence of short-chain fatty acids in the environment or as a means of recycling coenzyme A when the oxidation of acetyl-CoA is inhibited [66]. For example, it is known that the fungus *Phlebia radiata*, a Basidiomycete, produces 3-octanone (at microgram per liter levels) in this manner [24]. Significant amounts of 2-nonanone and 2-undecanone were obtained from various fungi grown on different triglycerides including tricaprin (synthetic) and palm kernel oil [67]. 2-Undecanone (rue ketone) possesses a dark, herbal odor and is important in cheese flavors. *Penicillium decumbens* particularly was promising; this fungus produced 8.22 mg of 2-nonanone from 50 mg of tricaprin and 6.54 mg of 2-undecanone from 50 mg of palm kernel oil. In another work, the same investigators optimized conditions for this organism and obtained a 46 percent conversion of palm kernel oil to 2-undecanone [68]. In still another study, these authors found that three *Trichoderma* strains will also convert tricaprin to 2-nonanone at a yield as high as 217 mg/100 mL [69].

It is sometimes preferable to use fungal spores rather than vegetative cells. Spores give high, reproducible conversion of substrate and minimal amounts of undesirable products. Furthermore, product recovery is easier without having to deal with mycelium, and spores are easier to store and transport [70]. Several studies have been conducted where immobilized *Penicillium roqueforti* spores produced 2-heptanone (blue cheese note) from octanoic acid. The productivity of the system was determined by noting the consumption of octanoic acid rather than by recovering 2-heptanone from the exhaust gas [71,72]. Creuly et al. [72] designed a fed-batch process for this system and found that the process was pH dependent. At 6.5 pH, the system was highly stable with a constant reaction rate over a long period of time. However, at the lower 5.5 pH, a constant rate of production was not achieved. This was attributed to the fact that at the lower pH acid altered the spore membrane permeability and caused the loss of intracellular metabolites. On the other hand, at the higher pH the acid in its undissociated form did not affect the spores in the same way. The drawback to this process is that it does not lend itself well to large-scale production given the complex manipulations involved.

Cell-free preparations of *Pseudomonas oleovorans* have been found to convert the allylic alcohols oct-1-enol and octan-3-ol to 3-octanone and oct-1-en-3-one (Table 3.6) [73]. In theory, one could supply an appropriate organism with 2-heptanol to obtain 2-heptanone in the same manner. However, the secondary alcohol precursors would have to be natural, which could lead to problems in finding sufficient quantities. Other natural precursors have been shown to yield substantial amounts of 2-heptanone, 2-nonanone, and 2-undecanone when added to cultures of *Aspergillus ruber* and *A. repens*. When grown on coconut oil as the sole carbon source, these organisms produced 15 g/L of 2-heptanone, 16 g/L of 2-nonanone, and 35 g/L of 2-undecanone [74].

Halim and Collins [59] found that *p*-methylacetophenone is a major aroma constituent of *Mycoacia uda*. They characterized this component as having a penetrating, sweet, fruity odor but failed to quantify amounts present.

There are not as many examples of phenol-derived ketones found in biological systems as there are of aliphatic ketones. Raspberry ketone was found in stationary liquid cultures of *Nidula niveo-tomentosa* (Bird's Nest Fungus), but levels were not quantified [75]. The authors presumed that niduloic acid, also produced by this fungus, is the biogenetic precursor of raspberry ketone in this system. With further manipulation of environmental conditions, this system has the potential to produce significant levels of raspberry ketone through *de novo* synthesis or through the biotransformation of niduloic acid.

It would appear that the groundwork has been laid for the commercial pro-

TABLE 3.6 Conversion Products of Allylic Alcohols and Methyl Ketones by *P. oleovorans*

Substrate	Product	Yield (mg/ml) versus Reaction Time		
		10 h	20 h	40 h
1-Octen-3-ol	3-Octanone	0.50	1.00	1.10
	1-Octen-3-one	—	0.50	0.10
1-Octen-3-one	3-Octanone	2.00	2.50	3.00
3-Octanol	3-Octanone	0.80	1.20	1.20

Source Adapted by permission of The Royal Society of Chemistry from May et al. [73].

duction of ketones. With a number of microorganisms already identified as potential producers of these compounds, the development of a natural source is a likely prospect.

3.3.6 Lactone Production by Whole Cells

Lactones are organoleptically important flavor compounds found in a large number of foods. Because they have low odor thresholds, they often have a high flavor value. In nature, they are found mainly as γ- and δ-lactones and more infrequently as macrocylic lactones [64], with γ-lactones found primarily in plants and δ-lactones primarily in animal products [76]. Both γ- and δ-lactones are intramolecular esters of corresponding hydroxy fatty acids (i.e., internal esters with an oxygen bridge); most of those found in nature have a *cis* configuration. Lactones are important contributors to the aromas of fruits (especially coconut) and dairy products (see Chapter 6). Another element of lactone odor is that of musk for which the macrocylic esters are responsible. A *trans* isomer, 15-pentadecanolide, gives angelica root oil its musklike odor. Maga [77] has done an extensive review article on lactones in foods and their sensory properties.

At present, lactones are made fairly inexpensively via chemical synthesis from keto acids. On the other hand, microbially produced lactones have the advantage of being pure optically and natural. There are numerous microorganisms that are known to synthesize lactones *de novo*; these are summarized in the review article by Welsh et al. [78]. Unfortunately, most of the organisms cited produce the compounds of interest only in trace amounts. Some of the studies mention the presence of lactones but fail to quantify them. However, from an industrial production viewpoint, there are several microorganisms that are promising for lactone production (Table 3.7) [79–83].

Berger et al. [79] supplemented *Polyporus durus* with Miglyol™, a synthetic triglyceride, and achieved a yield of 281 mg/L (sum of all lactones produced). It was determined that a pH below 4.5 was necessary for lactone production to be stimulated by the Miglyol™. This suggested that the fungus produced 4-hydroxy acids that were lactonized in the acidic extracellular environment.

In another study [80], it was found that *Bjerkandera adusta* also responded to the addition of Miglyol™ by producing several 4- and 5-olides. The addition of Miglyol™ to cultures of *Ischnoderma benzoinum*, on the other hand, inhibited growth but still led to the production of volatiles [17]. This inhibition was attributed to interaction of the fatty acids with the lipophilic portions of the cell membrane, causing leakage of cell contents and interference with cellular respiration. With *I. benzoinum*, the type of medium had an effect on which volatile compounds were produced. Yeast-malt medium, as opposed to the authors' synthetic medium, favored the production of γ-octalactone and 2-octen-4-olide when used with Miglyol™ in submerged culture (see Fig. 3.5).

A similar example of how media determine the type of lactone produced is found with *Trichoderma viride*. This organism produced 6-pentyl-α-pyrone

TABLE 3.7   Microorganisms Producing Lactones in Appreciable Amounts

Organism	Lactones Produced	Description	Synonyms	Ref.
Polyporus durus	γ-Butyrolactone[a]	Sweet, buttery; used in detergents, sunscreens, cosmetics, and air fresheners	1,4-Butanolide 4-Butanolide 4-Hydroxybutanoic acid lactone Lavender oil	[79]
	4-Hexanolactone[a]	Caramel, sweet, herbaceous; constituent of soft fruit	5-Ethyl dihydro-2(3H)-furanone 4-Hexanolide 4-Caprolactone 4-Ethyl butyro-lactone 4-Hydroxyhexanoic acid lactone γ-Hexalactone	
	5-Hexanolactone[a]	Found in coconut milk, strawberries, mango, and papaya juices	Tetrahydro-6-methyl-2H-pyran-2-one 5-Hexanolide 5-Caprolactone 5-Hydroxyhexanoic acid lactone δ-Hexalactone 5-methyl-δ-valerolactone	
	γ-Octalactone[b]	Fruity, coconut, very sweet; used in aroma compositions and heavy blossom perfumes	1, 4-Octanolide 4-Octanolide γ-*n*-Butyl-γ-butyrolactone 4-Hydroxyoctanoic acid γ-lactone	
Bjerkandera adusta	5-Hexanolactone[a]			[80]
Tyromyces sambuceus	γ-Decalactone[a]	Fruity, peachy; used in peach flavors and in perfumery for heavy, fruity flower odors	4-Decanolide 5-Hexyldihydro-2(3H)-furanone 4-Hydroxydecanoic acid lactone γ-Hexylbutyro-lactone	[80, 81]
Pleurotus euosmus	Coumarin[a]	Haylike, spicy; used in fine fragrance and soap perfumes for green, spicy notes	2H-1-Benzopyran-2-one Coumarinic anhydride	[82]
Trichoderma viride	6-Pentyl-α-pyrone[b]	Coconut, peachy.	5-Hydroxy-2, 4-decadienoic acid δ-lactone	[6]
	6-(Pent-1-enyl)-α-pyrone[a]			[83]

[a] 1–100 mg/L.   [b] > 100 mg/L.

67

when grown on potato dextrose medium, yet when grown on a modified Czapek Dox medium, 6-(pent-1-enyl)-α-pyrone was produced [83]. The former lactone was originally thought to be associated with spores rather than hyphae [6]. However, further investigation into 6-pentyl-α-pyrone synthesis [84] showed that the compound was an extracellular metabolite and sporulation was not necessary for its formation. In addition, agitation inhibited lactone formation when certain nitrogen sources were present; with other nitrogen sources, production was enhanced. *Ceratocystis moniliformis* was also responsive to media composition [15]. In a glycerol-urea medium, the organism produced γ- and δ-decalactones, whereas in a galactose-urea medium geranial and citronellol were the predominant products. In a dextrose-leucine medium, the major product was isoamyl acetate. Further investigation remains to be done as to how the carbon and nitrogen sources determine what is produced.

Lactone production in *Pityrosporum ovale* was stimulated by lecithin, oleic acid, and triolein [85]. Unfortunately, lactone levels were given in relative peak intensities, so they cannot be compared with those of other organisms for determination of commercial feasibility. The lactones found in all the *Pityrosporum* cultures were γ-octa-, γ-nona-, and γ-decalactones, the latter being the major component.

Tahara et al. [86] found that the yeast *Sporobolomyces odorus* produced γ-decalactone and *cis*-6-dodecen-4-olide, but did not quantify amounts. They found that γ-decalactone was most prevalent in a young culture; it peaked when growth did and it seemed unaffected by pH changes. The greatest yield was obtained using a mannitol-peptone medium. High yields of *cis*-6-dodecen-4-olide were obtained in a fructose-alanine medium. Further work on this organism by Tressl et al. [87] showed that decanoic acid is converted easily to γ-decalactone, indicating that saturated lactones are formed possibly by the γ- or δ-hydroxylation of saturated acids. From a similar derivation of 6-dodecen-4-olide from linoleic acid, it was confirmed that unsaturated lactones are derived from unsaturated fatty acids via β-oxidation.

Jourdain et al. [13] followed-up on Tahara's work and investigated further the production of γ-decalactone by *S. odorus* in batch culture. From their results (Fig. 3.8), they determined that there were two points in the growth curve where γ-decalactone was produced. The first was at 40 hours, during growth phase, and the second was around 100 hours, when nongrowing cells were synthesizing the lactone during biomass volume increase (due to reserve accumulation). When grown in a fermenter, the biomass and γ-decalactone production were maximized at 35 percent of the saturation level of oxygen, with continuous production for a month at a maximum productivity of 0.3 mg/L per hour.

A semicontinuous fermentation of *Tyromyces sambuceus* at the 5-L scale was done by Kapfer et al. [81] for the purpose of obtaining γ-decalactone. The culture was supplemented with ricinoleic acid (12-hydroxyoctadec-9-enoic acid) in the form of castor oil as a precursor of γ-decalactone. At 14 days, a peak value of 372 mg/L was reached. When the clumped mycelia in the headspace

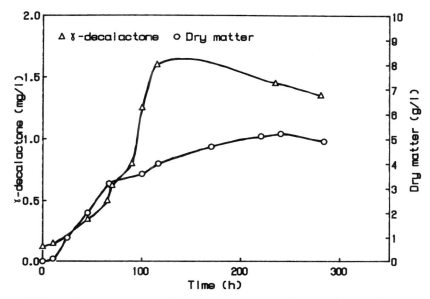

FIG. 3.8 Batch growth curve and γ-decalactone production for *Sporobolomyces odorus*. *Source* Reprinted by permission of the Editor, S. Nitz, from Jourdain et al. [13].

were washed back down into the medium, a level of 880 mg/L was recorded, leading to the conclusion that headspace conditions were favorable for lactone production. Okui et al. [88] found that various *Candida* strains convert ricinoleic acid via β-oxidation to 4-hydroxydecanoic acid. As described in a patent by Farbood and Willis [89], the 4-hydroxydecanoic acid produced by the yeast can be converted further to γ-decalactone by boiling the acidified fermentation broth, with a yield of solvent-extracted lactone of 6 g/L. Fritzsche Dodge & Olcott is reportedly working on Farbood's microbial biotransformation of castor oil to γ-decalactone [90].

Gervais and Battut [91], working with *Sporidiobolus salmonicolor*, found that the intracellular accumulation and extracellular excretion of γ-decalactone could be modified by altering the water activity (a_w) of the culture with glycerol. They found that the greater the dilution intensity (osmotic shock), the greater the release of γ-decalactone (Fig. 3.9). A scale-up was done in a 2-L fermenter, where aroma was produced at levels five to seven times greater than in the shake flasks due to improved aeration. It was determined that an a_w range of 0.97 to 0.99 was optimum. It is interesting to note that Lanza et al. [15] observed that *C. moniliformis* produces lactones when using glycerol as a carbon source. Gervais and Battut could have been supplying their organism with a needed precursor (glycerol) for lactone production when their only intent was to alter the a_w.

Muys et al. [92] demonstrated that certain yeasts, molds, and bacteria can reduce γ- and δ-keto acids to optically active γ- and δ-lactones. *Saccharomyces cerevisiae* converted δ-keto-capric acid to δ-decalactone with a conversion efficiency of 71 percent. Various other keto acids were also converted to their corresponding lactones using *S. cerevisiae* grown on glucose. This process has the advantage of using a generally-recognized-as-safe (GRAS) organism to produce a natural end product and also of replacing a multistep chemical synthetic route. Unilever is now using this organism to produce γ-lactones [93]. Tressl et al. [87] also studied lactone production by *S. cerevisiae* and demonstrated the conversion of certain 4- and 5-oxoacids to their corresponding lactones, but no further work on the topic has since been published.

It was mentioned earlier that one advantage of producing lactones microbially is that pure optical isomers can be obtained. The optical rotation of an isomer can be very important to its odor impact; for example, of the two diastereoisomers of whiskey lactone (3-methyl-4-octanolide), the *cis*-lactone has an odor threshold value approximately ten times greater than the *trans*-lactone (0.79 ppm vs. 0.067 ppm) and has a more pleasing aroma [94]. Microorganisms also can be used to convert optically active lactones to other forms. For example, washed cells of *Rhodococcus erythgropolis* were able to convert the L-(+)-isomer of pantoyl lactone to the D-(−)-isomer [95].

There are a few drawbacks with the microbial systems identified here. In most cases the organism must be grown for a long period of time (20 to 35 days) before the lactones appear in any appreciable amounts. Often the compounds are produced only from surface fungal cultures (i.e., shallow liquid culture). Many will not produce in submerged culture (i.e., shake flasks or fermenters) where the most efficient use of the volume/area ratio is obtained. Because most volatile production seems to be linked to secondary metabolism, a large amount of biomass is needed to produce appreciable levels of compounds. Studies performed in fermenters showed that greater biomass yields are obtained than in shake flasks and that the fungi are amenable to environmental manipulation.

As an alternative to whole cell systems, there are a few enzyme systems that can be used to produce lactones. For example, Gatfield [96] used a lipase-esterase preparation from *Mucor miehei* in a nonaqueous environment to convert 4-hydroxybutyric acid to γ-butyrolactone. The future may see more lactone production via other such nonconventional approaches.

3.3.7 Green Aldehydes and Alcohols—Possibilities for Bioprocessing

Some good examples of high-value–low volume products are leaf aldehyde (*trans*-2-hexenal) and leaf alcohol (*cis*-3-hexenol), which are responsible for the "green" flavors and aromas of fruits and vegetables. *trans*-2-Hexenal has a sharp, herbal-green aroma that also makes it desirable for use in perfumes and for providing a green nuance in fruity flavors. *cis*-3-Hexenol is used in perfumes and flavors to obtain green top notes [64]. Currently, synthetic com-

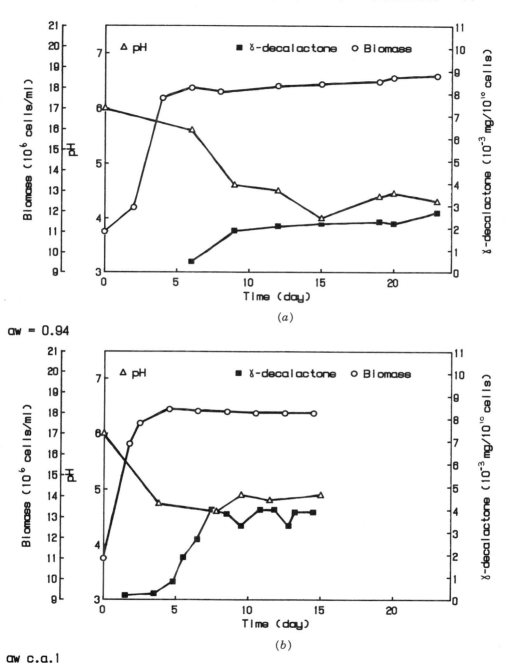

FIG. 3.9 Influence of water activity on growth of *Sporidiobolus salmonicolor,* pH of medium, and extracellular production of γ-decalactone. *Source*   Reprinted by permission of The American Society for Microbiology from Gervais and Battut [91].

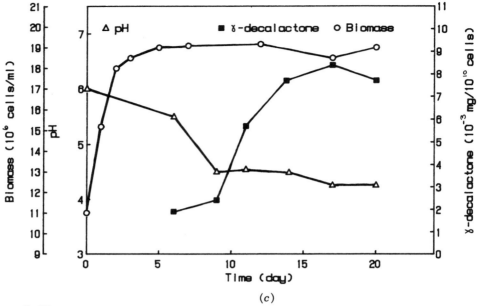

$$(c)$$

$$aw = 0.99$$

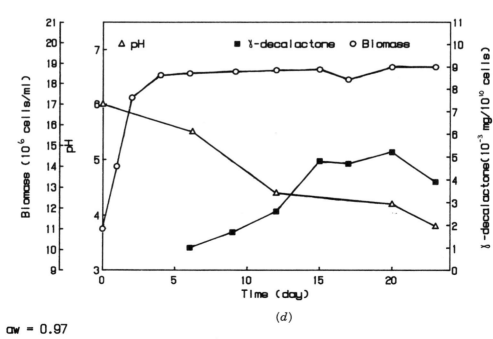

$$(d)$$

$$aw = 0.97$$

FIG. 3.9   *(Continued)*

pounds are used extensively; however, a natural product would be worth $250 to $350/kg [35].

The natural compounds are obtained primarily from plant tissues that have been disrupted in some fashion. In general the unsaturated fatty acids, linoleic and linolenic, are degraded via a lipoxygenase-catalyzed formation of hydroperoxides and a subsequent cleavage by a hydroperoxide lyase to form aliphatic C_6 compounds such as hexanal, *cis*-3-hexenal, and *trans*-2-hexenal (Fig. 3.10) [97]. However, these lipoxygenases and hydroperoxide lyases differ considerably in their enzymatic characteristics depending on their source. For example, lipoxygenases, such as those found in tomatoes, catalyze the production of 9-hydroperoxides from linoleic and linolenic acid. These 9-hydroperoxides are in turn cleaved by hydroperoxide lyase to form the volatile products *cis*-3-nonenal and 2-*trans*-6-*cis*-nonadienal (violet leaf aldehyde), which are found in cucumber aroma. *trans*-2-*cis*-6-Nonadienal is of commercial importance because it is one of the most potent fragrance and flavoring substances known. Lipoxygenases found in plants like soybeans catalyze the formation of 13-hydroperoxides, which in turn are converted to hexanal and *cis*-3-Hexenal. *trans*-2-Hexenal is generated through the isomerization of *cis*-3-Hexenal by an unidentified enzyme in the plant tissue [98].

Currently, there is no description of isolated enzymatic systems that could be utilized in the commercial production of green flavor components [99]. It would be impractical to utilize the endogenous enzymes in plant tissues for this purpose due to the low levels of the required enzymes and because of the processing difficulties involved with the use of mascerated tissue. A way around these difficulties is through plant cell cultures. In the presence of linolenic acid, callus homogenates of apple, pear, strawberry, and tomato are capable of producing hexanal, *cis*-3-hexenal, *trans*-2-hexenal, *cis*-3-hexenol, and *cis*-3-hexenyl acetate (a prototype for green odors used in combination with *cis*-3-hexenol [64]) [100]. An alternative to using plant cell culture is the use of a microbial system that has been found to contain lipoxygenase enzymes. *Fusarium oxysporum* lipoxygenase oxidizes methyl linoleate (the methyl ester of linoleic acid) and forms two isomeric 9- and 13-hydroperoxides [101]. The hydroperoxide lyase necessary for the conversion of these hydroperoxides to green flavor compounds could possibly be obtained from cultures of tobacco green cells [102]. This fungal/plant cell culture system might be further enhanced by the isolation of the lyase gene from the fungus and rDNA modification of a suitable microbial host obtain an organism proficient in lipoxygenase activity. Given the high value of natural leaf aldehydes and alcohols, further investigation into a successful biological system could be very profitable.

3.4 ENZYME-BASED TECHNOLOGIES

The use of enzymes on an industrial scale is common practice now. In the food industry, for example, amylase is used to hydrolyze starch and dextrins to

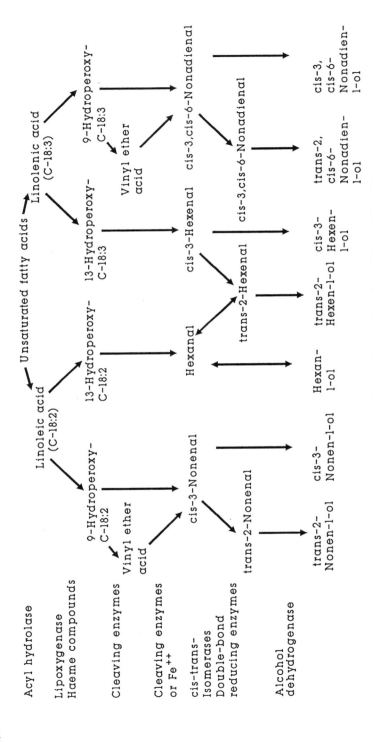

FIG. 3.10 Summary of proposed pathways for the formation of C_6 and C_9 aldehydes and alcohols from plant lipids. *Source* From Eriksson [97].

glucose. Glucose isomerase is then used to convert glucose to fructose, which is used in the production of high fructose syrups (HFS). However, as discussed in Section 3.3, the production of certain flavor substances using whole microbial cell systems is often preferred over using isolated enzymes, because the former contain many enzymes that can act in a multistep sequential manner. On the other hand, if the biotransformation of interest involves a limited number of steps, then the use of isolated enzymes can be beneficial. For instance, the use of isolated enzymes can result in extremely pure products by precluding the formation of side products. Enzymes can be important from an economic standpoint in flavor processing, because they are required only in catalytic rather than stoichiometric amounts and their activity can be very selective. Unfortunately, unlike whole cell systems where in certain cases it is possible to produce flavor substances *de novo* from simple sugars or precursors, enzyme bioconversions often require precursors or substrates that have severe availability problems. This section discusses several important enzyme conversions currently or potentially important in flavors production.

Pure enzymes can be obtained from a number of sources using conventional methods of protein purification. However, in many cases the enzyme content of these sources is very low, so that large quantities of plant materials would be required to obtain a useful amount. Some useful enzymes (e.g., lipase) can be obtained from animal cells, but again one encounters difficulties with availability and cost. Microbial sources are much more suitable for enzyme production, because, in general, they are not subject to such limitations. Microorganisms such as bacteria, yeasts, and fungi are capable of producing a large variety of diverse enzymes that have the potential to be isolated and used in industrial processes. The levels and types of enzymes produced by specific microbial systems depend on environmental factors such as nutrient availability, pH, and temperature, along with genetic adaptability. Apart from environmental manipulations, the use of classical genetic modifications or modern rDNA methods is another means of achieving elevated production of novel enzymes or the enhancement of existing systems.

It is useful to look at the number of enzymes known versus the number actually used commercially. Upward of 3000 enzymes have been described in the scientific literature [33]. Of these, there are probably only a few hundred that are commercially available, in quantities ranging from milligrams to several grams. Only approximately 20 are available in amounts suitable for use in commercial processes. Even though the ratio of the number of known enzymes to the number used in commercial applications is very high, a presentation of several important enzymatic bioconversions should show the tremendous utility and potential of this approach. Lipase-type enzymes will be discussed, because these represent an important emerging bioprocessing area having many different applications to production of flavors. Many of the approaches and limitations described for lipase also apply to other enzymes used in flavor bioprocessing.

3.4.1 Lipase Flavor Processing

The use of lipase for the modification of readily available fats (e.g., palm oil) can provide value-added products (e.g., cocoa butter). For an excellent review on this subject, see Macrae and Hammond [103]. Specific examples are discussed later in this section, but first it is useful to understand some of the physical and chemical characteristics of this class of enzyme. Lipases (glycerol ester hydrolases EC 3.1.1.3) comprise a group of enzymes of widespread occurrence in nature [104,105,106]. Lipase activity does not depend on cofactors; the chemical reagent is simply water. This feature simplifies process requirements and can have a significant impact on process economics. Lipases are active at the oil–water interface of heterogeneous reaction systems, which are difficult systems to study. As a result, relatively few studies have been done on lipases, unlike other industrial enzymes such as amylases and proteases [107]. However, the collective functional characteristics of lipases lend themselves well to commercial processes for flavor production.

In general, lipases are acidic glycoproteins of molecular weight ranging from 20,000 to 60,000 and with specific activity of 1000 to 10,000 units/mg of protein (one unit of lipase releases one μmole of fatty acid from a triacylglycerol per minute) [103]. Lipases are capable of the partial or even complete hydrolysis of triacylglycerols to provide free fatty acids (FFA), diacylglycerols, monoacylglycerols, and glycerol. Examples of FFAs that are produced by this reaction are propionic and butyric for cheese-type flavors [108]. These acids are now produced commercially is this manner.

It was believed previously that lipases contain a higher proportion of hydrophobic than hydrophilic amino acids, enabling closer interaction with hydrophobic substrates. However, a relatively recent survey of the amino acid profiles of various lipases has indicated that, in fact, as a group lipases are no more hydrophobic than other groups of enzymes [109]. Because lipases do not exhibit exceptionally high hydrophobicity, the strong interaction with hydrophobic substrates at an interface is most likely the result of hydrophobic patches on the surface of the enzyme [110]. This is also the probable explanation of the strong self-association that is noted with lipases in aqueous solution. It has been established recently that lipases, in general, have a catalytic site located on the bottom of a groove that is covered with a kind of lid [111]. The lid is important for the binding of the lipase to its substrate. In the groove, between the lid and the catalytic site, there is a secondary binding site for the substrate. It may be possible to increase significantly the efficiency of lipase by modifying these regions around the active site (modification of hydrophobicity, etc.). Recent advances in protein engineering should facilitate directed site modification.

Two types of important reversible reactions are catalyzed by lipases: (1) ester synthesis and (2) inter- and intraesterifications. Reactions favoring esterification can be encouraged by relatively simple adjustment of the hydrophobic nature of the reaction solvent surrounding the enzyme [112]. That is, the esterification reaction is favored if the water activity is low (Fig. 3.6), and this can be very useful for production of flavor esters [110]. In these nonaqueous systems the bulk solvent is generally an apolar solvent such as hexane.

It is becoming increasingly clear, at least with the majority of enzymes operating in nonaqueous reaction systems, that a small amount of water must be present to maintain conformational integrity and hence activity [113–115]. The presence of this small amount of water allows the enzyme conformational changes required to accommodate substrate entry. In a recent study [113], the presence of some water was deemed to be essential, and it also improved initial lipase activity, although excessive levels of water generated by the enzyme reaction were found to cause progressively a reduction in reaction rate. The latter phenomenon was attributed to phase-separation effects.

Control of water activity can be accomplished by the addition of certain agents, such as glycerol or hygroscopic polymers (i.e., Sepharose), although glycerol has the disadvantage of also being a reactant in triacylglycerol synthesis. Another means to control the a_w of the nonaqueous reaction is the use of hypobaric conditions (i.e., a vacuum) to pull off the water in a controlled manner as it is formed [52]. This approach removes excess water but still leaves the essential water required for functionality of the enzyme. Miller et al. [52] found that significant amounts of myristyl myristate (a model ester) could be produced with a commercially available lipase (Lipozyme™) in hexane under hypobaric conditions (0.5 bar) using the following process:

1. 3-Liter stirred reactor
2. 60°C/0.5 bar for 20 hours
3. 1 kg each of alcohol and acid reactants
4. 10 g Lipozyme™ charge
5. 12 repeated batch operations (using the same Lipozyme™ charge) were conducted at over 96 percent yield

The yield was a minimum of 1.2 tons of ester/kg of lipase, which is very significant. The purity of the formed ester was better than that of a chemically generated (low-pH) ester, because the formation of strong acid-catalyzed degradation products was avoided. It is likely that numerous flavor esters can be prepared using this system.

Gatfield [96] showed that lipase preparations from *Mucor miehei* are capable of synthesizing esters in nonaqueous reaction media. In a model system comprised of equimolar quantities of oleic acid and ethanol (no other solvents added) and 3 percent by weight lipase, a remarkable stability of the system (which produced ethyl oleate) was observed. That is, the system was stable for 36 days, with esterification efficiencies of 83 to 87 percent. Interestingly, certain acids such as propionic and acetic do not give rise at all to esterification products with this system. It was found that keto and hydroxy derivatives of propionic and butyric acids are converted more readily, relative to their parent acids, into their respective ethyl esters. For example, hydroxypropionic acid can be esterified to ethyl isovalerate (apple odor), whereas the parent acid, propionic, is esterified poorly. On the other hand, the structural derivates of the alcohol component are not as stringent, as demonstrated by the ease of both primary and secondary aliphatic alcohol esterification. However, tertiary alcohols such as *tert*-butanol do not esterify readily, and terpene alcohols such as geraniol, menthol, and linalool show a wide-ranging degree of esterification with oleic acid (90, 20, and 5 percent, respectively). It was recognized in this study that the removal of water is a key driving force in encouraging the esterification by lipase of these model compounds. It is important to recognize that, even under conditions of low initial water content, one mole of water is formed for every mole of ester synthesized during the condensation reaction. As a result it becomes important to continuously remove excess water from the reaction if quantitative yields of ester are required. In the *M. miehei* system, it was speculated that this water is adsorbed by the enzyme directly, because of the considerable change in physical appearance of the enzyme during the course of the reaction.

In a more recent study, biocatalytic production of natural flavor esters was examined in both batch and fed-batch immobilized enzyme reactors [110]. Lipase from *C. cylindracea*, immobilized on silica gel, was used as the catalyst, because it had been shown previously to possess broad substrate specificity [116]. Significant levels of ethyl butyrate were produced by this system along with various other important flavor esters (Table 3.8). It was also demonstrated

TABLE 3.8   Ester Production by Immobilized *C. cylindracea* Lipase Using a Range of Substrates[a]

Ester	Concentration (M)	Molar Conversion (%)
Ethyl propionate	0.19	76
Ethyl butyrate	0.25	100
Ethyl hexanoate	0.09	44
Ethyl heptanoate	0.21	84
Ethyl octanoate	0.26	100
Ethyl laurate	0.13	52
Ethyl isobutyrate	0.18	72
Ethyl isovalerate	0.01	3
Isobutyl acetate	0.06	25
Isoamyl acetate	0.06	24
Isoamyl butyrate	0.22	91

[a]The molar concentration is that of ester in heptane after 24 h. Percent molar conversion is that of initial acid to ester after 24 h.

Source Reprinted by permission of *Science & Technology Letters*, England from Gillies et al. [116].

that care must be taken in drawing conclusions from initial rates of reaction. Not only does the reaction efficiency depend upon the specificity of the lipase but also on the differences in substrate solubility. For example, ethyl lactate and ethyl cinnamate were not produced in the system outlined above, probably because of the low solubility of the acids in the bulk organic solvent (hexane) used. An important observation was the requirement for intermittent hydration of the immobilized enzyme reactor system, even though the reaction generated water [110]. It was speculated that certain nonpolar solvents have a tendency to strip away the water required for maintaining the structural integrity of the enzyme. For this particular lipase–silica-gel system, 10 to 20 percent water allowed the system to remain stable for 1 month while producing ethyl butyrate.

Various other studies have examined lipase-mediated esterifications and are listed in Table 3.9 [78]. One of these investigations has shown that acetic acid can cause the lipase from *C. cylindracea* to have a 12-fold lower efficiency of esterification of butyric acid and isoamyl alcohol [117]. It was suggested that this inhibition was the result of the disruption of the hydration-layer–protein interaction or even the overall protein conformation.

An important practical factor that could affect commercialization of lipase-based production of flavor esters is the initial water content of the substrates [118]. In a model system using anhydrous substrates (heptanol and oleic acid), as water was produced in the synthetic reaction, the amount of water in the enzyme microenvironment increased and the reaction rate also increased, giving rise to an autocatalytic process. However, the use of water-saturated substrates led to the slowing of esterification (i.e., heptyl oleate synthesis). It was suggested that this was not due to a hydrolysis reaction, because it occurred very slowly compared to esterification. A more likely explanation is the slower

TABLE 3.9 Lipase-mediated Esterification Reactions Described in the Literature[a]

Substrates	Lipase Source
Aliphatic acids and alcohols	Lactic acid and psychrotropic bacterial whole cells
	Mucor miehei
	Candida cylindracea
	Aspergillus niger
	Immobilized *C. cylindracea*
	Immobilized *Mucor, Candida,* and *Pseudomonas*
Terpene alcohols and aliphatic acids	Various fungi
	C. cylindracea on various supports
	Various lipases on porous glass beads

[a]Bulk oganic solvents (including ethanol, *n*-heptane, and hexane) used as reaction media.

Source Adapted with permission from *CRC Critical Reviews in Biotechnology.* Copyright CRC Press, Inc. Boca Raton, FL, 1989, from Welsh et al. [78].

diffusion of the hydrophobic substrates to the enzyme-active site as the a_w increased. Although heptyl oleate has no application in flavor production, the model does suggest that the use of hydrophobic flavor precursors in organic media could be problematic unless the a_w is controlled at all stages of the process.

Zaks and Klibanov [114] have used specific buffers to modify the pH of the enzyme hydration layer before removal of water in an attempt to modify activity. In a totally nonpolar solvent, pH phenomena are limited to the aqueous shell around the enzyme. It was suggested that certain enzymes might remember the last pH prior to drying, which could modify their activity when reconstituted (e.g., lipase in solvent). A more recent study [119], based on the work of Gillies et al. [110], has suggested that the pretreatment of an enzyme with various buffers of differing pH gives results similar to those found by Zaks and Klibanov [114] for the formation of C_2 to C_5 flavor esters.

Sonomoto and Tanaka [120] have demonstrated the use of lipase in organic solvent for the preparation of optically active terpene alcohol esters for fragrance applications. Terpenes have many double bonds in their structures and tend to be oxidized during conventional organic synthesis. Lipase-based esterifications involving terpenes were studied to examine the utility of this approach. The investigators reported that cyclohexane allowed increased solubility of the apolar substrates (terpene alcohols, including DL-menthol) over totally aqueous systems, thereby allowing more effective enzyme-substrate interaction. Furthermore, the lipase equilibrium was shifted from a hydrolytic to a synthetic direction in the organic media. Successful esterification of DL-menthol with 5-phenylvaleric acid was accomplished leading to optically pure L-menthyl-5-phenylvalerate and D-menthol. As demonstrated here, enzyme biotransformations have the potential to circumvent some of the problems related to synthetic organic chemistry approaches.

A problem related to the use of any enzyme in an organic solvent, including lipase, is limited solubility of the enzyme. Early attempts to overcome this problem involved the use of lipase modified by polyethylene glycol (PEG) [121]. Because PEG is amphipathic in nature, it was thought that it could introduce solubility and stability to enzymes in organic media. By allowing an association of lipase with an amphipathic molecule, thereby stabilizing surface interactions with the surrounding milieu, one can reduce the potential for disruption of internal noncovalent stabilizing molecular interactions. A comparison of native and PEG-modified lipase showed, in general, that the PEG-modified enzyme had somewhat lower hydrolytic and synthetic capability in aqueous media. However, in various organic media, the PEG-modified lipase showed 2.4 to almost 5 times more activity than the native enzyme. Another interesting observation was the change in substrate specificity with the polarity of the organic solvent. More specifically, native *C. cylindracea* lipase showed different substrate selectivity than the PEG-modified enzyme in polar methanol versus nonpolar hexane. This may have been caused by a subtle change in the active and/or binding site(s) resulting from the PEG derivation procedure. This environmental manipulation could be useful if one wished to use the same enzyme for different end products. With better solubilization of the lipase by PEG modification, it is expected that enhanced reaction efficiencies can be obtained, making possible more efficient use of reactor volume for flavor ester generation.

The PEG–lipase complex described above is essentially in suspension and would be lost in a continuous flow-through reactor system. As an alternative to PEG modification, the immobilization of lipase on a fixed matrix would maintain and, in the process, could enhance operational stability [122]. Immobilization of enzymes has seen commercial application on a very large scale (e.g., glucose isomerase). Different methods of immobilization can be used, ranging from those leading to a loose association of the enzyme with the matrix to those resulting in more permanent associations. In the former case, ionic or hydrophobic interactions can be used, whereas, in the latter case, other means such as glutaraldehyde-induced covalent linking can be employed.

Experience has shown that conventional immobilization methods applied to lipase have resulted in preparations of low activity [122,123]. A better understanding of how the enzyme interacts with the immobilization matrix and how this relates to alterations in activity, stability, and changes in substrate specificity is essential. Shaw et al. [122] evaluated the effect of such factors as matrix hydrophobicity, pore size, chain length of the hydrocarbon attached to the matrix on the immobilization efficiency, and the specific activity of lipase from *C. cylindracea*. They found that the amount of lipase immobilized on polyvinylchloride (PVC) was affected by the length of the hydrocarbon chain attached to the matrix but not by the matrix pore size. In general, enzymes immobilized with a spacer length of two carbons had a lower activity than those with a spacer length of 6 or 12 carbons, although the spacer length did not alter specific activity of lipase. In studies on the effect of matrix pore size on lipase immobilization, they found that Sepharose® of different porosities led to similar lipase coupling efficiencies, but there was a marked decline in biocatalyst activity with decreasing porosity. It was suggested that greater mass transfer resistance was responsible for this observation.

These researchers pointed out that one of the major challenges of successfully using immobilized lipase is the accessibility of substrate, as determined by the porosity of the support. Of course, esterification of lower molecular weight esters would be less of a problem, whereas some of the higher chain alcohols and acids or bulkier aromatic substances might be more difficult. Another problem related to lipase immobilization is product inhibition caused by localized elevated levels of product around the enzyme. In one study of lipase immobilized within a PVC membrane, it was shown that the activity and operational stability of the immobilized lipase improved when controlled removal of the end products from the membrane-enzyme region was used [123].

Various other factors affecting lipase-mediated esterification reactions have been explored, including structural attributes of the substrates, such as the presence of side chains and double bonds [52,124]. Miller et al. [52] found that esterifications catalyzed by the lipase derived from *Mucor miehei* had dramatically different rates depending upon the position of an alkyl substituent on a straight chain fatty acid ($R-CH_2CH_2CH_2-COOH$). These investigators were interested in the possibility of using this enzyme with branched-chain fatty acids; they found that branching decreased the rate of esterification in all cases. Another facet of this study examined the effect of size of the alkyl substituent by using a methyl and an ethyl group. Interestingly, the enzyme showed no activity toward the ethyl-substituted fatty acid. It is clear from this work that subtle changes in precursor substituents can have a dramatic effect on enzyme recognition and productivity.

To date, lipases have shown potential for use in stereo- and regiospecific hydrolysis and esterifications to yield pure, optically active aliphatic and aromatic esters [110,120,125–127]. There is the possibility that lipases could be used to purify mixed enantiomeric compounds to yield optically active flavor esters [125,128]. However, careful selection of different lipases must be done as they can have markedly different regiospecificities [103,126]. These types of studies reinforce the need to examine systematically the interaction between enzymes and different substrates.

The commercial use of lipases under more rigorous industrial conditions (elevated temperature, extremes of pH, etc.) is likely, and as such a number of studies have examined this aspect. Lipases are known to be reasonably stable in neutral aqueous solutions at room temperature. In general, pancreatic and many extracellular microbial lipases lose activity during storage at temperatures above 40°C, although a number of microbial lipases are more resistant to heat inactivation [103]. Therefore, lipases with extreme heat stability would be particularly useful for some industrial applications because higher operational temperatures (e.g., >65°C) could accelerate catalysis and also preclude microbial contamination. Contamination could result in, for example, potential enzyme degradation (proteolysis) and product contamination. Another important operational parameter that can allow for better processing (e.g., minimize chances of contamination) is pH. Lipases are known to have a broad pH-activity profile, with greatest activity normally seen in the range of pH 6 to 8 [103]. Exceptions include lipases produced by *Rhizopus arrhizus* [129], which have a very low pH optimum, and by *Pseudomonas nitroreductans* [130], which have a pH optimum of 11. Clearly, more detailed studies related to thermotolerance and other physiochemical extremes of enzymes are required to enhance bioprocessing of flavors.

Other important practical factors come into play when designing an industrial enzyme-based process. For example, a lipase from a given biological system can vary in performance from one supplier to another. Langrand et al. [126] have shown that the rates of formation of various esters from different acids and alcohols vary greatly depending on the biological source of the lipase

(Table 3.10). The commercialization of any such process should attempt to determine, as early as possible, the variability of the lipase supply from different enzyme lots.

The practical use of any enzyme for industrial application, apart from its operational characteristics, depends on its availability in substantial amounts. New biotechnology or rDNA approaches are powerful methods with potential for the augmentation of production of commercial quantities of important en-

TABLE 3.10   Ester Formation Using Lipase

| | Degree of Conversion (%)[a] | | | | | |
| | Isoamyl Esters | | | Geranyl Esters | | |
Lipase	Acetate	Propionate	Butyrate	Acetate	Propionate	Butyrate
Alcaligenes (Meito-Sangyo)	0.5	3	18	<0.5	4	24
Aspergillus (Rohm)	23[b]	33	40[b]	6[b]	15	6[b]
Fungal (Rohm)	0.5	40	65[b]	<0.5	48	61
A. niger (Palatase)	0.5[b]	22	98[b]	<0.5[b]	5	40[b]
C. rugosa (cylindracea) (MY)	3.5	95	98	<0.5	95	90
G. candidum (SNEA)	<0.5	2	18[b]	<0.5	2	5
M. miehei (Gist-brocades)	80[b]	92	97	14	96	96
M. miehei (Lipozyme)	15	42	98	<0.5	30	93
P. cyclopium A (SNEA)	0.5	4	48	<0.5	2.5	38
P. cyclopium B (SNEA)	18	3.5	45	14	3.5	38
Porcine pancreas (Rohm)	0.5	55	95	<0.5	94	85
Porcine pancreas (Kochlight)	0.5	0.5	35	0.5	<0.5	12
R. arrhizus (SNEA)	5	70	97	3.5	58	97

[a]Using 0.2 g lipase with analysis at 24 h.
[b]0.5 g of lipase was used instead of 0.2 g.

Source   Reprinted by permission of Science and Technology Letters, England from Langrand et al. [126].

zymes (e.g., lipases). Riisgaard [111] of Novo Nordisk A/S presented an interesting example of how a bench scale-level development could be made into a commercial reality by applying rDNA technology. He found that a lipase with suitable functional properties was produced by an unspecified organism isolated from the environment in a wide-ranging screening program, but major obstacles remained to be resolved. These obstacles included the use of an uncharacterized microorganism (giving rise to technical and regulatory issues) and slow growth rates of the microorganism. From the regulatory standpoint, process approval can be facilitated if one uses a well-characterized microbial system (ideally having had precedent for use in food-related processing). As such, it was decided to produce a strain through lipase gene transfer to a GRAS host suitable for industrial production. The fungus *Aspergillus oryzae*, which is used extensively for food enzyme production, was selected as the host. The initial cloning strategy is outlined in Figure 3.11.

Not only did the outlined procedure allow the development of a suitable, high enzyme-yielding, industrial strain but it also led to improvements in product purification. That is, the lipase could be isolated directly after enzyme crystallization, a result which clearly has a favorable impact on economics. Riisgaard [111] anticipates that in the future highly pure products will become the standard for industrial enzyme production. This example illustrates an approach that can be taken not only with lipase but with many other enzymes and opens the possibility of using more exotic but limited-availability enzymes

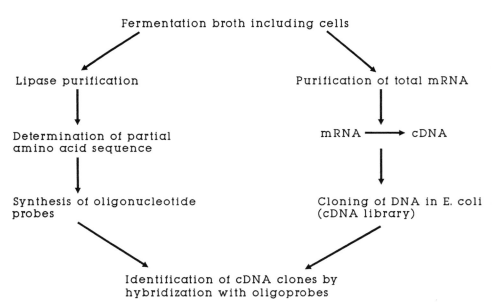

FIG. 3.11   Initial cloning studies for lipase from *Mucor miehei* with *Aspergillus oryzae* as the host. *Source*   From Riisgaard [111].

for production of flavors (e.g., lyase enzyme for ''green'' alcohols/aldehydes—
see Section 3.3.7).

3.4.2 Aldehyde Generation by Enzymes

A number of aliphatic and aromatic aldehydes can be produced using isolated,
well-known enzymes such as alcohol oxidase and alcohol dehydrogenase as
discussed below, although there are very few examples of commercial success.
However, there is an exciting possibility that novel reactions with atypical
substrates can occur with other enzymes. As an illustration, it has been shown
that galactose oxidase, which normally oxidizes D-galactose to give D-galacto-
hexodialdose, can also catalyze the reaction of certain atypical substrate aliphatic and aromatic alcohols to give the respective aldehydes [131]. For example, glycerol is converted to glyceraldehyde. Although these particular products are not related directly to flavors, the example demonstrates the need to
look beyond the obvious substrate and enzyme relationships. Many more such
examples will begin to surface as more is learned about molecular enzyme
mechanisms.

The market demand for acetaldehyde and the possibility of producing it
through a simple one-step biotransformation of ethanol has put much focus on
this particular product. One of the early attempts to develop a natural enzymatic
route to acetaldehyde involved alcohol dehydrogenase (ADH) in a complex
processing scheme (Fig. 3.12) [132]. The original intent of this study was to
generate acetaldehyde from ethanol *in situ* in orange essence to restore freshness. Because the process used a dehydrogenase, the requirement for its cofactor nicotinamide adenine dinucleotide (NAD^+) was a potential problem.
That is, the oxidation of ethanol to acetaldehyde results in the reduction of the
NAD^+ to NADH. If the NADH can not be reoxidized to NAD^+, a stoichiometric amount of an expensive cofactor is required, which would be impractical. One way to reoxidize the NADH as it is formed is to add a second
coupled substrate. For example, cinnamaldehyde could be added to regenerate
the cofactor to the oxidized state and also simultaneously generate cinnamyl
alcohol [133].

Alternatively, in the acetaldehyde process mentioned above [132], regeneration was achieved by light-catalyzed oxidation of the reduced cofactor with

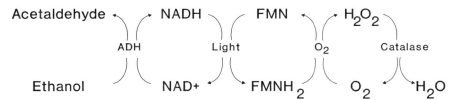

FIG. 3.12 Alchohol deyhdrogenase (ADH) enzyme system coupled with light-based
catalysis for cofactor regeneration.

flavin mononucleotide (FMN); fluorescent lighting could be used. The reduced FMN (i.e., $FMNH_2$) was in turn reconverted to FMN by oxidation with molecular oxygen. As a consequence, hydrogen peroxide was produced but decomposed into oxygen and water by virtue of the inclusion of the enzyme catalase. The formed acetaldehyde was complexed *in situ* by reacting with TRIS. Unfortunately, ADH is sensitive to both light- and oxygen-based regenerative processes, which cause its degradation, inactivation, and/or destruction. To prevent these problems, the ADH was isolated from the potentially destructive steps (i.e., light and oxygen regeneration) by use of an Amicon H1P10 hollow fiber semipermeable membrane cartridge. A steady-state concentration of 2.5 g/L of aldehyde was achieved in the effluent, using a continuous ethanol substrate infusion of 30 g/L. The aldehyde was trapped by the TRIS buffer as an addition product, which could be later regenerated by an alkaline pH shift for product recovery. Unfortunately, this step could present a problem because the use of an inorganic base might negate the natural designation of the final flavor (see Chapter 2). Presumably, gas stripping of the effluent to remove the acetaldehyde could be used (acetaldehyde b.p. 21°C), but then very efficient external traps would have to be incorporated for downstream aldehyde recovery. Even through the process is not used commercially, this example does serve to highlight the potential along with the problems associated with the use of isolated cofactor-requiring enzymes for flavor aldehyde production.

Another means of producing acetaldehyde by isolated enzymes involves the use of alcohol oxidase (AO) [134]. In Section 3.3.1, a whole cell, AO-based conversion of ethanol to acetaldehyde by *P. pastoris* was described. It is also possible to isolate the AO from the yeast and use it for the bioconversion, thereby avoiding other metabolites in the final product (Fig. 3.4). However, for catalysis to proceed the isolated AO would require the presence of a cofactor (e.g., FAD). Although the enzyme itself can be obtained quite readily at a reasonable cost, the cofactor is expensive so extensive reuse is necessary. Another problem with the use of isolated AO is limited resistance to end-product inhibition by acetaldehyde [134]. The mechanism for this inhibition was speculated to involve competition with ethanol for active sites on the enzyme. Alternatively, the aldehyde could inactivate the enzyme by destroying essential sulfhydryl groups. Others have shown that the latter mechanism is important with respect to hydrogen peroxide-mediated inactivation of AO [135]. The hydrogen peroxide is generated as a by-product during the oxidation of the alcohol to the aldehyde. The approach described earlier, involving the use of TRIS for complexing acetaldehyde as it is produced [132], was also used by Duff and Murray [134] with AO. The binding, however, led to the release of H^+ during the amine/aldehyde complexation, and the resultant drop in pH significantly reduced the AO activity and stability. However, other buffers in the reaction medium were used quite successfully to counter this problem.

Other means to continuously remove acetaldehyde have been studied. For example, Armstrong et al. used a continuous gas stripping approach with whole cell acetaldehyde-producing fermentations for the alleviation of product toxicity [136]. This latter method can also be applied to isolated enzyme systems. Because acetaldehyde has a boiling point of only 21°C, it is readily removed by air sparging, which is already present for the oxidative bioconversion to proceed. There are various ways of trapping the aldehyde externally (including adsorbents and cryogenic methods), but these could be cost-prohibitive on a large commercial scale.

The use of ADH and other cofactor-requiring enzymes holds much potential. However, the key bottleneck will be the ability to recycle the cofactors on the order of 10^3 to 10^4 times, which would be necessary to achieve an acceptable cost. At present NAD costs \$700/mole and NADPH costs \$200,000/mole. Obviously approaches to recycle and minimize cofactor usage and/or maximize stability are critical.

3.5 ENHANCED ENZYME EXTRACTION AND PROCESSING OF BOTANICALLY DERIVED FLAVOR COMPONENTS

In addition to biosynthetic reactions for the production of natural flavors and fragrances, there is great potential for other applications of enzyme-based extraction technologies. These are especially useful with complex flavor mixtures where it would be technically difficult to re-create the final sensory impression by blending individual natural flavor components. Enzymes such as pectinase, cellulase, β-glucanase, and hemicellulase can be used to loosen the structural integrity of botanical material (Fig. 3.13), thereby enhancing extraction of the desired flavor components. These enzymatic reactions are normally conducted at low temperatures (15°C to 25°C), which could be important if the process entails isolating thermolabile components [137]. Some examples follow to illustrate the utility of enzyme-enhanced extractions of botanical flavor compounds.

A commercial Japanese process uses fungal cellulases to enhance juice extraction from oranges [138]. This approach allows for a gentle extraction of flavor components from the fruit without disruption of the rind oil glands which contain off-flavors and bitter notes. Another Japanese report describes the use of genetic engineering to enhance enzymic extraction of flavors [139]. The researchers isolated the genes for pectinase from the bacterium *Erwinia carotovora*, the causative agent in root-rot diseases. Through rDNA technology, the group was able to express the gene in *Escherichia coli* but at only 25 percent of the original host organism. However, with the addition of a pectinase inducer, the enzyme production increased 20-fold, 90 percent of which was excreted into the growth medium. With large, inexpensive supplies of the aforementioned enzymes, the prospects of enzyme-mediated flavor release from plant tissue appear promising.

Apart from technical advances in the use of enzymes for enhanced extraction of flavors from botanical materials, regulatory issues could also influence this

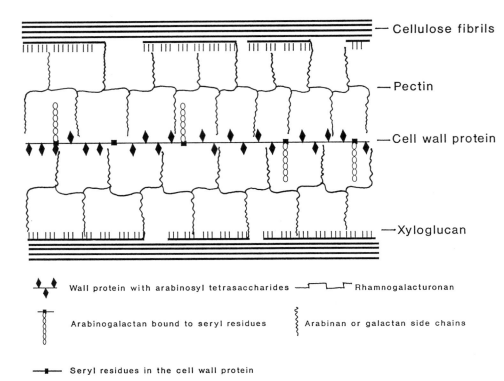

- Cellulose fibrils

- Pectin

- Cell wall protein

- Xyloglucan

Wall protein with arabinosyl tetrasaccharides ⌐‾⌐‾⌐ Rhamnogalacturonan

Arabinogalactan bound to seryl residues   Arabinan or galactan side chains

Seryl residues in the cell wall protein

FIG. 3.13   Model for the microstructure of the plant cell wall.

area. To illustrate, an important product that could be affected by these changes is benzaldehyde (useful in cherry-type flavorings). With the latest proposed changes in FDA/FEMA guidelines regarding the use of inorganic acids and bases to produce natural flavors, there could be problems with benzaldehyde extraction from fruit pits or bitter almonds. Amygdalin, which is a cyanogenic glycoside found in these sources, can be cleaved with inorganic acids to yield 1 mole of benzaldehyde, 2 moles of glucose, and 1 mole of hydrogen cyanide. Benzaldehyde is present in the bitter almond essential oil at levels up to 85 percent. However, the use of inorganic acids and bases could become a "gray" area with respect to natural processing requirements. FEMA and the FDA are currently addressing this area of concern, but clearly the use of biological approaches for hydrolysis of the glycoside would be considered natural. For example, the breakage of the glycosidic linkage with a β-glucosidase is possible [140]. This enzyme is normally found in emulsin, a preparation derived from almonds and other sources. It is possible that high levels of this enzyme could be produced by microbial systems to enhance the natural hydrolysis reaction that occurs in the botanical source.

The examples given above are illustrations of the opportunities for the biological use of enzymes for extraction of flavor components from botanical materials. Flavor-producing industries most will likely develop biotransformation approaches to flavor compound generation, but without question they also will look for novel, more efficient means to extract known botanical species. This latter area probably will see increased investigation in the near future.

3.6 CONCLUSION

Aliphatic, aromatic, and lactone flavor substances play a pivotal role in flavor formulations on their own or as precursors for production of other important compounds. It has been shown that biological processing has great potential for production of these compounds and will continue to form the basis of commercial processes.

REFERENCES

1. Omelianski, V.L., *J. Bacteriol.* **8,** 393 (1922).
2. Angold, R., Beech, G., and Taggart, J., *Cambridge Studies in Biotechnology*, Vol. 7. Cambridge University Press, New York, 1989.
3. Kaushik, S., *Chem. Mark. Rep.* **237,** SR30 (1990).
4. Armstrong, D.W., and Yamazaki, H., *Trends Biotechnol.* **4,** 264 (1986).
5. Newell, N., and Gordon, S., in *Biotechnology in Food Processing* (S. Harlander and T. Labuza, eds.). Noyes Publishers, Park Ridge, NJ, 1986.
6. Collins, R.P., and Halim, A.F., *J. Agric. Food Chem.* **20,** 437 (1972).
7. Armstrong, D.W., Gillies, B., and Yamazaki, H., in *Flavor Chemistry: Trends and Developments* (R. Teranishi, R. Buttery, and F. Shahidi, eds.). Am. Chem. Soc., Washington, DC, 1989.
8. Dhavlikar, R.S., and Albroscheit, G., *Dragoco Rep. (Engl. Ed.)* **12,** 251 (1973).
9. Nelson, J., McCormick & Strange, Hunt Valley, MD, personal communication (1990).
10. Topfer, K., *Chem. Mark. Rep.* **238,** 34 (1990).
11. Regenstein, J.M., and Regenstein, C.E., *Food Technol.* **44**(7), 90 (1990).
12. Regenstein, J.M., and Regenstein, C.E., *Food Technol.* **42**(6), 86 (1988).
13. Jourdain, N., Goli, T., Jallageas, J.C., Crouzet, C., Ghommidh, C., Navarro, J.M., and Crouzet, J., in *Topics in Flavour Research* (R. Berger, S. Nitz and P. Schreier, eds.). Eichhorn, Marzling-Hangenham, Germany, 1985.
14. Sastry, K.S.M., Singh, B.P., Manavalan, R., Singh, P., and Atal, C.K., *Indian J. Exp. Biol.* **18,** 836 (1980).
15. Lanza, E., Ko, K.H., and Palmer, J.K., *J. Agric. Food Chem.* **24,** 1247 (1976).
16. Hanssen, H.-P., Sprecher, E., and Klingenberg, A., *Z. Naturforsch., C: Biosci.* **39C,** 1030 (1984).
17. Berger, R.G., Neuhaeuser, K., and Drawert, F., *Biotechnol. Bioeng.* **30,** 987 (1987).

18. Hattori, S., Yamaguchi, Y., and Kanisawa, T., *Proc. Int. Conf. Food Sci. Technol.*, *4th*, *1974*, Vol. 2 (1974).
19. Tonon, F., and Odier, E., *Appl. Environ. Microbiol.* **54**, 466 (1988).
20. Moore, K., and Towers, G.H.N., *Can. J. Biochem.* **45**, 1659 (1967).
21. Lee, C.-W., and Richard, J., *J. Dairy Res.* **51**, 461 (1984).
22. Armstrong, D.W., Martin, S.M., and Yamazaki, H., *Biotechnol. Bioeng.* **26**, 1038 (1984).
23. Armstrong, D.W., and Yamazaki, H., *Biotechnol. Lett.* **6**, 819 (1984).
24. Gross, B., Gallois, A., Spinnler, H.-E., and Langlois, D., *J. Biotechnol.*, **10**, 303 (1989).
25. Buchanan, R., in *Biotechnology in Food Processing* (S. Harlander and T. Labuza, eds.). Noyes Publishers, Park Ridge, NJ, 1986.
26. Byrne, B., and Sherman, G., *Food Technol.* **38**(7), 57 (1984).
27. Murray, W.D., Duff, S.J., Lanthier, P.H., Armstrong, D.W., Welsh, F.W., and Williams, R.E., in *Frontiers of Flavor* (G. Charalambous ed.). Elsevier, Amsterdam, 1988.
28. Wekker, M.S., and Zall, R.R., *Appl. Environ. Microbiol.* **53**, 2815 (1987).
29. Duff, S.J., and Murray, W.D., *Biotechnol. Bioeng.* **31**, 790 (1988).
30. Ellis, S.B., Brust, P.F., Koutz, P.J., Waters, A.F., Harpold, M.M., and Gingeras, R.R., *Mol. Cell. Biol.* **5**, 1111 (1985).
31. Veenhuis, M., Zwart, K., and Harder, W., *FEMS Microbiol. Lett.* **3**, 21 (1978).
32. Bormann, C., and Sahm, H., in *Current Developments in Yeast Research: Advances in Biotechnology* (G. Stewart and I. Russell, eds.). Pergamon Press, Toronto, 1981.
33. Klibanov, A.M., in *Biotechnology Challenges for the Flavor and Food Industry* (R. Lindsay and B. Willis eds.), Elsevier Applied Science, London, 1989.
34. Schneider, H., Little, R., and Ross, N.W., *Biotechnol. Tech.* **4**, 85 (1990).
35. Lane, A., in *Biotechnology and the Food Industry* (P. Rogers and G. Fleet, eds.), Gordon & Breach, New York, 1987.
36. Boeing, A.J., Buck, K.T., and Dolfini, J.E., U.S. Patent 4,673,766 (1987).
37. MacKintosh, R.W., and Fewson, C.A., *Biochem. J.* **255**, 653 (1988).
38. Birkinshaw, J.H., Morgan, E.N., and Findlay, W.P., *Biochem. J.* **50**, 509 (1952).
39. Birkinshaw, J.H., Chaplon, P., and Findlay, W.P., *Biochem. J.* **66**, 188 (1957).
40. Chen, C.C., and Wu, C.M., *J. Food Sci.* **49**, 1208 (1984).
41. Williams, R.E., Armstrong, D.W., Murray, W.D., and Welsh, F.W., *Ann. N.Y. Acad. Sci.* **542**, 406 (1988).
42. Duff, S.J., and Murray, W.D., *Biotechnol. Bioeng.* **34**, 153 (1989).
43. Collins, R.P., and Halim, A.F., *Can. J. Microbiol.* **18**, 65 (1971).
44. Nordstrom, K., *J. Inst. Brew.* **69**, 142 (1963).
45. Morgan, M.E., *Biotechnol. Bioeng.* **18**, 953 (1976).
46. Yoshila, K., and Hashimoto, N., *Agric. Biol. Chem.* **45**, 2183 (1981).
47. Gray, W.D., *Am. J. Bot.* **36**, 475 (1949).
48. Armstrong, D.W., unpublished results (1985).
49. Farbood, M.I., Morris, J.A., and Seitz, E.W., U.S. Patent 4,657,862 (1987).

50. Stofberg, J., and Grundschober, F., *Perfum. Flavor.* **12,** 27 (1987).

51. Hosono, A., Elliot, J.A., and McGugan, W.A., *J. Dairy Sci.* **57,** 535 (1974).

52. Miller, C., Austin, H., Posorske, L., and Gonzlez, J., *J. Am. Oil Chem. Soc.* **65,** 927 (1988).

53. Bell, G., Blain, J.A., Patterson, J.D.E., Shaw, C.E.L., and Todd, R., *FEMS Microbiol. Lett.* **3,** 223 (1978).

54. Gancet, C., and Guignard, C., *Rev. Fr. Corps Gras* **33,** 423 (1986).

55. Kaminski, E., Stawicki, S., and Wasowicz, E., *Appl. Microbiol.* **27,** 1001 (1974).

56. Gracheva, I.M., in *Microbiology* (L.S. Smirnova, ed.), Vol. 1. G.K. Hall & Co., Boston 1974.

57. Berry, D.R., and Watson, D.C., in *Yeast Biotechnology* (D.R. Berry, I. Russell, and G. Stewart, eds.). Allen & Unwin, London, 1987.

58. Nordstrom, K., and Carlsson, B., *J. Inst. Brew.* **71,** 171 (1965).

59. Halim, A.F., and Collins, R.P., *Lloydia* **38,** 87 (1975).

60. Crueger, W., and Crueger, A., in: *Biotechnology: A Textbook of Industrial Microbiology* (T.D. Brock, ed.). Science Tech, Madison, WI, 1982.

61. Playne, M.J., in: *Comprehensive Biotechnology* (H.W. Blanch, S. Drew, and D.I.C. Wang, eds.), Vol. 3, Pergamon Press, Toronto, 1985.

62. Clausen, E.C., and Gaddy, J.L., in *Advances in Biotechnology* (M. Moo-Young and C.W. Robinson eds.), Vol. 2. Pergamon Press, Toronto, 1981.

63. Sharpell, F., and Stegmann, C., *Adv. Biotechnol.* **2,** 71 (1988).

64. Bauer, K., Garbe, D., and Surburg, H., *Common Fragrance and Flavor Materials*, 2nd ed. VCH Verlagsgesellschaft, Weinheim, Germany, 1990.

65. *Flavor and Fragrance Materials.* Allured Publ. Co., Wheaton, IL, 1989.

66. Lawrence, R.C., and Hawke, J.C., *J. Gen. Microbiol.* **51,** 289 (1968).

67. Yagi, T., Kawaguchi, M., Hatano, T., Fukui, F., and Fukui, S., *J. Ferment. Bioeng.* **70,** 94 (1990).

68. Yagi, T., Hatano, A., Kawaguchi, M., Hatano, T., Fukui, F., and Fukui, S., *J. Ferment. Bioeng.* **70,** 100 (1990).

69. Yagi, T., Kawaguchi, M., Hatano, T., Fukui, F., and Fukui, S., *J. Ferment. Bioeng.* **68,** 188 (1989).

70. Vezina, C., in *Basic Biotechnology* (J. Bu'lock and B. Kristiansen, eds.). Academic Press, London, 1987.

71. Larroche, C., Arpah, M., and Gros, J.-B., *Enzyme Microb. Technol.* **11,** 106 (1989).

72. Creuly, C., Larroche, C., and Gros, J.-B., *Appl. Microbiol. Biotechnol.* **34,** 20 (1990).

73. May, S.W., Steltenkamp, M.S., Borah, K.R., Katopodis, A.G., and Thowsen, J.R., *J. Chem. Soc., Chem. Commun.* **19,** 845 (1979).

74. Kinderlerer, J.L., *Phytochemistry* **26,** 1417 (1987).

75. Ayer, W.A., and Singer, P.P., *Phytochemistry.* **19,** 2717 (1980).

76. Ohloff, G., *Prog. Chem. Org. Nat. Prod.* **35,** 431–526 (1978).

77. Maga, J.A., *Crit. Rev. Food Sci. Nutr.* **8,** 1 (1976).

78. Welsh, F.W., Murray, W.D., and Williams, R.E., *Crit. Rev. Biotechnol.* **9,** 105 (1989).

79. Berger, R.G., Newhauser, K., and Drawert, F., Z. *Naturforsch.*, C: Biosci. **41C**, 963 (1986).

80. Berger, R.G., Newhauser, K., and Drawert, F., *Flavour Fragrance J.* **1**, 181 (1986).

81. Kapfer, G., Berger, R.G., and Drawert, F., *Biotechnol. Lett.* **11**, 561 (1989).

82. Drawert, F., Berger, R.G., and Newhauser, K., *Eur. J. Appl. Microbiol. Biotechnol.* **18**, 124 (1983).

83. Moss, M.O., Jackson, R.M., and Rogers, D., *Phytochemistry.* **14**, 2706 (1975).

84. Yong, F.M., Wong, H.A., and Lim, G., *Appl. Microbiol. Biotechnol.* **22**, 146 (1985).

85. Labows, J., McGinley, K., Leyden, J., and Webster, G., *Appl. Environ. Microbiol.* **38**, 412 (1979).

86. Tahara, S., Fujiwara, K., and Mizutani, J., *Agr. Biol. Chem.* **37**, 2855 (1973).

87. Tressl, R., Apetz, M., Arrieta, R., and Grunewald, K.G., in *Flavor of Foods and Beverages*, (G. Charalambous, ed.). Academic Press, New York, 1978.

88. Okui, S., Uchiyama, M., and Mizigaki, M., *J. (Tokyo) Biochem.* **54**, 536 (1963).

89. Farbood, H., and Willis, B., European Patent PCT 1072 (1983).

90. Dziezak, J., *Food Technol.* **40**(4), 108 (1986).

91. Gervais, P., and Battut, G., *Appl. Environ. Microbiol.* **55**, 2939 (1989).

92. Muys, G.T., van der Ven, B., and de Jonge, A.P., *Nature (London)* **194**, 995 (1962).

93. Unilever, European Patent 0371568 (1990).

94. Belitz, H.-D., and Grosch, W., *Food Chemistry.* Springer-Verlag, Berlin, 1987.

95. Shimizu, S., Hattori, S., Hata, H., and Yamada, H., *Appl. Environ. Microbiol.* **53**, 519 (1987).

96. Gatfield, I.L., *Ann. N.Y. Acad. Sci.* **434**, 569 (1984).

97. Eriksson, C.E., in *Progress in Flavour Research* (D.G. Land and H.E. Nursten, eds.). Applied Science, London, 1979.

98. Gatfield, I.L., in *Biogeneration of Aromas* (T.H. Parliment and R. Croteau, eds.). Am. Chem. Soc., Washington, DC, 1986.

99. Gatfield, I.L., *Food Technol.* **42**(10), 110 (1988).

100. Berger, R.G., Kler, A., and Drawert, F., *Plant Cell, Tissue Organ Cult.* **8**, 147 (1987).

101. Beppu, T., Shoun, H., Sudo, Y., and Seto, Y., in *Oxygenases and Oxygen Metabolism* (M. Nozaki, S. Yamamoto, Y. Ishimura, M. Coon, L. Ernster, and R. Estabrook, eds.). Academic Press, New York, 1982.

102. Sekiya, J., Tanigawa, S., Kajiwara, T., and Hatanaka, A., *Phytochemistry* **23**, 13 (1982).

103. Macrae, A.R., and Hammond, R.C., *Biotechnol. Genet. Eng. Rev.* **3**, 193 (1985).

104. Lilly, M.D., and Woodley, J.M., *Stud. Org. Chem.* **22**, 179 (1985).

105. Borgstrom, B., and Brockman, H.L., *Lipases.* Elsevier, Amsterdam, 1984.

106. Semeriva, M., and Desnuelle, P., *Adv. Enzymol.* **48**, 319 (1979).

107. Kloosterman, M., Elferink, V.H.M., van Lersel, J., Roskam, J.-H., Meijer, E.M., Hulshof, L.A., and Sheldon, R., *Trends Biotechnol.* **6**, 251 (1988).

108. Lee, K.-C.M., Shi, H., Huang, A.-S., Carlin, J.T., Ho, C.-T., and Chang, S.S., in *Biogeneration of Aromas* (T.H. Parliment and R. Croteau, eds.). Am. Chem. Soc., Washington, DC, 1986.

109. Macrae, A.R., in *Microbial Enzymes and Technology* (W.M. Fogarty, ed.). Applied Science, London, 1983.

110. Gillies, B., Yamazaki, H., and Armstrong, D.W., in *Biocatalysis in Organic Media* (C. Laane, J. Tramper, and M.D. Lilly, eds.). Elsevier, Amsterdam, 1987.

111. Riisgaard, S., *Genet. Eng. Biotechnol.* **10,** 11 (1990).

112. Iwai, M., Tsujisaka, Y., and Fukumoto, J., *J. Gen. Appl. Microbiol.* **10,** 13 (1964).

113. Boyer, J.L., Gilot, B., and Guiraud, R., *Appl. Microb. Biotechnol.* **33,** 372 (1990).

114. Zaks, A., and Klibanov, A., *Proc. Natl. Acad. Sci. U.S.A.* **82,** 3192 (1985).

115. Waks, M., *Proteins: Struct., Funct. Genet.* **1,** 4 (1986).

116. Gillies, B., Yamazaki, H., and Armstrong, D.W., *Biotechnol. Lett.* **9,** 709 (1987).

117. Welsh, F.W., and Williams, R.W., *J. Food Sci.* **54,** 1565 (1989).

118. Goldberg, M., Thomas, D., and Legoy, M.-D., *Enzyme Microb. Technol.* **12,** 976 (1990).

119. Welsh, F.W., and Williams, R.W., unpublished data (1988).

120. Sonomoto, K., and Tanaka, A., *Ann. N.Y. Acad. Sci.* **542,** 235 (1988).

121. Baillargeon, M.W., and Sonnet, P.E., *Ann. N.Y. Acad. Sci.* **542,** 244 (1988).

122. Shaw, J.-F., Chang, R.-C., Wang, F.F., and Wang, Y.J., *Biotechnol. Bioeng.* **35,** 132 (1990).

123. Rucka, M., and Turkiewicz, B., *Enzyme Microb. Technol.* **12,** 52 (1990).

124. Macrae, A.R., in: *Biotechnology for the Oils and Fats Industry* (C. Ratledge, P. Dawson, and J. Rattray, eds.). Am. Oil Chem. Soc., Champaign, IL, 1984.

125. Triantaphylides, C., Langrand, G., Millet, H., Rangheard, M.S., and Buono, G., *Bioflavour'87: Proc. Int. Conf., 1987*, 531–542 (1988).

126. Langrand, G., Triantaphylides, C., and Baratti, J., *Biotechnol. Lett.* **10,** 549 (1988).

127. Tressl, R., Heidlas, J., Albrecht, W., and Engel, K.H, *Bioflavour '87: Proc. Int. Conf., 1987*, 221–236 (1988).

128. Cambou, B., and Klibanov, A.M., *J. Am. Chem. Soc.* **106,** 2687 (1984).

129. Yamaguchi, T., Muroya, N., Isobe, M., and Sugiura, H., *Agric. Biol. Chem.* **37,** 999 (1973).

130. Watanabe, N., Ota, Y., Minoda, Y., and Yamada, K., *Agric. Biol. Chem.* **41,** 1353 (1977).

131. Klibanov, A.M., Alberti, B.N., and Marletta, M.A., *Biochem. Biophys. Res. Commun.* **108,** 804 (1982).

132. Raymond, W.R., U.S. Patent 4,481,292 (1984).

133. Deetz, J.S., and Rozzell, J.D., *Ann. N.Y. Acad. Sci.* **542,** 230 (1988).

134. Duff, S.J.B., and Murray, W.D., *Ann. N.Y. Acad. Sci.* **542,** 428 (1988).

135. Couderc, R., and Baratti, J., *J. Agric. Biol. Chem.* **44,** 2279 (1980).

136. Armstrong, D.W., Martin, S.M., and Yamazaki, H., U.S. Patent 4,720,457 (1988).

137. Ehlers, G.M., *Food Technol. N.Z.* **23,** 31 (1988).

138. Godfrey, T., in: *Industrial Enzymology: The Application of Enzymes in Industry* (T. Godfrey and J. Riechelt, eds.). Nature Press, New York, 1983.

139. Anonymous, *Biotechnol. Newswatch* **5,** 7 (1985).

140. Haisman, D.R., and Knight, D.J., *Biochem. J.* **103,** 528 (1967).

Fermentation Production
Of Pyrazines and Terpenoids
For Flavors and Fragrances

EUGENE W. SEITZ

Chr. Hansen's Laboratory Inc.,
Milwaukee, Wisconsin

INTRODUCTION

The first section of the chapter reviews the fermentation production of important flavor molecules known as *pyrazines*. The focus is on pyrazines produced in foods by various microorganisms and in broth systems by actinomycetes, Gram negative, and Gram positive bacteria and filamentous fungi. Only one organism, a Gram negative mutant *Pseudomonas*, was shown able to yield over 1.0 percent of a useful potato-like pyrazine.

The second section reviews production of valuable flavor and fragrance terpenoids by microorganisms. The emphasis is on molecules produced by actinomycetes, bacteria, yeasts, and filamentous fungi. Bacteria and filamentous fungi appear to be the most versatile in the variety of terponoids produced. Examples are presented showing both microbial forms able to transform acyclic, monocyclic, and bicyclic monoterpenoids. However, γ-irone transformations were reserved for bacteria while filamentous fungi were portrayed in transformation of ionones and sesquiterpenoids.

The highest yielding process involving a yeast (*Cryptococcus*) provides a 10.0 percent yield of a diterpenoid called *sclareolide*. This compound is of value in tobacco and food for flavoring and as a synthetic substrate for ambra derivatives useful in fragrances.

Bioprocess Production of Flavor, Fragrance, and Color Ingredients, Edited by Alan Gabelman, ISBN 0-471-03821-0 © 1994 John Wiley & Sons, Inc.

4.1 FERMENTATION PRODUCTION OF PYRAZINES

4.1.1 INTRODUCTION

Pyrazines are heterocyclic nitrogen-containing compounds, which since the mid-1960s, have been shown to contribute significantly to the unique taste and aroma of roasted or toasted foods [1]. They are important flavor molecules because of their diverse organoleptic properties. The FEMA GRAS list [2] of substituted pyrazines shows that 34 such compounds are approved for use in flavoring. For example (see Fig. 4.1), 2-isobutyl-3-methoxy pyrazine provides bell pepper odor, 5-ethyl-2,3-dimethyl pyrazine contributes to the roasted or nutty character of certain heated foods, tetramethyl pyrazine is described as pungent, and 2,6-dimethyl-3-methoxy pyrazine as musty [3].

The first evidence that microorganisms were able to produce pyrazines was provided by Kosuge et al. [4]. They showed that tetramethyl pyrazine (Fig. 4.1) is present in natto, a Japanese fermented soybean food. The compound is responsible for the characteristic odor of this food and a follow-up study [5] showed that *Bacillus subtilis* grown on a sucrose asparagine medium produced tetramethyl pyrazine. Studies by Reineccius et al. [6] implicated microorganisms in the formation of the same pyrazine during the fermentation of cocoa beans. More recently Barel et al. [7] confirmed the role of microorganisms in producing pyrazines in chocolate.

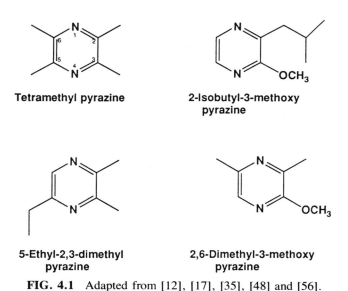

Tetramethyl pyrazine

2-Isobutyl-3-methoxy
pyrazine

5-Ethyl-2,3-dimethyl
pyrazine

2,6-Dimethyl-3-methoxy
pyrazine

FIG. 4.1 Adapted from [12], [17], [35], [48] and [56].

4.1.2 Pyrazines Produced in Foods by Microorganisms

Cheeses such as Camembert and Brie are ripened by a surface organism called *Penicillium caseicolum*. This organism, when grown in a defined medium [8], produced 2-methoxy-3-isopropyl pyrazine (Fig. 4.2), which imparted an earthy, nutty, potato-like flavor. Furthermore, rehydrated sterilized (autoclaved or gamma-irradiated) coconut supported luxuriant growth of a *Bacillus subtilis* culture. The product assumed a pungent odor, which was due to two alkyl pyrazines, namely, tetramethyl (Fig. 4.1) and 2,3,5-trimethyl pyrazine (Fig. 4.2).

The studies in Japan by Kosuge et al. [9] established that foods such as miso, soy sauce, natto, vinegar, and sake all contain tetramethyl pyrazine. The highest level (265 μg/kg) was found in miso stored for 1 year at room temperature. Only 29 μg/kg was found in 1-month old miso. Natto is made by fermenting cooked soybeans with a variant of *Bacillus subtilis* and was found to contain 22 μg/kg of this pyrazine compound. Sugawara et al. [10] showed that cooked soybeans which were beany in odor due to 2-pentyl furan and 1-octen- 3-ol (Fig. 4.3) contained no pyrazines. However, the fermented natto product made from the same beans contained both pyrazines and sulfur compounds that contributed significantly to the odor of the fermented food and masked completely the beany, cooked soybean odor.

Volatile compounds produced by *Pseudomonas perolens* ATCC# 10757 in sterile fish muscle (*Sebastes melanops*) included 2-methoxy-3-isopropyl pyrazine (Fig. 4.2), which was primarily responsible for the musty, potatolike odor in the fish tissue [11]. The same pyrazine was produced by *Pseudomonas*

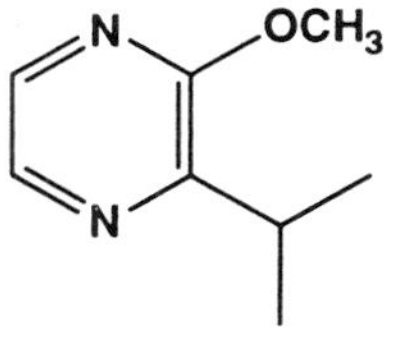

2-Methoxy-3-isopropyl pyrazine

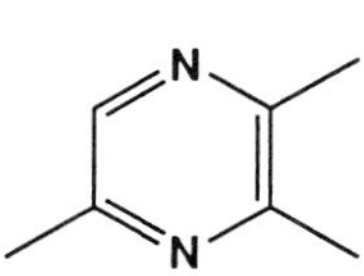

2,3,5-Trimethyl pyrazine

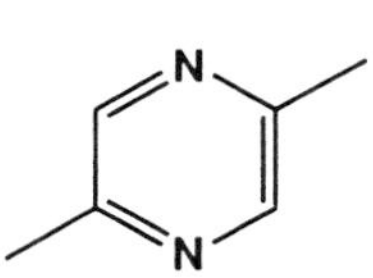

2,5-Dimethyl pyrazine

FIG. 4.2 Adapted from [13], [17], [19], [29], [34], [35] and [59].

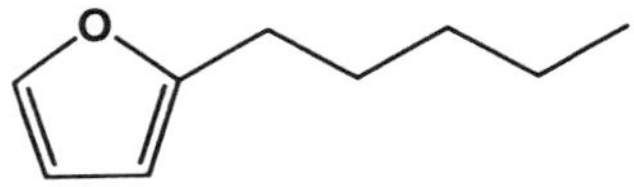

2-Pentyl furan

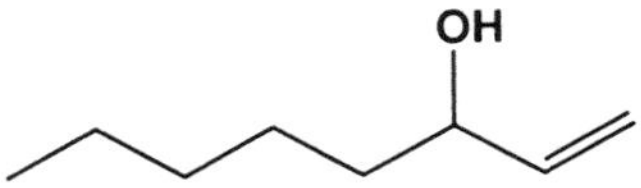

OH

FIG. 4.3 Adapted from [65].   **1-Octen-3-ol**

taetrolens [12] when grown in milk. Similarly, the musty potatolike pyrazine was shown by Dumond et al. [13] to be produced in smear-coated, surface-ripened cheese by *Pseudomonas perolens*.

Studies by Kowalewska et al. [14] showed that *Lactobacillus helveticus* produced various pyrazines when grown on whey permeate supplemented with filter-sterilized amino acids. These included a pyrazine with formaldehyde in the second position, alkyl pyrazines, and alkoxy pyrazines. However, the amounts formed were too small and their mass spectra too impure to ascertain which isomers were present. This work reinforces the theory that pyrazines are formed by cheese microorganisms.

4.1.3 Pyrazines by Actinomycetes

Several odorous substances, including 2-methoxy-3-isopropyl pyrazine (Fig. 4.2) were obtained by Gerber [15] from various actinomycetes grown in submerged culture. She acknowledged the 2-methoxy-3-isopropyl pyrazine to be the same musty, earthy, compound found by others, and reaffirmed its importance in the aroma of peas and potatoes. Gerber's organism, a streptomycete isolated from a Florida soil sample, produced the highest level of this pyrazine in Bennett's broth medium, which was found superior to Czapeck's plus yeast medium for production of the odorant.

4.1.4 Pyrazines by Gram Negative Bacteria

Potato-like odors produced by *Serratia* and *Cedecea* were found by Gallois [16] to be caused by alkyl methoxy pyrazines. The major compound produced by *S. odorifera* was 2-methoxy-3-isopropyl-5-methyl pyrazine, and that produced by *C. davisae* was 2-methoxy-3-*sec*-butyl pyrazine (Fig. 4.4). Other pyrazines produced by certain strains of these bacteria include 2-methoxy-3-isopropyl pyrazine (Fig. 4.2), 3-methoxy-2-isobutyl pyrazine (Fig. 4.1),

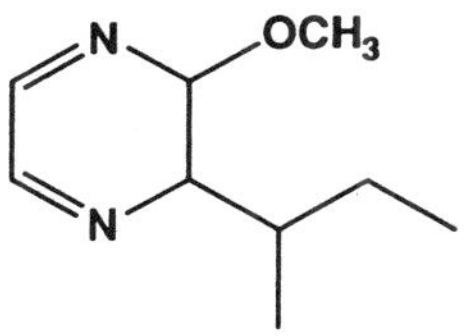

**2-Methoxy-3-sec-butyl
pyrazine**

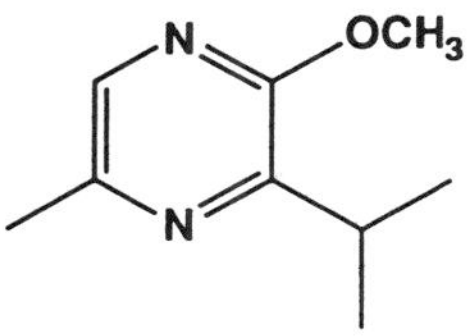

**2-Methoxy-3-isopropyl–
5-methyl pyrazine**

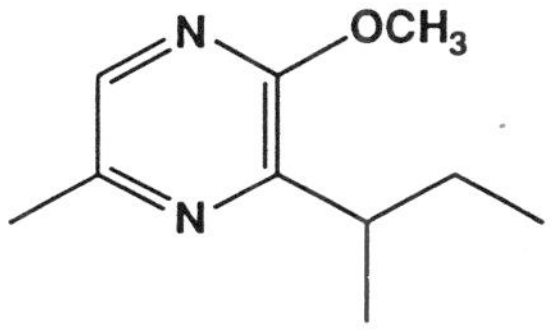

**2-Methoxy-3-sec-butyl–
5-methyl pyrazine**

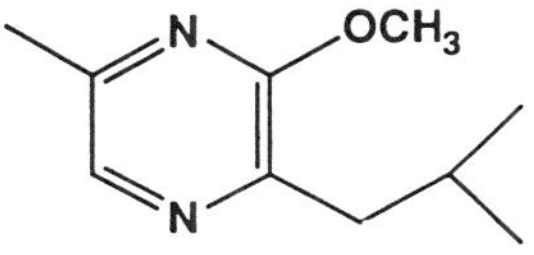

**2-Methoxy-3-isobutyl–
6-methyl pyrazine**

FIG. 4.4 Adapted from [17].

2-methoxy-3-*sec*-butyl-5-methyl pyrazine, and 2-methoxy-3-isobutyl-6-methyl pyrazine (Figure 4.4).

A gram negative bacterium (unidentified) was shown to produce 2,6-dimethyl-3-methoxy pyrazine [17] (Fig. 4.1). The organism was deposited under the accession no. 11802 with the National Collection of Industrial Bacteria (Torry Research Station, Aberdeen, Scotland).

McIver and Reineccius [18] showed that 2-isobutyl-3-methoxy pyrazine (Fig. 4.1) was produced with a yield of 12.5 g/L when a mutant *Pseudomonas perolens* was cultured in nutrient broth. The parent organism (ATCC# 10757) grown in the same medium produced only 46 mg/L of this musty, green-potato pyrazine compound; the production level was influenced slightly by various nitrogen source compounds, but was increased to 133 mg/L when the level of phosphate was increased to between 0.4 and 1.2 mM. The mutant also was able to produce 150 mg/L of 2-methoxy-3-*sec*-butyl pyrazine (Fig. 4.4) in the nutrient broth medium. This pyrazine imparted a green, pea podlike odor. Production of tetramethyl pyrazine (Fig. 4.1) and diacetyl also was demonstrated for this mutant. The tetra compound began to be produced in media containing either 2 percent pyruvate or 2 percent lactate after cessation of methoxy pyrazine biosynthesis. However, pyrazine synthesis using pyruvate (1 percent), lactate (1 percent) or nutrient broth as a carbon source did not commence in high-density cultures until they were well into their stationary phase of growth.

4.1.5 Pyrazines by Gram Positive Bacteria

A mutant of *Corynebacterium glutamicum* was found [19] to accumulate 3 g/L of tetramethyl pyrazine (Fig. 4.1) after 5 days; the pyrazine crystallized upon cooling of the broth. Isoleucine, valine, leucine, and pantothenate were required for growth of the organism deficient in a single enzyme in the isoleucine–valine pathway. Accumulation of this substituted pyrazine was also dependent on the addition of thiamine. A strain of *Bacillus cereus* #147 was shown [20] to produce only 0.3 g/L of this same tetra compound after a 7-day fermentation. The liquid synthetic medium required 4.0 percent glucose and 0.2 percent yeast autolysate for maximal yields.

When the natto organism, *Bacillus natto*, was cultured in a basal medium [21] of cooked soybeans, pyrazine production could be stimulated by the addition of 1 percent glucose and 1 percent sodium glutamate. Subsequent work in this laboratory [22] showed that when the organism was grown in liquid medium the addition of L-threonine was most stimulatory for pyrazine production and L-serine ranked second. The main product of the L-threonine-containing culture was 2,5-dimethyl pyrazine (Figure 4.2), while the one containing L-serine produced mainly tetramethyl (Fig. 4.1) and trimethyl pyrazine (Fig. 4.2).

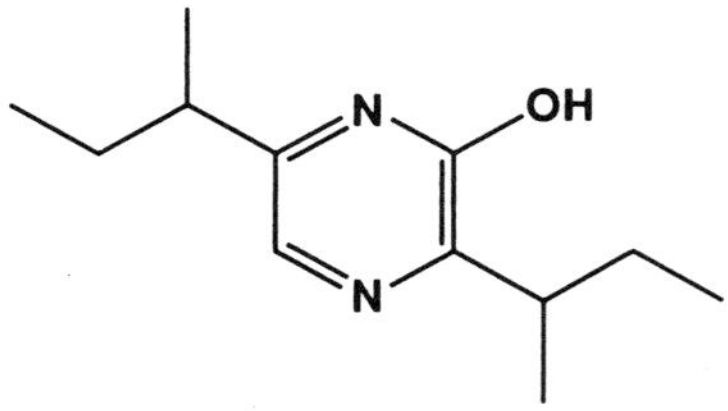

2-Hydroxy-3,6-di-sec-butyl pyrazine

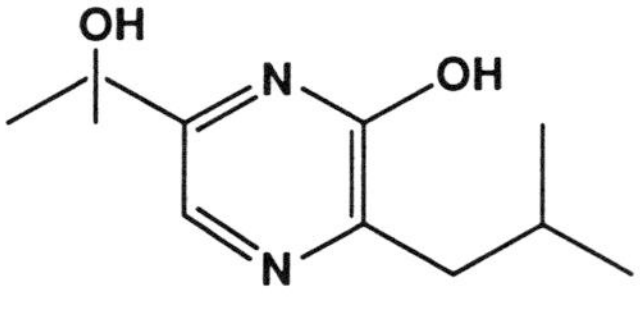

2-Hydroxy-3-isobutyl-6-(1-hydroxy-1-methylethyl) pyrazine

FIG. 4.5 Adapted from [60].

4.1.6 Pyrazines by Filamentous Fungi

Researchers in Japan [23–25], out of concern that food aspergilli (*A. sojae* strain X-1) could produce aflatoxins, showed for the first time that nontoxic (mice studies) fluorescent pyrazines, but no aflotoxins, were present in cultures of this organism. These included 2-hydroxy-3-6-di-*sec*-butyl pyrazine and 2-hydroxy-3-isobutyl-6-(1-hydroxy-1-methylethyl) pyrazine (Figure 4.5), deoxyaspergillic acid, deoxymutaaspergillic acid and 2-hydroxy-3-*sec*-butyl-6-(1-hydroxy-1-methylpropyl) pyrazine (Fig. 4.6).

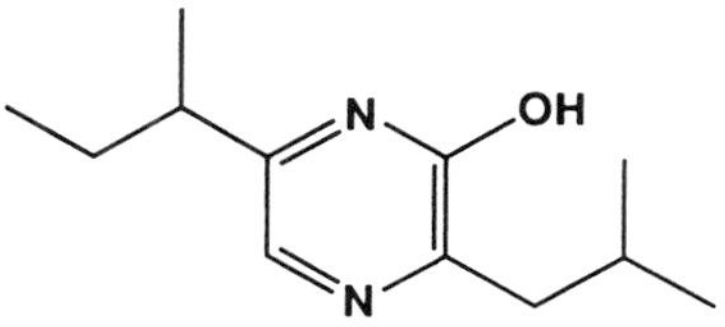

Deoxyaspergillic acid
(2-Hydroxy-3-isobutyl-6-sec-butyl pyrazine)

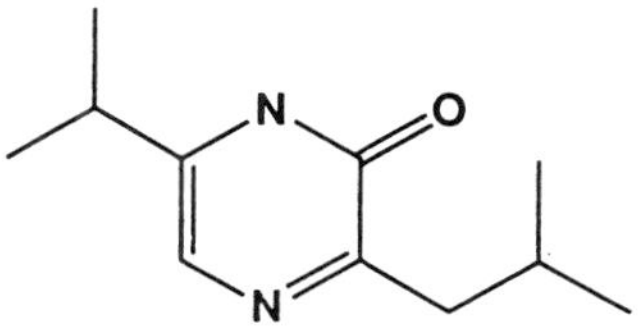

Deoxymutaspergillic acid (2-(1H)-pyrazinone-3-isobutyl-6-isopropyl pyrazine)

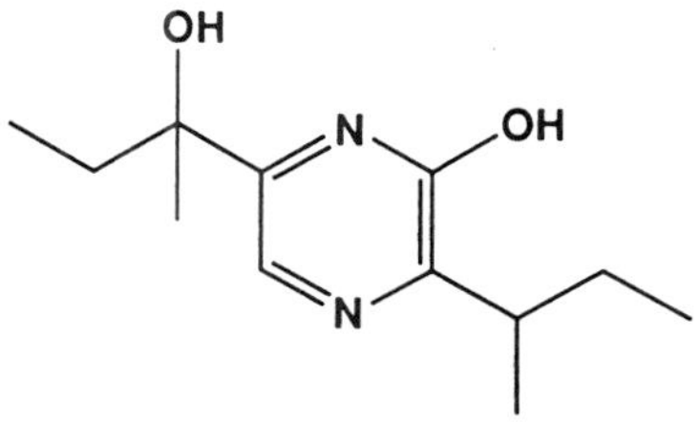

2-Hydroxy-3-sec-butyl-6-(1-hydroxy-1-methyl-propyl) pyrazine

FIG. 4.6 Adapted from [61].

These studies, however, did reveal the presence of toxic (i.e., to mice on I.P. injection) nonfluorescent pyrazine compounds in culture filtrates of the same *A. sojae* strain X-1. Three compounds were isolated, two of which were identified as 2-hydroxy-3,6-di-*sec*-butyl pyrazine-1-oxide and aspergillic acid (Fig. 4.7). In addition, three nonfluorescent toxic (to mice) pyrazines were identified as 2-hydroxy-3-isobutyl-6-isopropyl pyrazine-1-oxide, 2-hydroxy-3-*sec*-butyl-6-(1-hydroxy-1-methylpropyl) pyrazine-1-oxide and hydroxy aspergillic acid (Fig. 4.8). All of these, unlike the nontoxic, fluorescent pyrazines identified, possessed the "N-oxide" structural feature. This indicated that the oxide was important in bestowing toxicity by these pyrazines in mice (LD_{50} of

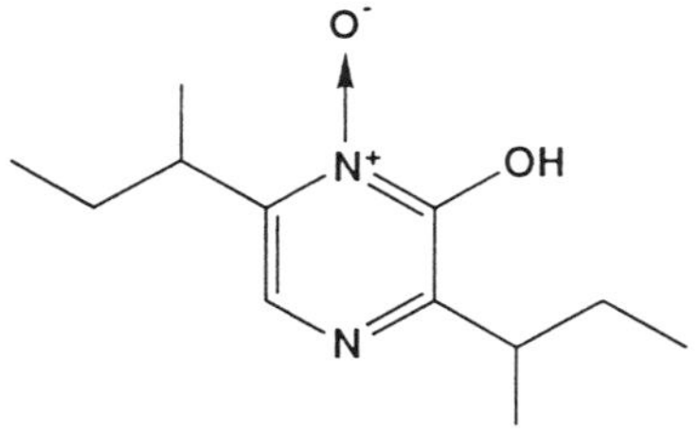

**2-Hydroxy-3-isobutyl-6-isopropyl
pyrazine-1-oxide**

**2-Hydroxy-3,6-di-sec-butyl
pyrazine-1-oxide**

**2-Hydroxy-3-sec-butyl-6-(1-hydroxy-
1-methylpropyl) pyrazine-1-oxide**

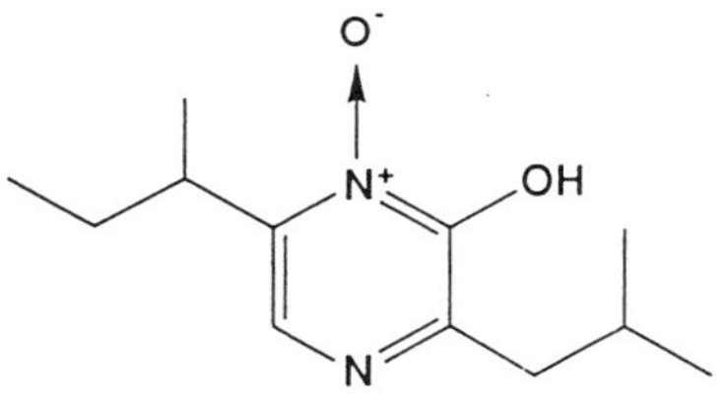

Aspergillic acid
(2-Hydroxy-3-isobutyl-6-sec-
butyl pyrazine-1-oxide)

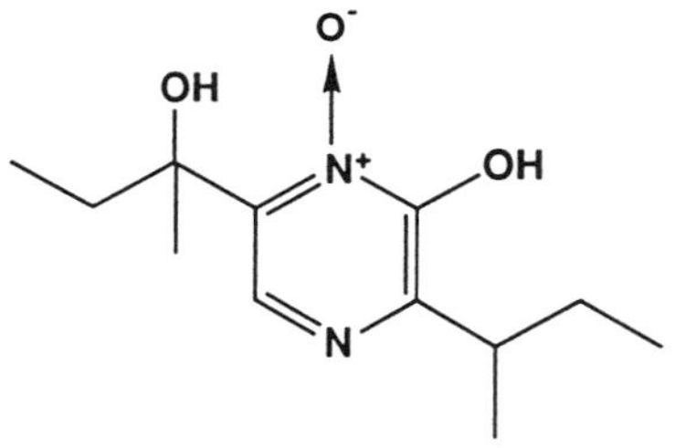

FIG. 4.7 Adapted from [72].

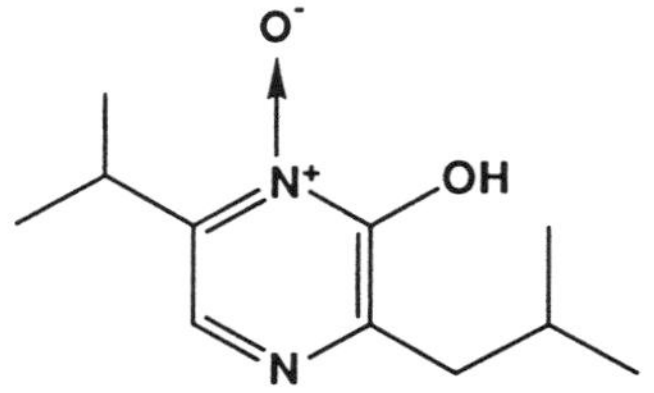

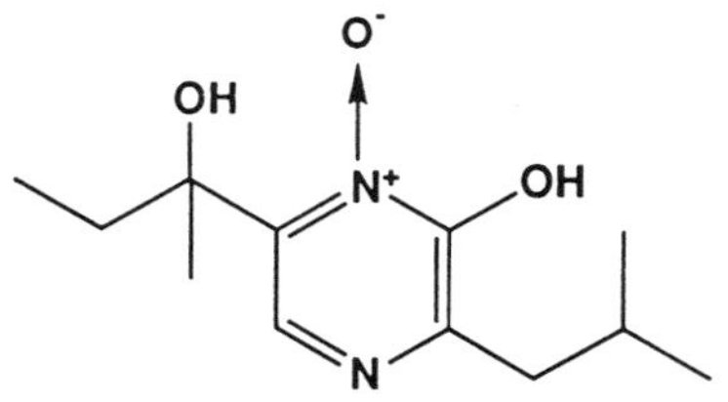

Hydroxy aspergillic acid (see Figure 4.7)

FIG. 4.8 Adapted from [61].

62.5–125 mg/kg) and that the structure negated or removed the ability of such pyrazines to fluoresce under UV light.

Aspergillus parasiticus NRRL 2999 accumulated four major, blue fluorescent pyrazines [26] identified as deoxyaspergillic acid (Fig. 4.6), flavocol (Fig. 4.9), 2-hydroxy-3,6-di-*sec*-butyl pyrazine (Fig. 4.5), and deoxyhydroxy mutaaspergillic acid (Fig. 4.9). All of these 2-hydroxy-3,6 substituted pyrazines interfere with aflatoxin analysis in this aflotoxigenic organism due to their fluorescence. Buchanan and Houston [26] showed that pyrazine production on peptone mineral salts medium was inversely related to carbohydrate content of the medium. That is, the pyrazine accumulation was greatest when the microorganism was grown on media rich in protein but low in carbohydrate (i.e., 1–2 percent glucose).

Aspergillus oryzae grown on soybeans and wheat flour in solid-state fermentations was shown [27] to be responsible for the production of 19 substituted pyrazine compounds. Hydrolysis and fermentation of the fermented soya cake (FSC) in brine resulted in a marked decrease in the total pyrazine concentration (i.e., 3312 versus 166 nanomoles/kg of FSC). Pyrazines were not formed during the Maillard reaction (nonenzymatic browning) that occurred in the brined, slightly acidic, hydrolyzate stored at room temperature. Those pyrazines found in levels exceeding 100 nanomoles/kg FSC [27] were 2-methyl, 2,5-dimethyl, 2,6-dimethyl, 2,3-dimethyl, trimethyl, and tetramethyl pyrazines. In addition, the nonbrined, nonyeast fermented FSC intermediate contained considerable quantities of 2-ethyl-3-methyl, 3-ethyl, and 2-ethyl-3,5,6 trimethyl pyrazines (Figs. 4.1 and 4.2).

Flavocol
2-Hydroxy-3,6-di-isobutyl pyrazine

Deoxyhydroxy mutaspergillic acid
(see Figure 4.6)   **FIG. 4.9**   Adapted from [7].

4.1.7 Summary

A Gram negative organism described as a mutant of *Pseudomonas perolens* [18], shown to produce 12.5 g/L of 2-isobutyl-3-methoxy pyrazine (Fig. 4.1), could be of value in commercial fermentations. The potatolike odor character attributed to this compound could play an important role in the creation of potato-flavored snacks, dips, pancakes, soups, crackers, breads, and stews. Its green musty qualities might allow its use in vegetable, mushroom, and cheese flavors.

Certain *Aspergillus oryzae* and *A. sojae* strains were shown to be prolific pyrazine producers under certain conditions. The wide variety of 2-hydroxy-3,6 substituted pyrazines produced by *A. sojae* undoubtedly contributes to the complex flavor qualities of soy sauce.

4.2 TERPENOIDS FOR FLAVORS AND FRAGRANCES

4.2.1 Introduction

Terpenoids are the primary flavor and fragrance impact molecules found in the essential oils of higher plants. Significant among these are the mono- and sesquiterpenoids. It is accepted today that certain higher fungi, though devoid of the storage organelles found in higher plants, are able to produce [28] an array of such volatile compounds, especially the acyclic monoterpene alcohols as well as a broad array of variously structured sesquiterpenes. The best studied terpene-producing fungi include species of the ascomycete genus *Ceratocystis*. The work of Colins [29] showed that many acyclic monoterpene alcohols, aldehydes, and esters were produced by cultures of *C. variospora*. A comprehensive review by Krasnobajew [30] focused on many microbial transformations of terpenoids of interest to the flavor and fragrance industry. The review provided examples of transformations of terpenoids: acyclic monoterpenoids, monocyclic monoterpenoids, bicyclic monoterpenoids, ionones, and related compounds, sesquiterpenoids, diterpenoids, and triterpenoids.

The microbial transformation of selected terpenoid substrates leads to compounds of value to the flavor and fragrance industry. The transformation reactions occur with high selectivity in terms of functional groups (chemoselectivity), site of the reaction (regioselectivity) and purity of stereoisomer or enantiomer (stereoselectivity). Such selectivity makes it possible to provide desired molecular forms such as *cis* and *trans* diols and pure optical isomers with the required odor and taste characteristics. However, one difficulty with this approach is the toxicity of both substrates and products to microorganisms, which requires product recovery from dilute aqueous solutions. In most cases, due to their toxicity, terpenoid substrates are added to a microbial culture after the growth rate or cell mass has been maximized. Often stepwise addition of the substrate is carried out to prevent growth inhibition of the organism and to avoid cell lysis. Care must be taken to prevent loss of volatile terpenoid compounds while the biological system is being vigorously aerated and agitated in aseptic bioreactor vessels.

Microorganisms for industrially successful processes [31, 32] must be selected by enrichment culture techniques from soils taken from all possible locations. The soil is incubated in a growth medium containing the substrate terpenoid (see Section 4.2.4 for more details) as the sole carbon source. Only those isolates which produce useful catabolites are retained, while those able to more completely degrade the terpenoid must be either mutated or discarded. The mutation ideally allows the organism to carry out a desired transformation, but due to the loss of certain metabolic enzymes is unable to produce unwanted, low-yield metabolites.

This review focuses on selected terpenoid transformation systems and provides examples for different microbial forms. Only those transformations perceived to be based on natural (i.e., unaltered by enzymatic or chemical catalysis) essential oil extracts or preparations were included. For example, the industrial process involving chemically prepared, mixed isomers of menthol [33] subsequently resolved microbially, was judged inappropriate. Similarly, the systems dependent on terpenoid glycosides [34], prepared using yeast enzymes, were categorized as nonmicrobial and hence not included.

4.2.2 Terpenoids by Actinomycetes

An earthy, musty-odored sesquiterpenoid (neutral oil) bearing the trivial name ''geosmin'' (Fig. 4.10) was isolated by Gerber and Lechevalier [35]. They isolated it from broth cultures of several *Actinomycetales* but particularly *Streptomyces griseus* LP-16, which produced 1 mg/L in shake flasks. Gerber [36] believed the molecule to be a biogenetic sesquiterpeniod missing an isopropyl group. Gerber and Lechevalier [37] described a method for growing the LP-16 strain in 5-liter fermenters and for purifying geosmin using a hydrophobic resin (XAD-2, Rohm and Haas Co., Philadelphia, PA). When their strain was grown in an aerated, enriched soybean broth, it produced 6 mg/L of broth after a 3-day incubation.

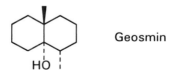

Geosmin

(Octahydro - 4, 8a - dimethyl - 4a [2H] - naphthalenol)

FIG. 4.10 Adapted from [35] and [62].

Two sesquiterpene alcohols were isolated by Gerber [38] from geosmin-producing actinomycetes (i.e., several *Streptomyces* species). The first, cadin-4-ene-1-ol (Fig. 4.11, left side) was identical with epicubenol from cubeb oil, except it was opposite (dextra-rotatory) in optical rotation. This microbial cadinol had an ''earthy'' or ''woody'' odor, while the nonmicrobial enantiomer was ''sweet–spicy'' and was a more intense odorant. The second sesquiterpene alcohol (Fig. 4.11, right side), (+)-selina-4(14)-7(11)-diene-9-ol, was produced by *S. fradiae* #3535. Again its optical rotation was opposite to that of the plant derived natural selina diene isolated from vetiver oil. This work substantiated that microbial sesquiterpenes can be enantiomers of the molecules derived from plant sources and as such they possess quite different aroma qualities.

Several Actinomycetes, mostly *Streptomyces* species were shown by Gerber [38] to produce a bicyclic monoterpenoid identified as 2-methyl isoborneol (Fig. 4.12). Yields by these microorganisms were in the 1–5 mg/L range. The compound had a camphoraceous and menthol like odor. It was shown to be levo-rotatory at $-14.8°$, which was in good agreement with an authentic sample synthesized from d-camphor.

4.2.3   Terpenoids by Bacteria

4.2.3.1   *Acyclic Monoterpenoid Degradations*   Several recombinant *Pseudomonas* organisms were evaluated for the transformation of linalool, geraniol, and citronellol, as described by Vandenbergh [39]. No claims of products from linalool or citronellol were made; however, a novel product (Fig. 4.13) described as 6-methyl-5-heptene-2-one was formed from geraniol using *Pseudomonas putida* NRRL-B-18040 (PP U 2.9). The compound was described as having a pleasant, fruity odor and suitable for use in foods.

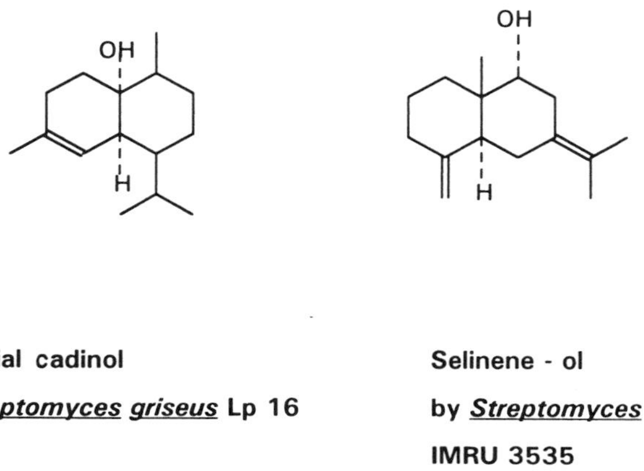

Microbial cadinol

by *Streptomyces griseus* Lp 16

Selinene - ol

by *Streptomyces fradiae*

IMRU 3535

FIG. 4.11   Adapted from [38].

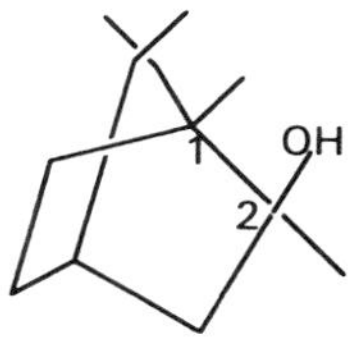

(-) 2 - Methyl isoborneol

1, 2, 7, 7 - tetramethyl bicyclo [2,2,1] heptan - 2 - ol

FIG. 4.12 This compound was produced by *Streptomyces lavendulae* IMRU 3440-1Y [20]. and *Penicillium caseicoleum* [34].

4.2.3.2 Monocyclic Terpenoid-Limonene Transformation

A *Corynebacterium* species grown by Japanese workers [40] on limonene was shown to produce about 10 mg/L of 99 percent pure L-carvone (Figure 4.14) in 24–48 hours. The optical rotation reported was $[\alpha]_D^{20} = -48°$ (literature $-62°$ for the 99 percent optically pure L isomer).

Not only has limonene served as a substrate for the microbial formation of L-carvone, but also to form (+)-α-terpineol together with (+)-perillic acid (Fig. 4.15). The organism used [41] was *Pseudomonas gladioli*. α-Terpineol was resistant to further degradation and accumulated to over 0.6 g/L in a 4-day-old broth culture. Perillic acid levels reached 1.8 g/L after 4 days and could be recovered at this yield if the process was terminated at this time to prevent its degradation to an unidentified major acid.

4.2.3.3 Bicyclic–Monoterpenoid Transformations

Pinene Degradation.   The studies by Gibbon et al. [42] will only be mentioned briefly. One *Pseudomonas* soil isolate, PX1 (NCIB 10684), transformed pure α-pinene into (+)-*cis*-thujone (1/3) and (+)-*trans*-carveol (2/3) (Fig. 4.16) at a yield of 0.4 g/L. Another *Pseudomonas* strain (NCIB No. 11671 from sewage) was shown [43] capable of transforming α- or β-pinene at low concentrations (6-13 mg/L) into L-carvone (see Fig. 4.17). Either one or the

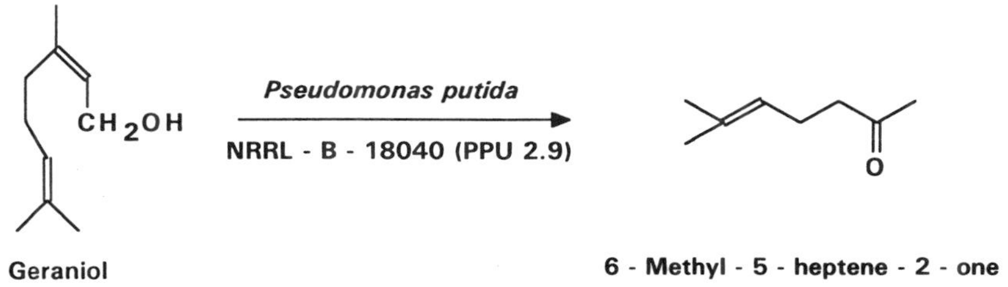

Geraniol

6 - Methyl - 5 - heptene - 2 - one

FIG. 4.13 Adapted from [39].

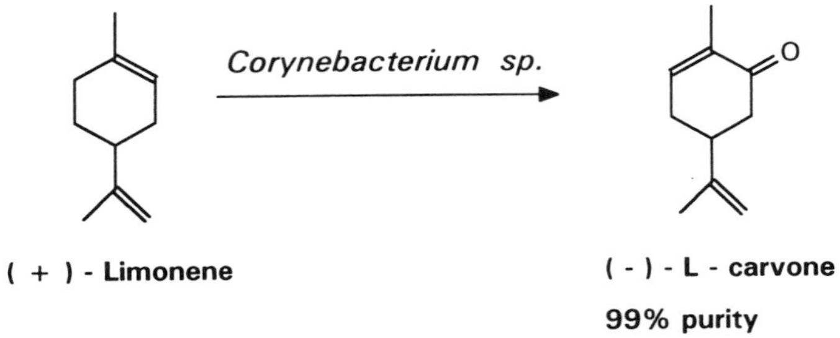

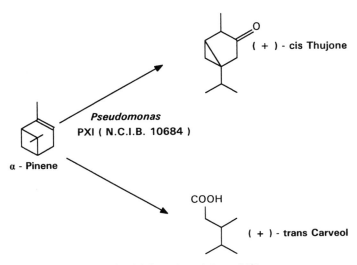

FIG. 4.14 Adapted from [40].

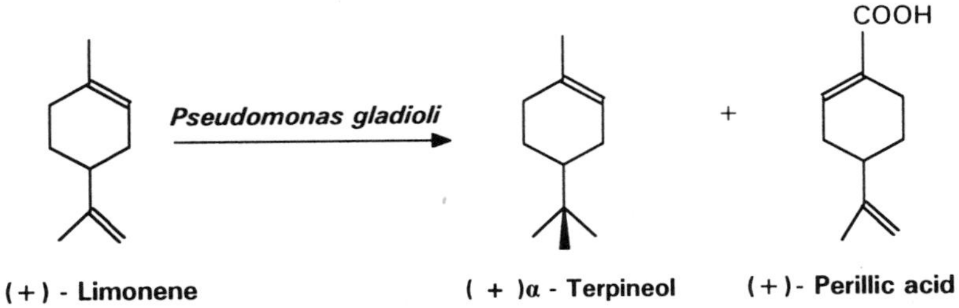

FIG. 4.15 Adapted from [41].

FIG. 4.16 Adapted from [42].

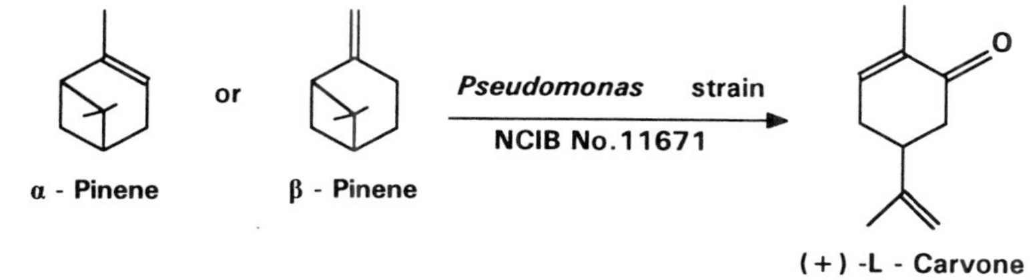

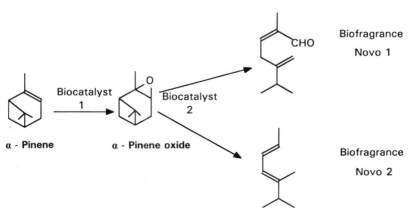

FIG. 4.17 Adapted from [43].

other substrate was added stepwise as a sole carbon source in a 1-to-3-day process involving nutrient broth media.

A commercially interesting microbial transformation of α-pinene was reported [44] in which the double bond in the primary ring structure was oxidized to form α-pinene oxide (Fig. 4.18). The latter is transformed in turn to acyclic monoterpenoid aldehydes, referred to as *biofragrance Novo 1 and 2* respectively. These products were recommended for use in fragrances and may be produced in tens of kilos in a 750 L fermenter using selected microorganisms. This development appears to be based on work of Graham et al. [45]. Their patent described a process involving a mutant *Pseudomonas fluorescens* and one dependent on a prior chemical treatment to assist in the decylization of the pinene structure.

Eucalyptol Transformation.   Eucalyptol (1,8-cineole), used in the flavor and fragrance industry, is a main constituent of the essential oil from *Eucalyptus polybractea*. A bacterium identified as *Pseudomonas flava* UQM 1742 isolated from eucalyptus leaves was found [46] to utilize 1,8-cineole provided in a mineral salts medium as a sole carbon source. Levels above 0.5 g/L of this

FIG. 4.18 Adapted from [44].

substrate proved to be toxic. When the organism was incubated for less than 15 hours to prevent complete 1,8-cineole consumption, it was found that 0.16 g/L of substrate was converted to 0.07 g/L (43 percent) and 0.05 g/L (31 percent) respectively of 2-oxocineole and ketolactone respectively (Fig. 4.19). The configuration of these major products indicated that they were derived by microbial oxidation at the same prochiral carbon atom of 1,8-cineole.

When 1,8-cineole was extracted from the essential oil of *Eucalyptus radiata* var. *australiana* [47] it was shown that *Bacillus cereus* (UI-1477) was able to carry out a stereospecific hydroxylation of this major component. The product (see Fig. 4.20) was 6-R-exo-hydroxy-1,8-cineole. The rate and yield of 1,8-cineole hydroxylation by the organism were determined by GC analysis of methylene chloride extracts of microbial broth samples taken at different incubation-time periods. A maximum yield of the hydroxylation product in a medium containing 1.0 g/L of 1,8-cineole was 0.74 g/L (74 percent), achieved in 24 hours. Levels of the product remained constant for an additional 48 hours, indicating that the product was not utilized for growth or energy requirements of the organism.

4.2.3.4 *Formation of γ-Irones*   Two strains of *Enterobacteriaceae* classified as *Serratia liquefaciens* and *Pseudomonas maltophilia* were isolated [48] from contaminated tissue cultures of rhizomes from *Iris pallida* or *germanica*. Sterilized (120°C, 30 min) rhizomes of this plant were shown to serve as a substrate for these organisms in the production of several irone isomers. The *S. liquefaciens* culture attained a maximum yield of 1.2 g/kg dry weight of rhizomes after 3–4 days of incubation. The major natural isomer produced was *cis-γ*-irone (Fig. 4.21), which represented 61 percent of the total quantity of four isomers identified. In one example provided, this isomer was increased

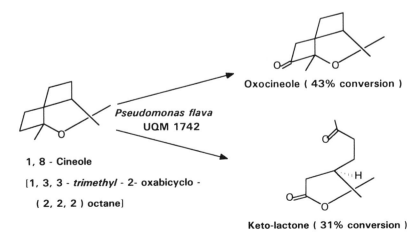

FIG. 4.19 Adapted from [46].

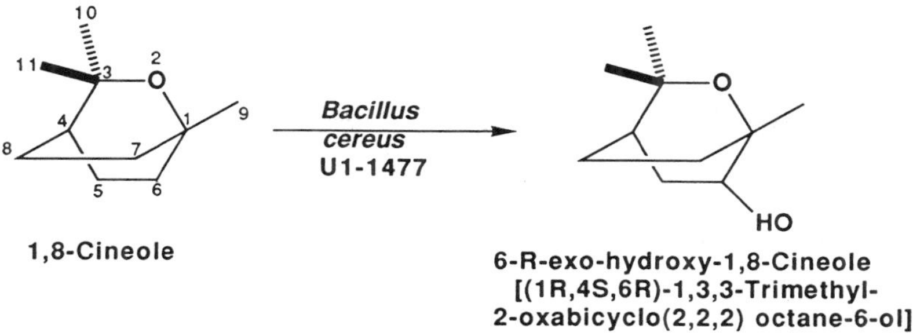
10
11
3
2
O
4
1
9
8
7
5
6
1,8-Cineole
Bacillus
cereus
U1-1477
O
HO
6-R-exo-hydroxy-1,8-Cineole
[(1R,4S,6R)-1,3,3-Trimethyl-
2-oxabicyclo(2,2,2) octane-6-ol]

FIG. 4.20 Adapted from [47].

by a factor of 24, from 2.8 mg/g to 67 mg/g of fresh rhizomes. The other quantitatively significant irone formed was *cis*-α-irone (Fig. 4.21), which made up 33 percent by weight of the total irone isomers in the transformed product.

4.2.3.5 Diterpene Transformations A study by Hieda et al. [49] in Japan showed that *cis*-abienol could be transformed into four compounds (see Fig. 4.22) by an unidentified soil bacterium JTS-162. These new compounds were

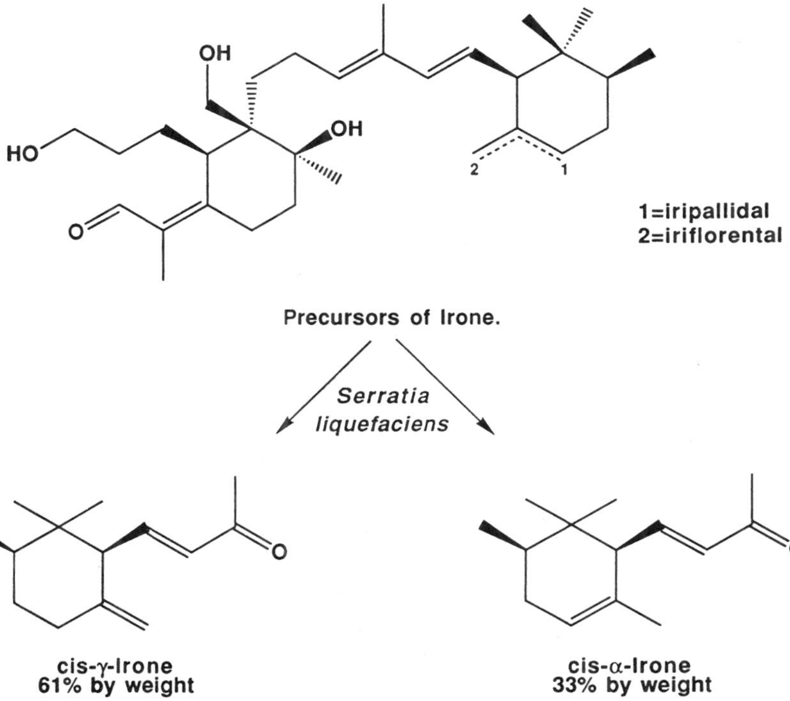
OH
OH
HO
O
2
1
1=iripallidal
2=iriflorental
Precursors of Irone.
Serratia
liquefaciens
O
O
cis-γ-Irone
61% by weight
cis-α-Irone
33% by weight

FIG. 4.21 Adapted from [48].

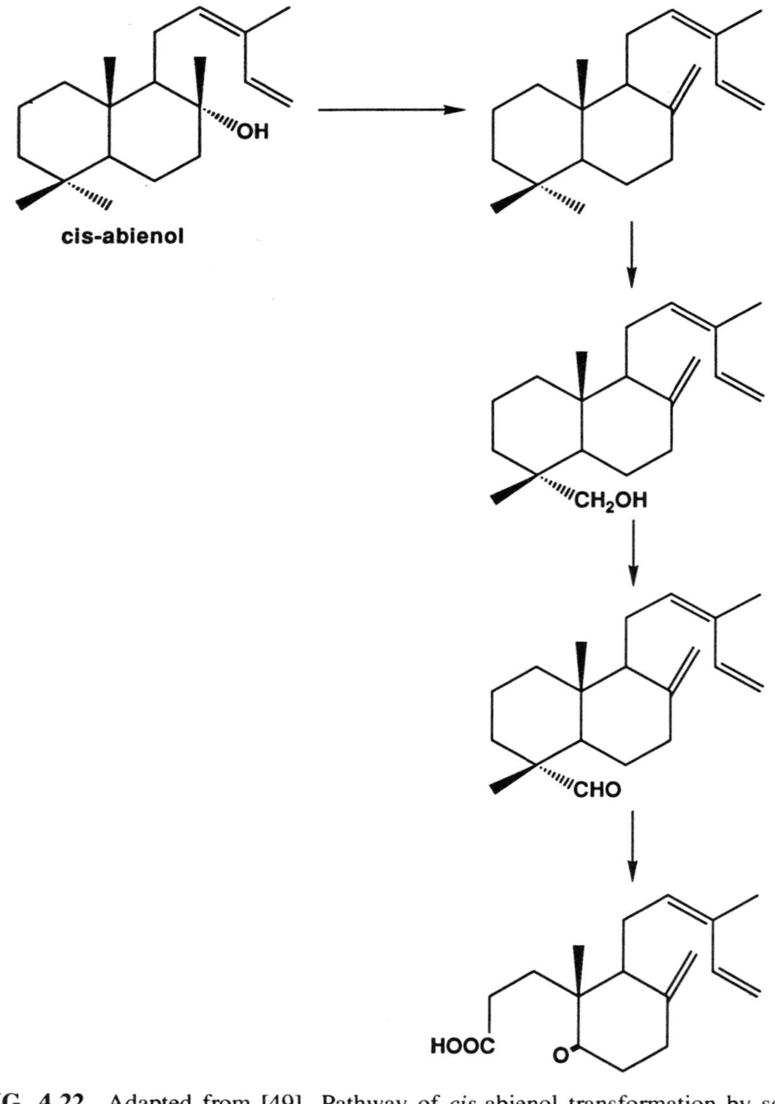

FIG. 4.22 Adapted from [49]. Pathway of *cis*-abienol transformation by soil bacterium JTS-162.

determined to be (12Z)-labda-8(17),12,14-triene (I), (12Z)-labda-8(17),12,14-trien-18-ol (II), (12Z)-labda-8(17),12,14-trien-18-al (III) and 4,18,19-tri-nor-3,4-seco-5-oxo-(12Z)-labda-8(17),12,14-trien-3-oic acid (IV). It was of interest that these conversion products were useful for flavoring tobacco. *Cis*-abienol is a known precursor in tobacco leaf wax for tobacco odorants formed by oxidative and leaf-enzyme tobacco curing processes. The best preparation, functional in tobacco at the ppm level, was obtained when the organism was incubated for 60 to 72 hours. If the fermentation was extended beyond 72 hours, over 90 percent of the *cis*-abienol was consumed but there was too much loss of compounds I and II when these were transformed sequentially into compounds III and IV respectively.

Sclareol, another tobacco (diterpenoid) labdenoid, was found [50] to be transformed by the same soil isolate JTS-162. Three transformation products were identified (see Fig. 4.23): labda-8(17),14-dien-13-β-ol (manool), C-8 methylene derivative of sclareol (I), 8,13-epoxylabd-14-ene, manoly oxide (II) and 14,15-dinorlabd-8(17)-en-13-one (III). Compounds I and II were the major products and the maximum yields of each compound was 20 and 24 percent (w/w) respectively (i.e., 0.2 to 0.24 g/L).

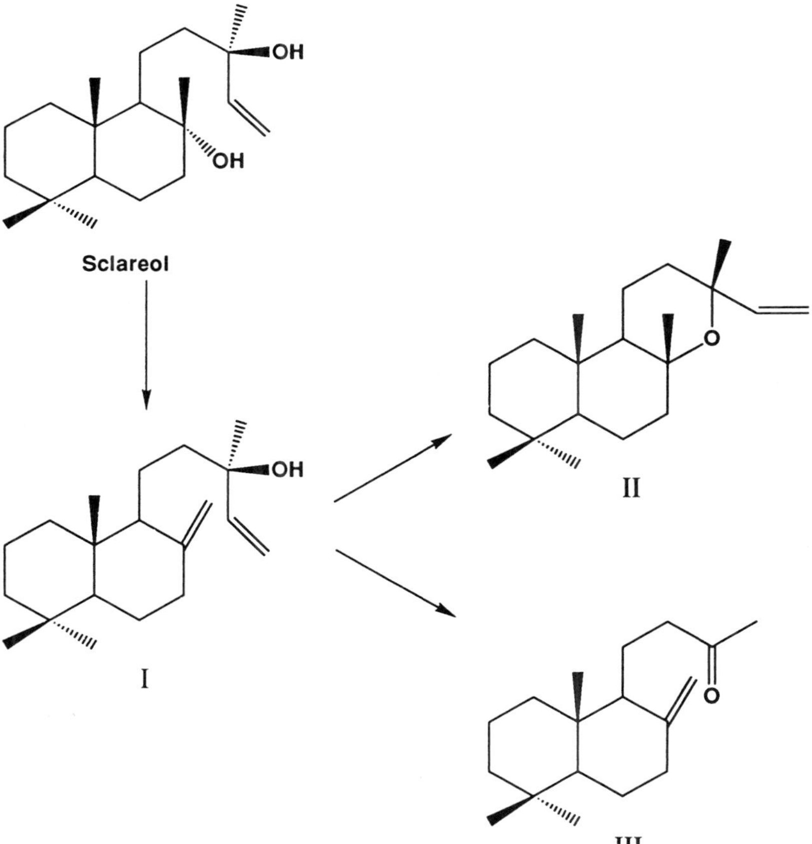

FIG. 4.23 Transformation pathway of sclareol by soil bacterium JTS-162. Adapted from [50].

4.2.4 Terpenoids by Yeasts

4.2.4.1 Acyclic and Monocyclic Terpenoids

In a review [51] it was shown that several yeasts, including *Candida reukaufii* AHU 3032 and *Rhodotorula minuta* are able to sterioselectively reduce racemic mixtures of β-(−)citronellal to β-(−)citronellol (see Fig. 4.24). This compound had 80 percent optical purity. The β-(+)-citronellal was left unchanged by these yeasts.

The process patented by Karst and Vladescu [52] involved mutants of *Saccharomyces cerevisae* blocked in the ergosterol synthetic pathway, which secreted geraniol, farnesol, and linalool (Fig. 4.25). Another ergosterol-requiring mutant (erg-9) produced mainly farnesol. The products of these organisms were applied to fruit juices and other drinks to enhance their flavor.

The predominant compounds reported by Klingenberger and Hansen [53] were geraniol, citronellol and linalool (Fig. 4.26). The yeast *Ambrosiozyma monospora* CBS #5514 was grown in still cultures on defined, liquid media. Yields of these volatile alcohols could be raised fourfold by placing the polystyrene adsorbent XAD-2 (Rohm and Haas Co., Philadelphia, PA), enclosed in a small polyester bag, directly into the culture medium. However, even after a 30-day culturing time only 80 mg/L of total alcohols were accumulated. The products were recovered by steam distillation, which normally leads to partial degradation of the acyclic monoterpene alcohols. The addition of XAD-2 prevented loss of citronellol and geraniol (200 and 16 mg/L respectively) and enhanced the amount of linalool and α-terpineol (9 and 2 mg/L respectively) recovered by steam distillation.

4.2.4.2 Diterpene Transformations

Sclareol, a labdane diterpene, ditertiary alcohol, is widely distributed in nature. According to Kouzi and McChesney [54], it was first isolated from the essential oil of *Salvia sclarea* L. (*Labiatae*). It has commercial application as a fixative in perfumery, as a tobacco flavorant, and as a synthetic substrate for the preparation of a series of ambra odorants in perfumery.

Microbial systems to prepare commercial flavor and fragrance compounds

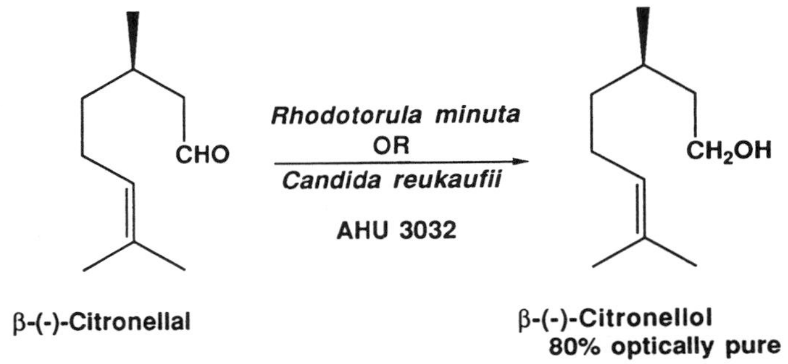

β-(-)-Citronellal   *Rhodotorula minuta* OR *Candida reukaufii* AHU 3032   β-(-)-Citronellol 80% optically pure

FIG. 4.24 Adapted from [51].

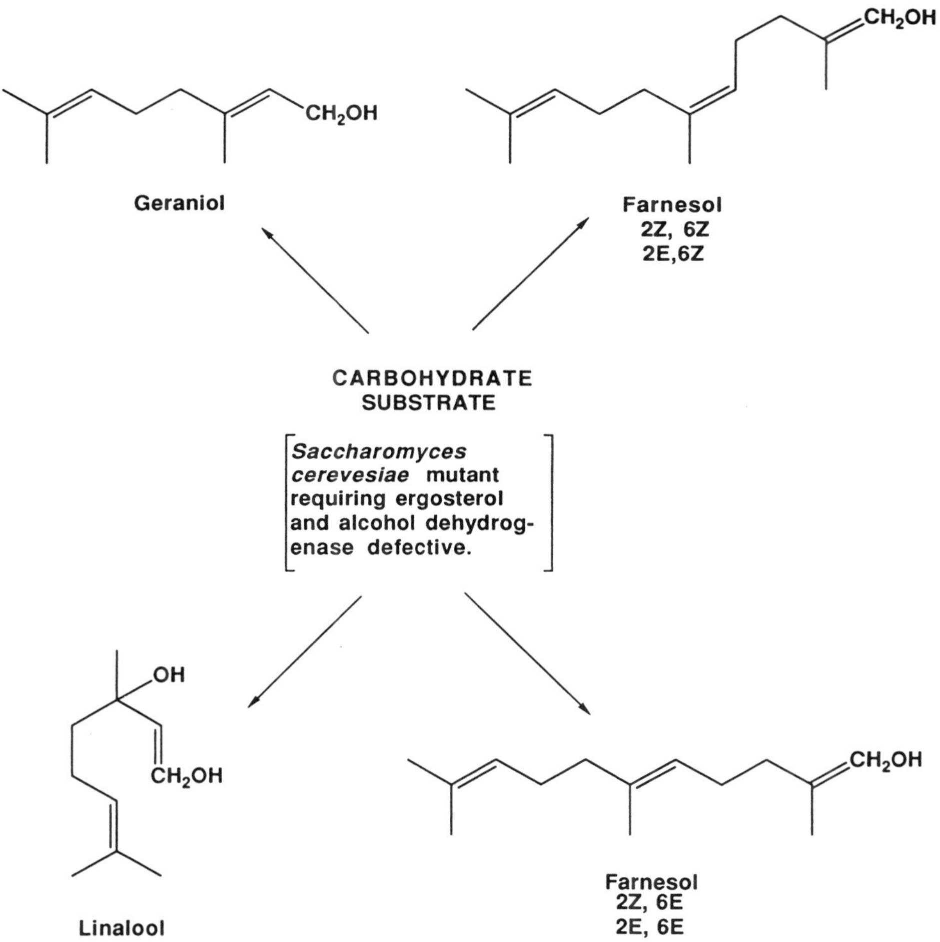

Geraniol

Farnesol
2Z, 6Z
2E, 6Z

CARBOHYDRATE
SUBSTRATE

Saccharomyces
cerevesiae mutant
requiring ergosterol
and alcohol dehydrog-
enase defective.

Linalool

Farnesol
2Z, 6E
2E, 6E

FIG. 4.25   Adapted from [52].

using sclareol as substrate have been described [31, 32]. Special filamentous yeasts were used biocatalytically for carrying out high-yielding, stereo- and regioselective transformations. The patented process [31] provided a method by which sclareol (0.1 g/L) was added to a 3-day-old culture of *Hyphozyma roseoniger* CBC 214.83 (ATCC 20624). Yeast extract was an important source of trace minerals and vitamins. The aerated, stirred cultures were incubated for 3 or 4 days, then a mixture of sclareol dissolved in Tween 80 (polyoxyethylene sorbitan monoleate, obtained from Sigma Inc., St. Louis, MO) was added incrementally during the next 5 days. The process was terminated after 4 more days of incubation. The recommended amount of sclareol to be added was 30 g/L of media. It was found beneficial to recrystallize the sclareol from hexane and provide it as a fine powder (i.e., 50 mesh sieve) mixed with an equal

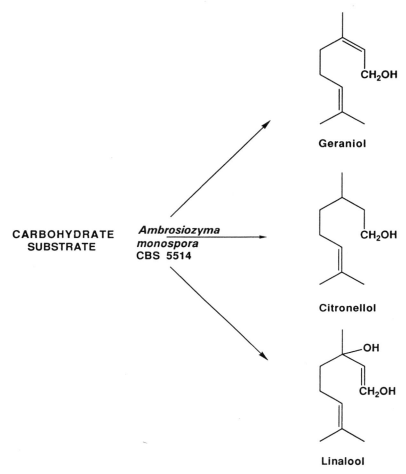

CARBOHYDRATE
SUBSTRATE

Ambrosiozyma monospora CBS 5514

Geraniol

Citronellol

Linalool

FIG. 4.26 Adapted from [53].

weight of Tween 80. The substrate was regularly converted to a diol compound (see Fig. 4.27) at yields better than 75 percent. Recovery was performed by extracting the broth using ethyl acetate.

In a related process Farbood et al. [32] provided a method by which a sclareol (powdered)/Tween 80 (2 to 1) emulsion was added stepwise to an aerated broth (salts–yeast extract–glucose) containing a mature culture of *Cryptococcus albidus* ATCC 20918. The powdered sclareol preparation which passed through a 100-mesh screen was more efficiently transformed than less finely dispersed substrate. The organism isolated from soil obtained from Mexico City, under optimized conditions, quantitatively transformed one or more isomers of substrate into a lactonized diterpenoid called sclaerolide (Fig. 4.28)

OH
OH
Hyphozyma
roseoniger
CBC 214.83
ATCC 20624
OH
OH

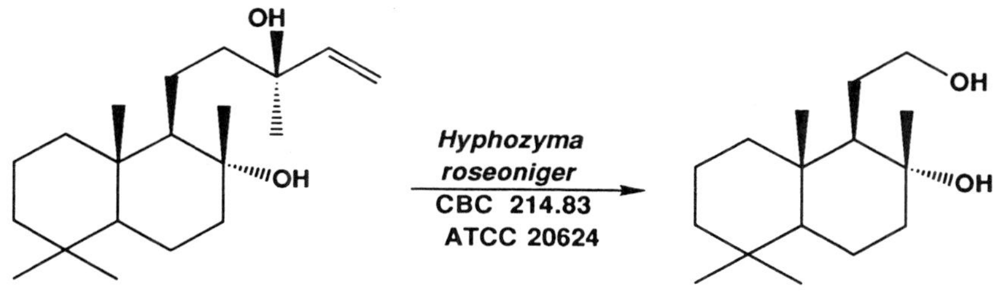

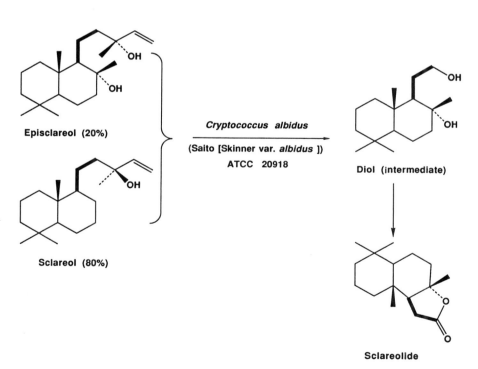

Sclareol (2-ethenyldecahydro-2-hydroxy-α-2,5,5,8a-pentamethyl-1-napthalene propanol)

Diol (Decahydro-2-hydroxy-α-2,5,5,8a-tetramethyl-napthalene ethanol)

FIG. 4.27 Adapted from [31].

OH
OH
Episclareol (20%)
OH
Sclareol (80%)
Cryptococcus albidus
(Saito [Skinner var. albidus])
ATCC 20918
OH
OH
Diol (intermediate)
O
O
Sclareolide

FIG. 4.28 Adapted from [32].

by going first to a diol intermediate. The most preferred level of substrate was in the range of 2.5 to 30.0 g/L but 150 g/L was successfully used. The transformation to solid product recovered under ambient conditions was completed in a maximum incubation time of 48 hours.

4.2.5 Terpenoids by Filamentous Fungi

4.2.5.1 Acyclic Monoterpenoid Transformation
Research at the Henkel Company in Germany by Schindler and Bruns [55] established a method for production of a mixture of monoterpenes by *Ceratocystis variospora* ATCC# 12866. The organism was grown in a salts–corn steep liquor–soybean meal medium or malt extract–glucose (oversupplied at 12 percent) medium with shaking. After 120 hours, 0.5 percent of Amberlite XAD-2 resin (lipophilic adsorbent) was added and incubation continued for another 48 hours. The yield of total terpenoids was increased from 60 mg/L without XAD-2 to 2700 mg/L with XAD-2. These terpenoids consisted of hexane-extracted geraniol (about 50 percent), nerol, neral, citronellol, linalool, and geranial.

A filamentous fungus classified as an *Actinomycetales* and named *Eremethecium ashbyi* BKM F3009B was shown by Russian microbiologists [56] to produce 200–280 mg/L of an essential oil-like product, three times more than a known reference strain of the organism. The improved isolate was grown on 2 percent soy flour and 1 percent sucrose medium in submerged (shaken) culture for 48 hours at 28°C. The composition of the aromatic components on a weight percent basis was geraniol 75.6, citronellol 8.3, nerol 3.1, β-phenyl ethanol 11.2, linalool 0.3, citral 0.5, and so on.

Studies in Germany by Abraham et al. [57] indicated that microbial ω-hydroxylation of acyclic terpenes such as myrcene, *trans-* and *cis-*nerolidol and farnesol could be carried out by filamentous fungi. The best conversion was achieved with *Aspergillus niger* ATCC# 9142 on the substrate *trans-*nerolidol; 20 percent of the primary alcohol was recovered (see Fig. 4.29). A bacterium, *Rhodococcus rubropertinctus* DSM 43197 was shown initially to produce the same primary alcohol by the ω-hydroxylation of *trans-*nerolidol. However, 16 percent of the end product was the corresponding carboxylic acid formed by a slower oxidation of the intermediary ω-hydroxy-*trans*-nerolidol.

The famous fungus *Botrytis cinera*, which grows in late grape harvests on fully ripened fruit, is of major interest in wine making. It was found under certain conditions able to carry out a controlled reductive metabolism on citral [58]. This substrate was completely broken down in grape must media, but in a pH 3.5 synthetic medium (not given) the organism formed 0.32 g/L of geraniol from citral supplied as the only carbon source (1 g/L; see Fig. 4.30). The same strain also produced 0.24 g/L of nerol in the synthetic medium.

4.2.5.2 Monocyclic Terpenoid Transformations
A patented process was described by Stumpf et al. [59] to make (+)-α-terpineol via *Penicillium digitatum* using either dl-limonene or R-(+)-limonene as substrate (see Fig. 4.31).

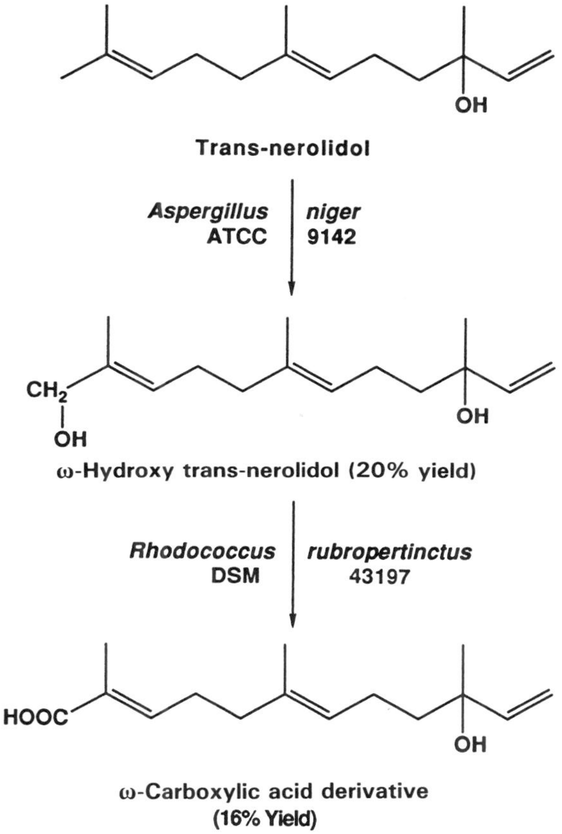

Trans-nerolidol

Aspergillus | *niger*
ATCC | 9142

ω-Hydroxy trans-nerolidol (20% yield)

Rhodococcus | *rubropertinctus*
DSM | 43197

ω-Carboxylic acid derivative
(16% Yield)

FIG. 4.29   Adapted from [57].

The organism was grown in a peptone–malt extract–yeast extract medium containing 1 percent dextrose. When the dextrose was consumed after a 72-hour incubation at 100 rpm, the substrate was added in incremental (stepwise) quantities up to 1.0 g/L. The stationary phase culture was shown to produce about 0.46 g/L of (+)-α-terpineol from R-(+)-limonene and 0.20 g/L from dl-limonene. The resulting α-terpineol (hydroxylated) compound is important to the perfumery industry for lilac, lavender, and other floral qualities.

Jasmonic acid (Fig. 4.32) may be isolated in the form of its methyl ester from the essential oil of jasmine made from jasmine flowers known in India as Chameli or Chambeli. Methyl jasmonate is an important ingredient in perfumery [60]. A patented process was described by Broadbent et al. [61] whereby the acid can be made using the organism *Lasiodiplodia theobromae* S22L (Forest Products Research Laboratory, England). The organism was grown as a liquid surface culture in Thomson bottles, each containing 1 L of medium. The medium contained various salts including sodium nitrate (0.2 percent),

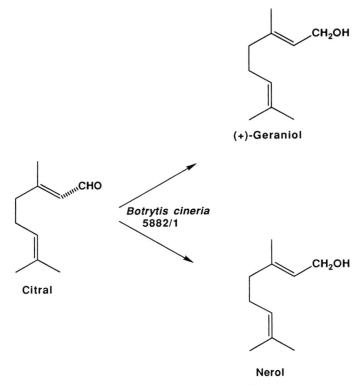

FIG. 4.30   Adapted from [58].

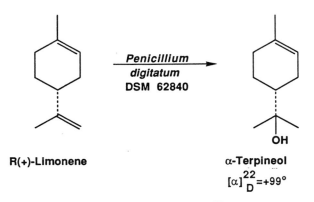

FIG. 4.31   Adapted from [59].

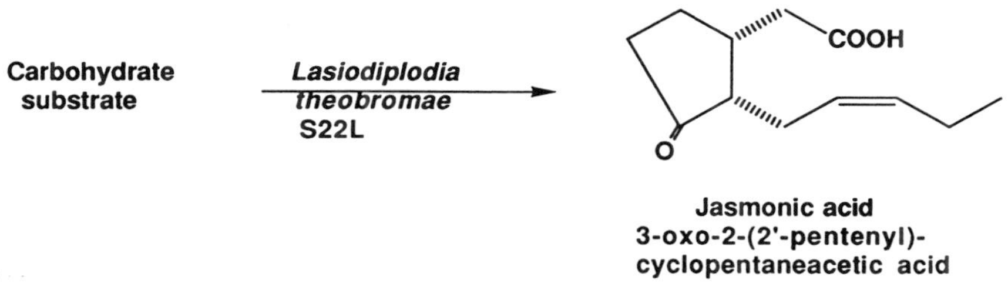

Carbohydrate substrate	*Lasiodiplodia theobromae* S22L

Jasmonic acid
3-oxo-2-(2'-pentenyl)-
cyclopentaneacetic acid

FIG. 4.32 Adapted from [61].

yeast extract and Cerelose (5.0 percent). Up to 0.5 percent elementary nitrogen and up to 15 percent by weight of carbon sources (i.e., sucrose, glucose, or glycerol) were found beneficial. When the organism was grown for about 13 days on the liquid medium a yield of 0.48 g/L of culture filtrate was recovered as crude jasmonic acid. The pure jasmonic acid recovered by distillation was recommended for use in perfumery as esters of the acid.

4.2.5.3 Bicyclic Monoterpenoid Transformations Fenchone occurs in fennel seed essential oil and has a camphoraceous odor [62]. The Latin taxanomic name of the source plant is *Lavandula stoechas* L. (*Labiatae*). Japanese workers [63] succeeded in carrying out a 32 percent bioconversion of (+)-fenchone to (+)-6-endo-hydroxy fenchone using a stationary, 3-day surface (mat) culture of *Aspergillus niger.* The organism was grown on a broth medium (mineral salts, peptone, and 1.5 percent glucose), received 500 mg/L of substrate, then was incubated for an additional 5 days. The mycelial mat was removed and the major metabolite, recovered at a yield of about 0.16 g/L, was hydroxy fenchone (Fig. 4.33).

4.2.5.4 Ionone Transformations Many bacteria and yeast strains were found [64] incapable of accumulating significant quantities of aroma products derived from α- and β-ionones. However, the ability to do so was prevalent among many filamentous fungi especially *Aspergillus* and *Rhizopus* cultures.

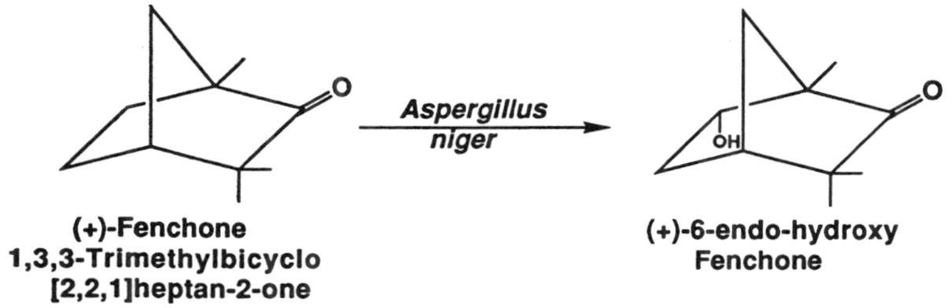

(+)-Fenchone
1,3,3-Trimethylbicyclo
[2,2,1]heptan-2-one

Aspergillus niger

(+)-6-endo-hydroxy
Fenchone

FIG. 4.33 Adapted from [63].

The best isolate was *Aspergillus niger* JTS 191, which accumulated neutral compounds imparting floral, violet or roselike odors. These when applied at 1 to 5 ppm concentration caused low-grade tobacco to assume desirable smoking qualities described as sweet, fresh, and mild. When β-ionone and β-methyl ionone was provided as substrate, the major products formed after 6 to 8 days of incubation with myclelial pellets were the four hydroxylated products as depicted in Fig. 4.34. When hydroxylation occurred at position 4 of the cyclic ionone structure, it assumed only the (R) configuration, but the hydroxyl group formed at position 2 consistently assumed the opposite stereochemical config-uration. Both β-ionone (A) and β-methyl ionone (B) were added at about 1 g/L and were shown to be transformed in yields of 26.6 percent and 18.4% to compounds 1A and 2A respectively, while the yields were 24.4 percent and 18.0 percent for compounds shown as 1B and 2B respectively (Fig. 4.34).

In a more recent effort Yamazaki et al. [65] showed that the same *A. niger* JTS 191 was able to transform (hydroxylate) α-ionone, α-methyl ionone and α-isomethyl ionone to useful tobacco aroma compounds. These neutral-fraction

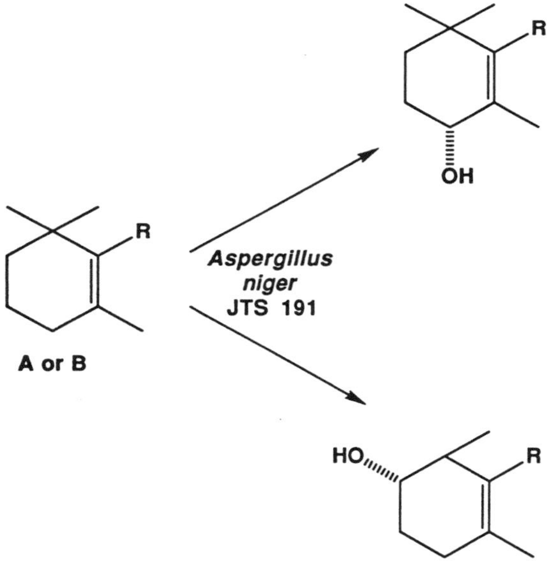

A: R is CH=CH-CO-CH$_3$ (β-ionone)
B: R is CH=CH-CO-CH$_2$-CH$_3$ (β-methyl ionone)

1A: (R)-4-Hydroxy-β-ionone
2A: (S)-2-Hydroxy-β-ionone
1B: (R)-4-Hydroxy-β-methyl ionone
2B: (S)-2-Hydroxy-β-methyl ionone

FIG. 4.34 Adapted from [64].

components were effective at 2.5 to 6.5 ppm. The type of aroma varied with the incubation time period of 48-hour-old mycelial pellets; the optimum was 3 to 4 days. The conversion products from α-methyl ionone were more effective than were those from α-ionone.

The major products from α-ionone added at the rate of 1 g/L were depicted as four structural isomers (Fig. 4.35). The *trans* stereoisomers were recovered at 43.1 percent and the *cis* stereoisomers at 25.3% (wt/wt). All were hydroxylated at position 3 on the ionone ring structure and were present at these yields after 72 hours, when the substrate had completely disappeared. Similarly, when the substrate (1 g/L) was α-methyl ionone, the products shown in Figure 4.36 were formed. The *trans* and *cis* isomers were recovered in yields of 51.3 percent and 32.1 percent (wt/wt) respectively. This bioconversion was completed more rapidly than for α-ionone.

The conversion of α-isomethyl ionone (see insert Fig. 4.36) is not shown. It was completed in 60 hours, the most rapid of the three substrate conversions. Again all four isomers were found. The yield approximated that of the other two systems and was reported [65] as 36.5 percent and 35.7 percent (wt/wt) for the *cis* and *trans* isomers respectively.

All the substrates were used as racemic mixtures. The conversion products initially exhibited levorotatory powers. However, with the passage of incubation time, when the dextrorotatory substrate enantiomer was transformed and all the substrate was consumed, the specific rotations of the conversion products approached zero.

The damascones are ionone-type compounds and an important group of carotenoid-derived compounds. In his review, Krasnobajew [66] described the microbial transformation of β-damascone by fungi, including *Aspergillus*, *Botryosphaeria* and *Lasiodiplodia*. All three genera were shown able to hydroxylate the ionone ring structure at position 4 and position 2 (Fig. 4.37), and no degradation of the substrate was observed. The *A. niger* transformation produced the former compound with a positive $[\alpha]_D^{22} = +24.3°$ optical rotation, whereas *B. rhodina* CBC 175.25 and *L. theobromae* ATCC 28570 yielded this compound with a negative rotation of $[\alpha]_D^{22} = -30.8°$ and $-42.2°$ respectively.

4.2.5.5 Sesquiterpenoid Transformations.

In his review, Krasnobajew [66] described the microbial transformation of patchouli alcohol, a member of this class of terpenoids. The alcohol is a major constituent (30–45 percent) in steam distillates of dried leaves of *Pogostemon cablin* Benth; more than 500 metric tons of essential oil are produced worldwide per annum [67]. The primary fragrance molecule in the essential oil is norpatchoulenol [68], which is present at 0.3 percent or less. A method was published for preparation of norpatchoulenol from patchoulol [69]; the first step involved a microbial process to convert patchoulol to 10-hydroxy patchoulol (see Fig. 4.38). The organism, a soil isolate, was able to catalyze the regio-selective hydroxylation reaction with a recovered product yield of up to 1.2 g/L in 5L fermentations.

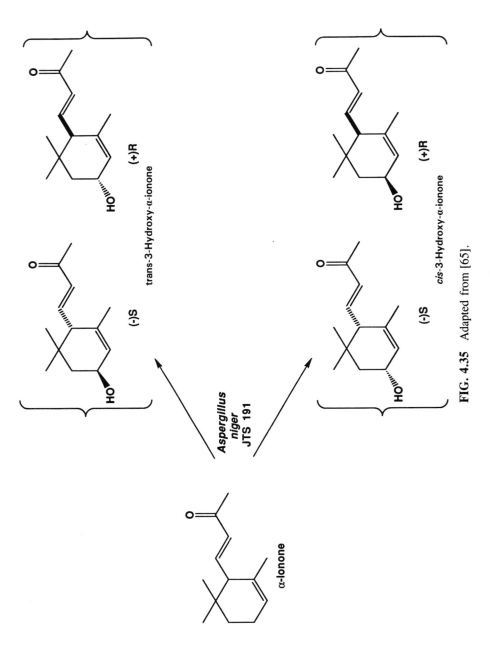

FIG. 4.35 Adapted from [65].

trans-3-Hydroxy-α-ionone

cis-3-Hydroxy-α-ionone

(+)R

(−)S

Aspergillus niger JTS 191

α-Ionone

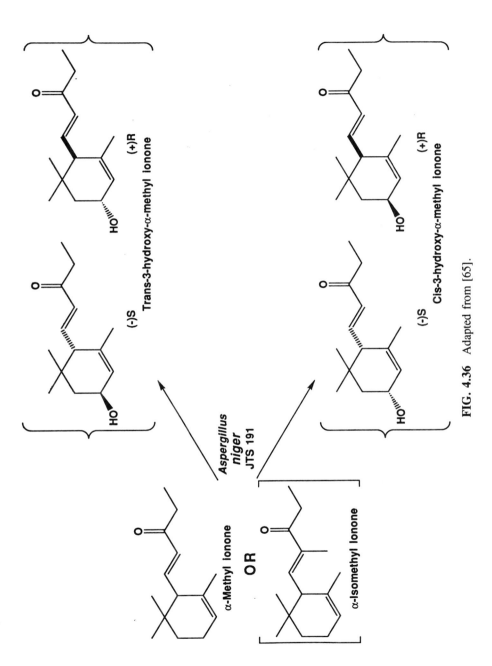

FIG. 4.36 Adapted from [65].

(+)R
Trans-3-hydroxy-α-methyl Ionone

(-)S

(+)R
Cis-3-hydroxy-α-methyl Ionone

(-)S

Aspergillus niger JTS 191

α-Methyl Ionone

OR

α-Isomethyl Ionone

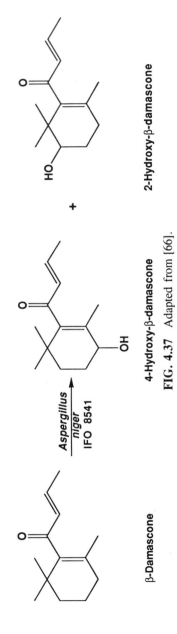

β-Damascone

Aspergillus niger IFO 8541

4-Hydroxy-β-damascone

+

2-Hydroxy-β-damascone

FIG. 4.37 Adapted from [66].

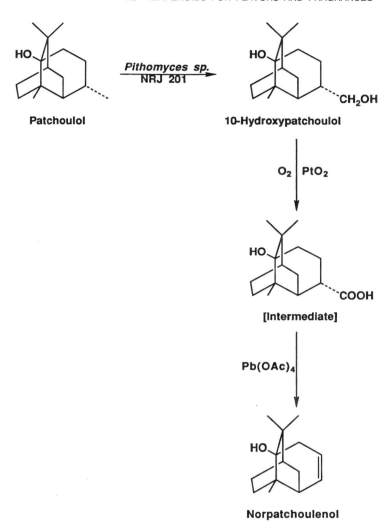

FIG. 4.38 Adapted from [69].

The 10-hydroxy compound was readily converted chemically via a two-step process to norpatchoulenol.

In a study of the odor structure relationship of patchoulol and norpatchoulenol [67], the most remarkable and balanced odorous impression was displayed by the norpatchoulol analog (±) A and the patchoulol analog (±) B (Fig. 4.39). Both were intensely woody and weakly camphoraceous but the A compound was more earthy and imparted a more intense overall odor. B was significantly weaker, yet very pleasant with a warm woody note.

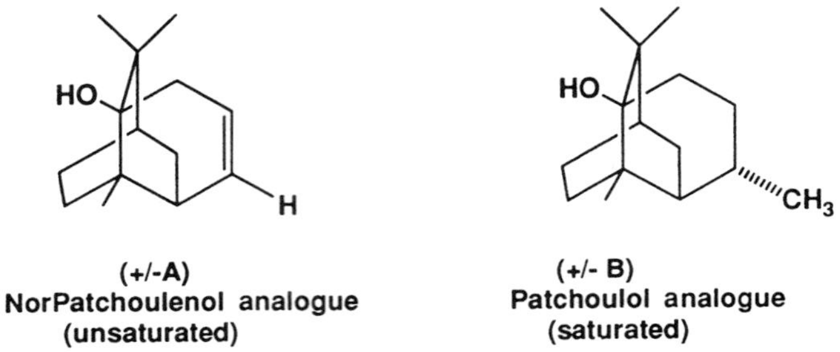

(+/-A)
NorPatchoulenol analogue
(unsaturated)

(+/- B)
Patchoulol analogue
(saturated)

FIG. 4.39   Adapted from [67].

The microbial transformation of the sesquiterpene alcohol (−)-α-bisabolol was investigated by Miyazawa et al. [70]. This terpenoid is found in a plant essential oil of value to the flavor and fragrance industry. The transformation to (−)-α-bisabololoxide B, the major metabolite (see Fig. 4.40), was carried out by *Aspergillus niger* grown in glucose–peptone broth as mycelial (surface) mats. The conversion approximated 14 percent with a yield of about 0.05 g/L.

Studies in England at the British American Tobacco Company [71] established that ambroxide was transformed by *Cladosporium oxysporum* (CBS-548.80) to three major oxygenated derivatives. The organism was cultivated aerobically in a malt extract broth for 4 days and was then supplied with 1 g/L of ambroxide dissolved in Tween 80 (polyoxyethylene sorbitan monoleate

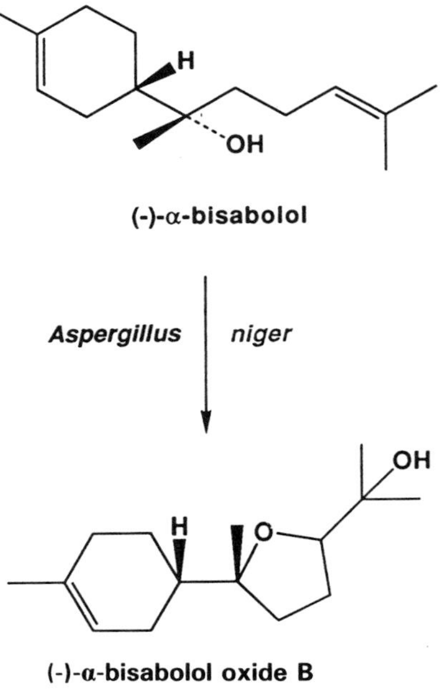

(-)-α-bisabolol

Aspergillus | niger

(-)-α-bisabolol oxide B

FIG. 4.40   Adapted from [70].

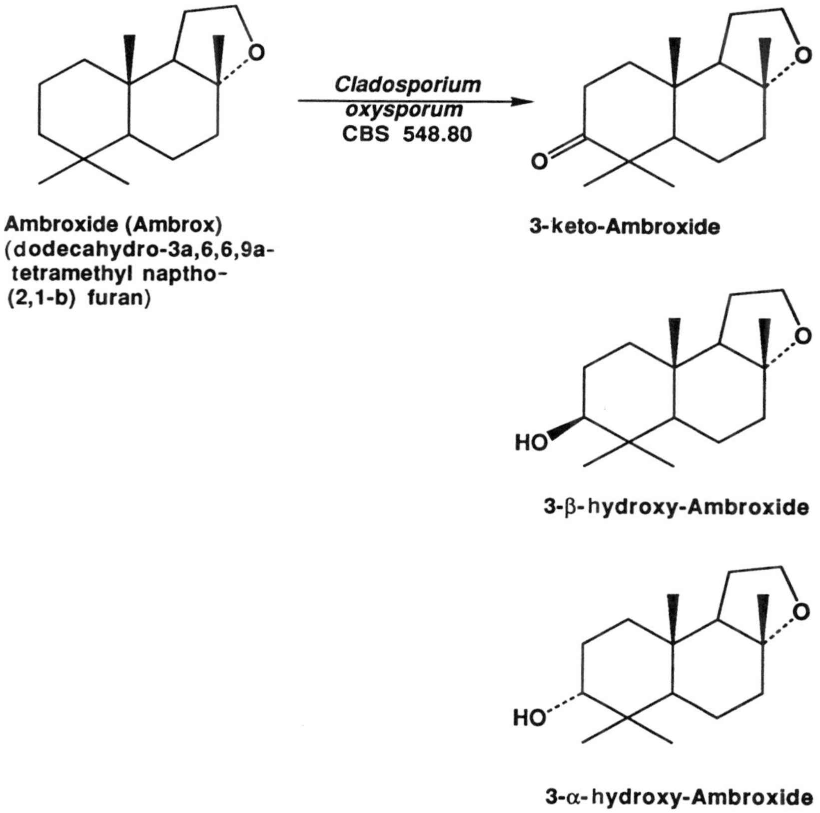

Ambroxide (Ambrox)
(dodecahydro-3a,6,6,9a-
tetramethyl naptho-
(2,1-b) furan)

Cladosporium
oxysporum
CBS 548.80

3-keto-Ambroxide

3-β-hydroxy-Ambroxide

3-α-hydroxy-Ambroxide

FIG. 4.41 Adapted from [71].

from Sigma Chemicals, Poole, England). After a 7-day incubation the broth was extracted with 1,1,1-trichloroethane. Customarily 30 percent of the ambroxide (0.3 g/L) was transformed (see Fig. 4.41) to a mixture of 3-keto ambroxide (0.06 g/L), 3-β-hydroxy ambroxide (0.2 g/L) and 3-α-hydroxy ambroxide (0.04 g/L). 3-Keto-ambroxide was the most significant as a smoke-flavor modifier without imparting undesirable attributes when used at 0.5 to 10 ppm in tobacco.

4.2.6 Summary

The microbial transformations of selected natural terpenoid substrates lead to modified terpenoids of value to the flavor and fragrance industry. While the actinomycetes thus far studied achieved disappointingly low yields, they are of interest in being able to elaborate two sesquiterpenoids, namely cadinol (earthy, woody) and selina diene-ol. These two compounds displayed optical rotations opposite, and possessed quite unique and different aroma qualities from the respective plant-derived enantiomers.

The yields of simple terpenoids in bacterial transformation were more promising, approaching gram quantities per liter in some cases. A process with commercial potential involved a *Pseudomonas fluorescens* in the transformation of α-pinene to a cyclic monoterpene aldehydes [44, 45]. These were given the trivial names of novalal 1 (lemonlike) and isonovalal 2 (citrus, woody-spicy).

Bacterial and filamentous fungi were shown to be most versatile in the variety of terpenoids produced from acyclic, monocyclic, and bicyclic monoterpenoids. However, γ-irone transformations were limited to *Enterobacteriaceae*, while filamentous fungi, especially *Aspergillus*, *Rhizopus* and *Cladosporium* species, were shown capable of significantly transforming ionones and sesquiterpenoids.

The highest yielding process involving a yeast (*Cryptococcus*) was provided by Farbood et al. [32]; in this process sclareol from clary sage was transformed into a lactonized diterpenoid called *sclareolide*. Up to 150 g/L of substrate was transformed over a maximum incubation time of 48 hours. The product is of value in tobacco and food flavors and as a synthetic substrate for various ambra derivatives useful as fixatives in fragrances.

REFERENCES

1. Maga, J. A., and Sizer, C. E., Pyrazines in foods. A review. *J. Agric. Food Chem* **21,** 22–50 (1973).

2. FEMA—Flavor and Extract Manufacturers Association, *Scientific Literature Review of Substituted Pyrazines in Flavor Usage*, Vol. I. Natl. Tech. Inf. Serv., U.S. Department of Commerce, Springfield, VA, 1984.

3. Kinderlerer, J. L., and Kellard, B., Alkylpyrazines produced by bacterial spoilage of heat-treated and gamma-irradiated coconut. *Chem. Ind. (London)* **16,** 567–568 (1987).

4. Kosuge, T., Kamiya, H., and Adachi, T., Odorous component of natto-fermented soybeans. *Yakugaku Zasshi* **82,** 190 (1962).

5. Kosuge, T., and Kamaiya, H., Discovery of a pyrazine in a natural product: Tetra methyl pyrazine from cultures of a strain of *Bacillus subtilis*. *Nature (London)* **193,** 776 (1962).

6. Reineccius, G. A., Kenney, P. G., and Weissberger, W., Factors affecting the concentration of pyrazines in cocoa beans. *J. Agric. Food Chem.* **20,** 202–206 (1972).

7. Barel, M., Leon, D., and Vincent, J.-C., Influence of cocoa fermentation time on the production of pyrazines in chocolate. *Cafe, Cacao, The* **29**(4), 277–286 (1985).

8. Kinderlerer, J. L., Volatile metabolites of filamentous fungi and their role in food flavour. *J. Appl. Bacteriol. Symp., Suppl.*, pp. 1335–1445 (1989).

9. Kosuge, T., Zenda, H., Tsuji, K., Yamamoto, T., and Narita, H., Studies in flavor components of foodstuffs. Part I. Distribution of tetra methyl pyrazine in fermented foodstuffs. *Agric. Biol. Chem.* **35**(5), 693–696 (1971).

10. Sugawara, E., Ito, T., Odagiri, S., Kubota, K., and Kobayashi, A., Comparison of compositions of odor components of natto and cooked soybeans. *Agric. Biol. Chem.* **49**(2), 311–318 (1985).

11. Miller, A., III, Scanlan, R. A., Lee, J. S., Libbey, L. M., and Morgan, M. E., Volatile compounds produced in sterile fish muscle (*Sebastes melanops*) by *Pseudomonas perolens*. *Appl. Microbiol.* **25,** 257–261 (1973).

12. Morgan, M. E., Libbey, L. M., and Scanlan, R. A., Identity of the musty potato aroma compound in milk cultures of *Pseudomonas taetrolens*. *J. Dairy Sci.* **55,** 666–669 (1972).

13. Dumond, J. P., Mourgues, R., and Adda, J., Potato-like off-flavour in smear coated cheese: A defect induced by bacteria. In *Sensory Quality in Foods and Beverages: Definition, Measurement and Control* (A. R. Williams and R. K. Atkin, eds.), pp. 424–428. Ellis Horwood, Chichester, UK, 1983.

14. Kowalewska, J., Zelazowska, H., Babuchowski, A., Hammond, E. G., Glatz, B. A., and Ross, F., Isolation of aroma-bearing material from *Lactobacillus helveticus* culture and cheese. *J. Dairy Sci.* **68**(9), 2165–2171 (1985).

15. Gerber, N. N., Three highly odorous metabolites from an actinomycete: 2-isopropyl-3-methoxy pyrazine, methylisoborneoli and geosmin. *J. Chem. Ecol.* **3**(4), 475–482 (1977).

16. Gallois, A., and Grimont P. A. D., Pyrazines responsible for the potato-like odor produced by some *Serratia* and *Cedecea* strains. *Appl. Environ. Microbiol.* **50**(4), 1048–1051 (1985).

17. Mottram, D. S., Patterson, R. L. S., and Warrilow, E., 2,6-Dimethyl-3-methoxy-pyrazine: A microbiologically produced compound with an obnoxious musty odour. *Chem. Ind. (London)* **12,** 448–449 (1984).

18. McIver, R. C., and Reineccuis, G. A., Synthesis of 2-methoxy-4-alkyl pyrazines by *Pseudomonas perolens*. *ACS Symp. Ser.* **317,** 266–274 (1986).

19. Demain, A. L., Jackson, M., and Trenner, N. R., Thiamine-dependent accumulation of tetramethyl pyrazine accompanying a mutation in the isoleucine valine pathway. *J. Appl. Bacteriol.* **94**(2), 323–326 (1967).

20. Ogai, D. K., Zunnundzhanov A., Urunbaeva, D. Y. U., Musaev, Sh. M., Shckerbakova, V. K., and Bessonova, I. A., *Bacillus cereus* strain 147 producer of tetramethyl pyrazine. *Appl. Biochem. Microbiol (Eng. Transl.)* **18**(5), 522–525 (1983).

21. Ito, T., Sugawara, E., Sukurai, Y., Takeyoma, S., Uchizawa, H., and Odagiri, S., Culture media for pyrazine production by a commercial natto bacillus (*Bacillus natto*). *J. Agric. Chem. Soc. Jpn.* **61**(8), 963–965 (1987).

22. Ito, T., Sugawara, E., Miyanohara, J.-I., Sukurai, Y., and Odagiri, S., Effect of amino acids as nitrogen sources on microbiological formation of pyrazines. *J. Jpn. Soc. Food Sci. Technol.* **36**(9), 762–764 (1989).

23. Sasaki, M., Kikuchi, T., Asao, Y., and Yokotsuka, T., Compounds produced by molds. II. Fluorescent compounds produced by Japanese industrial molds. *Nippon Nogei Kagaku Kaishi* **41**(4), 154–158 (1967).

24. Sasaki, M., Asao, Y., and Yokotsuka, T., Compounds produced by molds. III. Fluorescent compounds produced by Japanese commercial molds. *Nippon Nogei Kagaku Kaishi* **42**(5), 288–293 (1968).

25. Yokotsuka, T., Asao, Y., and Sasaki, M., Compounds produced by molds. IV. Isolation of non-fluorescent pyrazines. *Nippon Nogei Kagaku Kaishi* **42**(6), 346–350 (1968).

26. Buchanan, R. L., and Houston, W. M., Production of blue fluorescent pyrazines by *Aspergillus parasiticus*. *J. Food Sci.* **47,** 779–782 (1982).

27. Liardon, R., and Ledermann, S., Volatile components of fermented soy hydroly-sate. 2. Composition of the basic fraction. *Z. Lebensm.-Unters. Forsch.* **170**(3), 208–213 (1980).
28. Hanssen, H. P., Fermentative Gewinnung von Duft und Aroma Stoffen aus Pilz-kulturen. *Pilzaromastoffe* **33**, 996–1004 (1989).
29. Colins, R. P., and Halim, A. F., Production of monoterpenes by the filamentous fungus *Ceratocystis variospora. Lloydia* **33**, 481–482 (1970).
30. Krasnobajew, V., and Terpenoids. *Bio Technology*, Chapter 4 (Ed. Klaus Kieslich, Verlag Chemie, Weinberger, Fed Rep. Ger.) 97–125 (1984).
31. Farbood, M. I., and Willis, B. J., Process for producing diol and furan and microorganism capable of same. U.S. Pat. 4,798,799 (1989).
32. Farbood, M. I., Morris, J. A., and Downey, A. E., Process for producing diol and lactone and microorganisms capable of same. U.S. Pat. 4,970,163 (1990).
33. Yamaguchi, Y., Komatsu, A., and Moroe, T., Asymmetric hydrolysis of dl-men-thyl acetate by *Rhodotorula mucilaginosa*. (Optical resolution of menthols and related compounds. Part IV). *J. Agric. Chem. Soc. Jpn.* **51**, 411–416 (1977).
34. Gunata, Y. Z., Baynove, C. L., Cordonnier, R. E., and Arnaud, A., and Galzy, P., Hydrolysis of grape monoterpenyl glycosides by *Candida molischiana* and *Candida wickerhamii* β-glucosidases. *J. Sci. Food Agric.* **50**, 449–506 (1990).
35. Gerber, N. N., and Lechevalier, H. A., Geosmin, an earthy-smelling substance isolated from actinomycetes. *Appl. Microbiol.* **13**(6), 935–938 (1965).
36. Gerber, N. N., A volatile metabolite of actinomycetes, 2-methylisoborneol. *J. Antibiot.* **22**, 508–509 (1969).
37. Gerber, N. N., and Lechevalier, H. A., Production of geosmin in fermentors and extraction with an ion-exchange resin. *Appl. Environ. Microbiol* **34**(6), 857–858 (1977).
38. Gerber, N. N., Odorous substances from actinomycetes. *Dev. Ind. Microbiol.* **20**, 225–238 (1979).
39. Vandenbergh, P. A., Bacterial method and compositions for linalool degradation. U.S. Pat. 4,800,158 (1989).
40. Japanese Patent, Production of L-carvone from limonene using a *Corynebacterium* strain. Jpn. Pat. 47-38,998 (1972).
41. Cadwallander, K. R., Braddock, R. J., Parish, M. E., and Higgins, D. P., Bio-conversion of (+)-limonene by *Pseudomonas gladioli. J. Food Sci.* **54**(5), 1241–1245 (1989).
42. Gibbon, G. H., Millis, N. F., and Pirt, S. J., Degradation of α-pinene by bacteria. *Ferment. Technol. Today, Proc. Int. Ferment. Symp., 4th, 1972,* pp. 609–612 (1972).
43. Rhodes, P. M., and Winskill, N., Microbiological process for the preparation of L-carvone U.S. Pat. 4,495,284 (1985).
44. Davis, K. J., *Biofragrances* (personal communication). Novalal PLC, Plant Tech-nology Centre, The Merks Estate, Great Dunmow, Essex, UK, 1991.
45. Graham, B. A., Best, D. J., and Davis, K. J., Production of 2-methyl-5-isopropyl hexa-2,5-dien-1 ol and of 2-methyl-5-isopropyl-hexa-2,4-dien-1 al in microorga-nisms. Eur. Pat. Appl. EP 304,318 (1989).

46. MacRae, I. C., Alberts, V., Carman, R. M., and Shaw, I. M., Products of 1,8-Cineole oxidation by a pseudomonad. *Aust. J. Chem.* **32,** 917–922 (1979).

47. Liu, W. G., and Rosazza, J. P. N., Stereospecific hydroxylation of 1,8-cineole using a microbial biocatalyst. *Tetrahedron Lett.* **31**(20), 2833–2836 (1990).

48. Belcour, B., Courtois, D., and Ehret, C., Process for the preparation of γ-irone. U.S. Pat 4,963,480 (1990).

49. Hieda, T., Mikami, Y., Obi, Y., and Kisaki, T., Microbial transformation of cis-abienol. *Agric. Biol. Chem.* **46**(9), 2249–2255 (1982).

50. Hieda, T., Mikami, Y., Obi, Y., and Kisaki, T., Microbial transformation of sclareol. *Agric. Biol. Chem.* **46**(10), 2477–2484 (1982).

51. Kieslich, K., Abraham, W. R., Stumpf, B., and Washausen, P., Microbial transformation of terpenoids. *Top. Flavour Res., Proc. Int. Conf., 1985,* pp. 405–425 (1985).

52. Karst, F., and Vladescu, B. D. V., Procédé d'obtention d'aromes terpéniques par un procédé microbiologique. Fr. Pat 2,622.208 (1987).

53. Klingenberger, A., and Hanssen, H. P., Enhanced production of volatile flavour compounds from yeasts by adsorber techniques. I. Model investigations. *Chem. Biochem. Eng. Q* **2**(4), 222–224 (1988).

54. Kouzi, S. A., and McChesney, J. D., Microbial metabolism of the diterpene sclareol: Oxidation of the A ring by *Septomyxa affinis. Helv. Chim. Acta* **73**(8), 2157–2164 (1990).

55. Schindler, J., and Brune, K., Verfahren zur Fermentativen Gewinnung Monoterpenhaltiger Riechstoffe. Ger. Pat. DE 2,840,143 A1 (1980).

56. Mironov, V. A., Tsibul'skaya, M. I., and Yanoshevskii, M. Ts., Formation of monoterpenes by the *Ascomycete eremothecium ashbyi.* Russ. Pat. SU 1,454,845 A1 (1980).

57. Abraham, W. R., Arfman, H. A., Stumpf, B., Washington, P., and Kieslich, K., Microbial transformation of some terpenoids and natural compounds. *Bioflavour '87: Proc. Int. Conf., 1987,* pp. 399–414 (1988).

58. Brunerii, P., Benda, I., Bock, G., and Schrier, P., Bioconversion of monoterpene alcohols and citral by *Botrytis cinerea. Bioflavour '87: Proc. Int. Conf., 1987,* pp. 435–444 (1988).

59. Stumpf, B., Abraham, W.-R., and Kieslich, K., Verfahren zur Herstellung von(+)-α- Terpineol durch Mikrobiologische Umwandlung von Limonen. Ger. Pat. DE 3,243,090-A1 (1984).

60. Vick, B. A., and Zimmerman, D. C., Biosynthesis of jasmonic acid by several plant species. *Plant Physiol.* **75,** 458–461 (1984).

61. Broadbent, D., Hemming, H. G., and Turner, W. B., Preparation of jasmonic acid. Br. Pat. 1,286,266 (1968).

62. Merck Index, *An Encyclopedia of Chemicals and Drugs,* 11th ed. Merck & Co., Rahway, NJ, 1989.

63. Miyazawa, M., Yamamoto, K., Noma, Y., and Kameoka, H., Bioconversion of (+)-fenchone to (+)-6-*endo*-hydroxyfenchone by *Aspergillus niger. Chem. Express* **5**(4), 237–240 (1990).

64. Mikami, Y., Fukunaga, Y., Arita, M., and Kisaki, T., Microbial transformation of β-ionone and β-methyl ionone. *Appl. Environ. Microbiol.* **41**(3), 610–617 (1981).

65. Yamazaki, Y., Hayashi, Y., Arita, M., Hieda, T., and Mikami, Y., Microbial conversion of α-ionone, α-methylionone and α-isomethyl-ionone. *Appl. Environ. Microbiol.* **54**(10), 2354–2360 (1988).

66. Krasnobajew, V., and Helminger, D., Fermentation of fragrances: Biotransformation of β-ionone by *Lasioplodia theobromae*. *Helv. Chim. Acta* **65**, 1590–1601 (1982).

67. Spreitzer, H., A study on the odour/structure relationship of patchoulol and nor-patchoulenol. *Helv. Chim. Acta* **73**, 1730–1733 (1990).

68. Teisseire, P., Chemistry of patchouli oil. *Riv. Ital. Essenze, Profumi, Piante Off., Aromi, Saponi, Cosmet., Aerosol* **55**, 572–592 (1973).

69. Suhara, Y., Itoh, S., Ogawa, M., Yokose, K., Sawada, T., Sano, T., Ninomiya, R., and Maruyama, H. B., Regio-selective 10-hydroxylation of patchoulol, a sesquiterpene, by *Pithomyces* species. *Appl. Environ. Microbiol.* **42**(2) 187–191 (1981).

70. Miyazawa, M., Funatsu, Y., and Kameoka, H., Biotransformation of (−)-α-bisabolol to (−)-α-bisabololoxid B by *Aspergillus niger*. *Chem. Express* **5**(8), 589–592 (1990).

71. Barnes, A. G., and Bevan, P. C., Preparation of oxygenated ambroxides. U.S. Pat. 4,585,737 (1986).

Savory Flavors

TILAK W. NAGODAWITHANA

Red Star Specialty Products, Universal Foods Corporation, Milwaukee, Wisconsin

Savory foods always have constituted an important part of our daily diet. The discovery of soy source centuries ago perhaps, can be considered as one of the earliest savory products made by man to improve palatability of foods. Recent developments in food research have resulted in a wide range of cost-effective savory flavoring ingredients. The best known are the hydrolyzed vegetable proteins and autolyzed yeast extracts. The amino acid–peptide mixtures produced by protein hydrolysis and the reactions that follow with components like sugar account for flavor generation. In addition, there are other compounds in foods that have little or no flavor or aroma themselves, yet are capable of enhancing the flavor of other foods. These are known as flavor enhancers; examples include monosodium glutamate (MSG), inosine 5'-monophosphate (5'-IMP), and guanosine 5'-monophosphate (5'-GMP). Traditionally, the Japanese have used natural foods rich in these compounds as condiments to improve the palatability of other foods. Technologies now available for the production of these flavor potentiators and other savory flavor ingredients are discussed in some detail in this chapter.

5.1 INTRODUCTION

Ever since *Homo sapiens* appeared on the surface of the earth, their food habits have been clearly different from those of their evolutionary ancestors. Their higher level of intelligence and the experience they gained with time permitted these primitive civilizations to process their food to make it flavorful, enhance its palatability, and often improve its nutritional value. For example, fermented foods have been prepared and extensively consumed by humans quite unwittingly for thousands of years. Fermentation not only offered ancient cultures

Bioprocess Production of Flavor, Fragrance, and Color Ingredients, Edited by Alan Gabelman, ISBN 0-471-03821-0 © 1994 John Wiley & Sons, Inc.

the possibility of enjoying better-tasting food products with greater shelf-life and stability, but also reduced the time and effort that was previously required to obtain the daily food supply. It even made stockpiling of necessary food products practical. The fermentation concept thus was accepted as one of the oldest and most economical methods of producing and preserving foods. Nearly every past civilization has relied on fermented food products; one or more of these were characteristic of each specific culture. These fermented foods were socioculturally bound, especially in rural household and village community traditions.

The term *savory* is described in the dictionary as that which is not sweet but piquant, pungent, or salty to the taste or appetizing to the taste and smell. It has a strong similarity to the term *flavorful* when applied to a food system. Foods derived from muscles of animals and fish (e.g., beef, chicken, pork, bacon, seafoods), prepared foods like soups, sauces, and gravies, certain snack foods and formulated meat products like sausages and meat pies are all known to exhibit savory impact.

The public has always been flavor conscious, so that bland, tasteless foods generally have minimal demand commercially. Among the considerable variety of food preparations containing meat, the most popular lines are undoubtedly those having a higher proportion of quality meats, which impart the best taste and flavor. However, the rising price of meat has caused two major variations in the production of meat products. One is the use of less meat and more bulking agents and filler. The other is the wider use of savory flavorings combined with appropriate bulking agents or carriers. Meat extracts are usually unsuitable as savory flavoring because of high cost. Synthetic meat flavors are also unsatisfactory because they impart only a beefy taste and are too salty. The consumer is constantly demanding more flavorful, nutritionally rich foods with low fat and low cholesterol, and, because of this, the need for savory flavorings for the formulation of meat products and meat analogs has increased significantly in recent years.

The discovery of soy-sauce fermentations by the Orientals centuries ago can be considered one of the outstanding developments made by man in the area of food science. In the production of soy sauce, the fungus *Aspergillus oryzae*, together with *Lactobacillus*, *Hansenula* and *Saccharomyces* species, grows on soy beans to produce a dark brown liquid with a distinct aroma capable of imparting a savory flavor when applied to many foods. Later studies on soy sauce manufacture, specifically those conducted during the last century, have led to the development of new areas of food science particularly important to the flavor industry. Based on these findings, for the first time in history it became possible to produce meatlike, savory flavors from proteins and lipids of vegetable origin.

Recent developments in food research have resulted in a wide range of cost effective savory flavoring ingredients that can be of immeasurable significance to those who process meat products and meat analogs. Because of the similarity of the free amino acid profile of meats, these products of either yeast or plant origin are often used in meat flavor applications. Those of plant origin are prepared by acid hydrolysis of soy beans, wheat, and other plant substrates; these products are commonly known as *hydrolyzed vegetable proteins* (HVPs). Unfortunately, despite the effectiveness of HVP as a replacement for meat extract, recent attention has focused on some serious drawbacks of these products, which deter many food processors from using them; these drawbacks will be described later in this chapter.

On the other hand, many food processors are now using yeast-derived products, which generally are considered safer than HVPs, to enhance the savory characters of foods. These products, whether they are made from Brewer's spent yeast, Baker's yeast, or Torula yeast, are known to create the impression of high meat content and impart a rich meaty flavor [1]. Additionally, they are capable of generating a wide range of flavor intensities in many food formulations, ranging from delicate, bland foods to pungent, meaty flavors. Indeed, these yeast extracts have now become popular among food processors as inexpensive replacements for meat extracts capable of generating a variety of savory flavors.

5.2 HYDROLYZED VEGETABLE PROTEINS

Vegetable protein hydrolysates have for a long time been used in the food industry as flavor donors, flavor enhancers, or flavor donor/enhancer combination products. These classifications have been offered by the manufacturers to assist customers in grade selection for their specific application. Flavor donors contribute a taste that becomes a definite feature of the final product. Flavor enhancers (e.g., glutamic acid) enrich the naturally occurring flavor but contribute little or no taste of their own. The refined HVP liquid contains amino acids and their sodium salts (notably MSG), a high concentration of salt as a result of the neutralization of the hydrochloric acid, and other flavor solids formed by the different chemical reactions occurring during the processing.

Hydrolyzed vegetable proteins are known to contribute meatiness to foods, enhance and intensify naturally occurring savory flavors, and generally round off and balance the savory characteristics of the food material. These hydrolysates are known for their versatility and ease of application. They are relatively cost effective and become unstable only when heated above 200°C. At these high temperatures HVPs begin to lose their typical flavor, with the development of caramelized and salty tastes upon prolonged baking. Due to the high salt concentration, protein hydrolysates have an indefinite shelf-life by normal standards.

5.2.1 Commercial Process

The most commonly used proteinaceous materials of plant origin are wheat gluten, corn gluten, defatted soy flour, defatted peanut flour, defatted cotton seed flour, and so on. Wheat gluten is considered an excellent material for the production of HVP because of its high content of glutamic acid. The process for the manufacture of HVP is described in the literature in very general terms [2]. In a typical process, the proteinaceous material is hydrolyzed with HCl either for 11 hours at 212°F or 1.5 hours at 250°F (Fig. 5.1). This obviously requires acid-resistant tanks; furthermore, hydrolysis at 250°F requires pressure vessels that have the capability of maintaining at least 15 psig during the processing operation. The resulting mixture is a concentrate containing 32 percent flavor solids rich in proteinaceous material, 23 percent water, 32 percent food grade hydrochloric acid of 35 percent strength, and 13 percent pure sodium hydroxide of food grade quality. Under the conditions used here the protein chains are broken down to their component amino acids. At the high temperatures used, some of these amino acids react with carbohydrates in the protein-rich substrate to produce a meaty character. The well-known Maillard reaction is responsible for such flavors. The significance to food processing of this reaction is outlined briefly later in this chapter. The relationship between amino nitrogen and total nitrogen usually provides a reliable indication of the extent of hydrolysis during the acid-mediated reaction.

When hydrolysis is complete, the charge is pumped into another vessel, where it is neutralized with sodium hydroxide or sodium carbonate to a pH in the range of 4.8 to 6.0. The liquid medium at this stage is intensely dark in color, primarily due to the formation of a sludge during acid hydrolysis. This

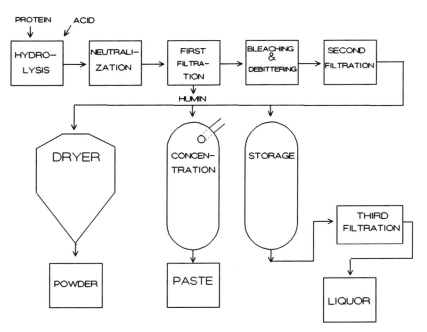

FIG. 5.1 Flowchart for the commercial production of protein hydrolysates.

sludge is referred to as *humin*, thought to be a condensation product of tryptophan and aldehydes present in the medium. The liquor containing this insoluble material is filtered, and the filtrate is then bleached and debittered using activated carbon. The latter adsorbs preferentially bitter-tasting phenylalanine and tyrosine and certain color substances. The decolorized liquor is then filtered and stored, concentrated to a paste by vacuum evaporation, or converted to a powder by spray drying. Some manufacturers produce HVP in granular form; in this case, after the storage of clarified liquor for almost 8 months, it is concentrated into a paste, oven dried under vacuum to produce a cake, and then ground into granules. These manufacturers claim that the granular form of HVP is less hygroscopic and more soluble than the spray-dried form.

5.2.2 Properties

Changes in the color and taste of HVPs can be made by making minor changes in process conditions or by varying the blend of vegetable proteins used in the process. The salt content, typically 40 to 45 percent, is directly related to the level of hydrochloric acid. To meet customer demands for savory flavor and to expand the application capabilities, development and refinement of the hydrolysis techniques have progressed considerably during the last few years. Moreover, salt, MSG, caramel color, and other food additives are often added to HVPs to generate the desired flavor profile.

Acid hydrolysis brings about the destruction of certain amino acids and vitamins, thereby reducing nutritive value. Tryptophan is one of the important amino acids destroyed by acid hydrolysis. Additionally, HVPs made using hydrogen chloride have been found to contain monochloropropanols (MCPs) and dichloropropanols (DCPs), which are known carcinogens. This has made many food processors severely restrict the use of HVP in new product development. The German government now requires HVP manufacturers to reduce chloropropanol levels below 50 ppb for DCPs and 1 ppm for MCPs.

5.3 REACTION FLAVORS

Flavor chemistry of raw and cooked meats has been investigated during the last three decades, and these studies have provided the flavor industry with invaluable information to help in the production of meatlike, savory flavorings from a number of basic precursor compounds (see Table 5.1). A considerable segment of the food-ingredient industry is still devoting its energies to finding new ways to enhance savory characteristics in soups, sauces, entrees, and soy-extended meat analogs through the use of reaction flavor technology.

The significance of Maillard reactions and their role in savory flavor development has occupied the attention of many researchers in the flavor industry. In general terms, *Maillard browning*, sometimes referred to as non-enzymatic browning, is a reaction that occurs between carbonyl compounds and amines

TABLE 5.1   Basic Precursors Used in the Development of Reaction Flavors

Amino acids	Cysteine, glutamic acid, valine, glycine, hydrolyzed vegetable protein (HVP), yeast extract, hydrolyzed animal protein, etc.
Reducing sugars	Glucose, xylose, ribose, ribose-5-phosphate
Vitamins	Thiamine
Sulfur compounds	Furanones, sulfides, thiols (cysteine, thiamine)
Nucleotides	Inosine 5'-monophosphate, guanosine 5'-monophosphate
Acids	Lactic acid, aliphatic carboxylic acids, acetic acid, etc.

during the cooking process. The outcome of the reaction is largely dependent on the concentration of reactants, moisture level, temperature, and pH of the medium. The variations and enormous complexity in the final flavor almost entirely are due to the large number of permutations arising from the interactions of a relatively large number of intermediate reactants characteristic of the Maillard reactions. Yet it is surprising how reproducible the flavor characteristics can be when the Maillard browning reaction is carried out under highly controlled conditions.

The term Maillard reaction in general terms refers to a complex series of reaction pathways that occurs following the initial reaction between the amino acids and reducing sugars. The extent to which these reactions proceed is dependent on the time and temperature conditions used. The characteristic taste and aroma compounds produced by these reactions largely are due to the thermal degradation of Amadori compounds of amines and amino acids (Fig. 5.2). It is beyond the scope of this review to present a more detailed description of the Maillard reaction; excellent reviews can be found elsewhere [3–5].

Much of the development work on savory flavors has been conducted in the last 30 years; the number of patents per year filed on reaction flavor technology peaked between 1966 and 1972. Food flavorists active in the field have been able to formulate several authentic, meatlike savory flavors by the direct controlled reaction between amino acids and carbohydrate sources in the presence of other ingredients, such as fatty acids, salt, certain vitamins, esters of amino acids, and sulfides. The role of the sulfur-containing amino acid L-cysteine in the development of meatlike, savory flavors is well established. May and his coworkers have filed several patents in which they claim that heating cysteine (or cystine residues in peptides) in the presence of furan or substituted furans, pentoses, or glyceraldehyde gives a meatlike, savory flavor [6–9]. Interestingly, methionine (also a sulfur-containing amino acid) does not produce the desired savory notes but instead produces undesirable flavors when present in the reaction mixture. Additionally, Bidmead et al. [10] and Giacino [11] have claimed

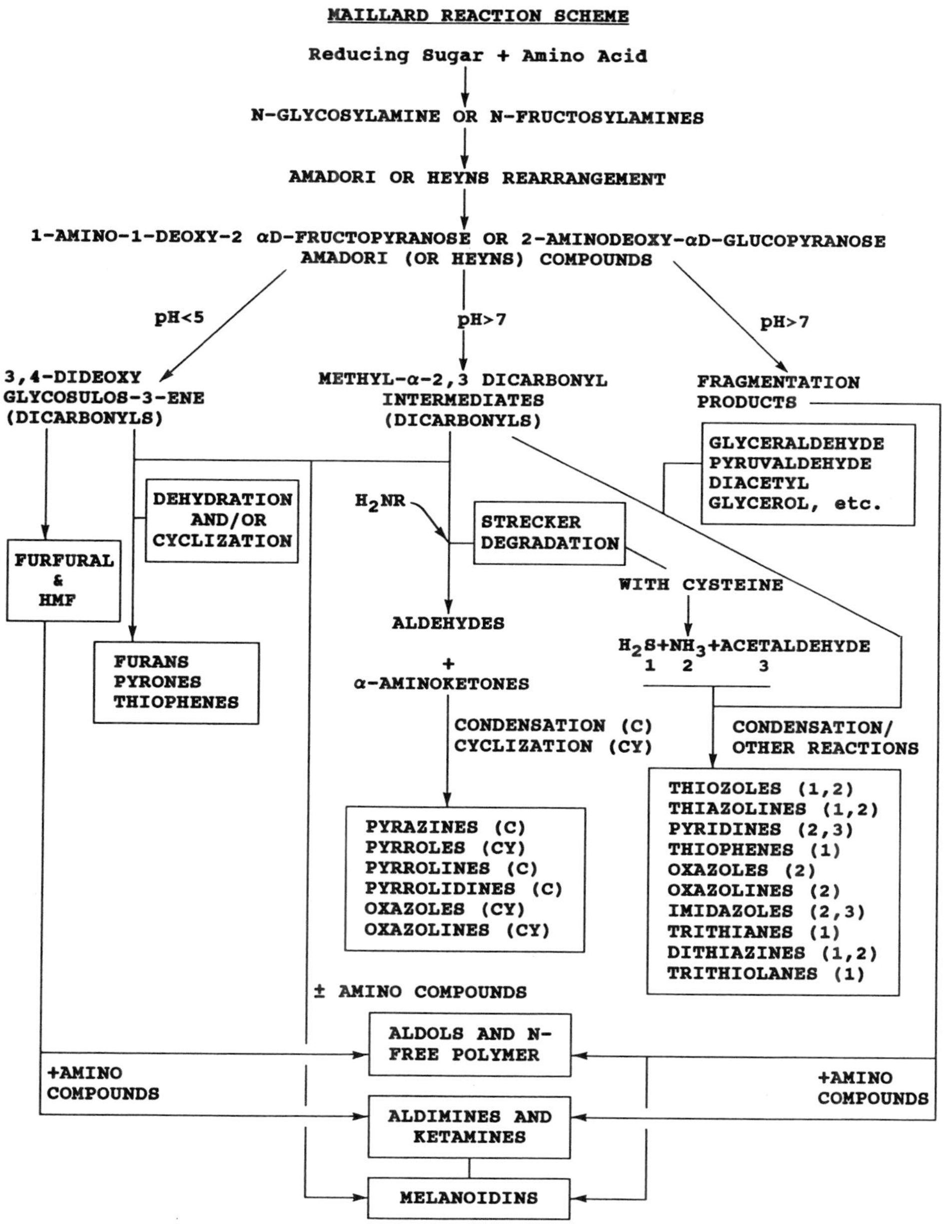

MAILLARD REACTION SCHEME

Reducing Sugar + Amino Acid

N-GLYCOSYLAMINE OR N-FRUCTOSYLAMINES

AMADORI OR HEYNS REARRANGEMENT

1-AMINO-1-DEOXY-2 αD-FRUCTOPYRANOSE OR 2-AMINODEOXY-αD-GLUCOPYRANOSE AMADORI (OR HEYNS) COMPOUNDS

pH<5   pH>7   pH>7

3,4-DIDEOXY
GLYCOSULOS-3-ENE
(DICARBONYLS)

METHYL-α-2,3 DICARBONYL
INTERMEDIATES
(DICARBONYLS)

FRAGMENTATION
PRODUCTS

GLYCERALDEHYDE
PYRUVALDEHYDE
DIACETYL
GLYCEROL, etc.

DEHYDRATION
AND/OR
CYCLIZATION

H_2NR

STRECKER
DEGRADATION

WITH CYSTEINE

FURFURAL
&
HMF

$H_2S+NH_3+ACETALDEHYDE$
1   2   3

ALDEHYDES

FURANS
PYRONES
THIOPHENES

+
α-AMINOKETONES

CONDENSATION (C)
CYCLIZATION (CY)

CONDENSATION/
OTHER REACTIONS

PYRAZINES (C)
PYRROLES (CY)
PYRROLINES (C)
PYRROLIDINES (C)
OXAZOLES (CY)
OXAZOLINES (CY)

THIOZOLES (1,2)
THIAZOLINES (1,2)
PYRIDINES (2,3)
THIOPHENES (1)
OXAZOLES (2)
OXAZOLINES (2)
IMIDAZOLES (2,3)
TRITHIANES (1)
DITHIAZINES (1,2)
TRITHIOLANES (1)

± AMINO COMPOUNDS

ALDOLS AND N-
FREE POLYMER

+AMINO
COMPOUNDS

+AMINO
COMPOUNDS

ALDIMINES AND
KETAMINES

MELANOIDINS

FIG. 5.2 Chemical pathways of the Maillard reaction.

that thiamine contributes to meatlike flavors when it is present in standard reaction mixtures.

It is common practice to heat semidry mixtures of HVPs and reducing sugars, such as glucose, to achieve the characteristic aroma and flavor of cooked meat. Cysteine also plays an important role in the production of these savory flavors, as mentioned above. Despite the presence of sulfur amino acids in HVPs, some formulations require the use of additional sulfur amino acids to further enhance the savory notes. Interestingly enough, no one has claimed yet the exact duplication of meaty aroma and taste by the heat processing of pure ingredients or chemicals.

As mentioned above, the development of savory flavors is extremely complex because of the large number of possible interactions and permutations that can occur between reaction intermediates. As we gain a better understanding of meat flavors and the reactions associated with their production, further improvements in the development of savory flavors will be possible.

5.4 YEAST-DERIVED FLAVOR PRODUCTS

Man's involvement with yeast extends from prehistoric times, and the term *yeast* has now become a household word because of its familiar application in bread, beer, wine, and so on. Considering its harmless nature, the food industry has long been interested in yeast and yeast-derived products in various food formulations. Yeast provides a means of imparting a rich, savory flavor to food products, and it is considered a relatively inexpensive alternative to meat extracts. Owing to its high flavor strength, often only a small amount of yeast extract is necessary to make a significant improvement in the taste of a finished product.

In addition to the range of dried yeast products that food processors have already become accustomed to, there are a number of yeast extracts and yeast-derived flavor enhancers in the present market in various grades and flavor characteristics to suit varying needs. The combined ranges of yeast extracts and dried yeasts now make it possible for food processors to choose flavoring ingredients of superior quality applicable to a large number of products. The wholesomeness of these yeast products makes their use absolutely safe when used within the recommended limits. Recent developments in the processing of Brewer's and Baker's yeast for the production of a wide range of yeast-derived products of high flavor strength have been reported [1]. It is therefore of significant interest to review the recent progress in the development of yeast-derived food ingredients and briefly describe the processes by which Baker's and Brewer's yeasts are converted into a wide variety of valuable flavor ingredients to meet the needs of the food industry.

Yeast-derived products for food use are available in three forms: (1) inactive dried yeast, (2) autolyzed yeast, and (3) yeast extract. In general, Baker's, Brewer's, and Torula yeast strains serve as the common substrates for the production of these products following the removal of undesirable constituents. Inactive dry yeast is produced by drying the yeast cream after heat inactivation. Autolysis involves self-digestion of the yeast cells, which occurs when environmental conditions are set to activate certain endogenous enzymes. The total content following autolysis is termed the autolysate. The soluble portion of the autolysate after the removal of the particulate fraction (cell wall) is termed the yeast extract. The specific characteristics of these three products are discussed in greater detail in the following sections.

5.4.1 Dry Yeast

Dried Brewer's, Baker's, and Torula yeasts contain approximately 50 percent protein and have the ability to enhance the flavor of rich, savory foods. Although dry Torula and Baker's yeast generally have a bland flavor, the slightly bitter Brewer's yeast blends well with other flavors present in the food formulation. Surprisingly, it is only in the last three or four decades that the immense potential of dried Brewer's yeast has been realized. Over these years, many technological advances have taken place in the development of food flavors from yeast. The rich savory taste of dry yeast products and their remarkable flavor strengths make their use highly cost effective. These products are known to improve and enhance the appearance, texture, nutritional properties, and savory flavor of many foods such as processed meats, dips, cheese spreads, dressings, and chili preparations. They also provide an ideal nutritional complement to the amino acid profile of breakfast and snack foods, cereal grain products, and bakery goods.

The types of dried yeast products vary considerably depending on the method of manufacture used to develop different flavor profiles. In general, a well-washed Baker's yeast produces a yeast product that has a cleaner flavor profile than any product derived from Brewer's spent yeast. The dried primary grown yeasts, *Saccharomyces cerevisiae* and Torula yeast grown exclusively for food use, are superior because of their higher and more uniform content of B-complex vitamins, protein, certain minerals, enzymes, RNA, and important flavor compounds. In addition, primary grown yeast that is propagated exclusively for use as a flavor or for extract production has a natural, pleasant, toasted, nutty flavor preferred by most food processors to the characteristic bitter flavor of Brewer's yeast. Primary grown yeasts generate a bready, savory flavor in food products and hence are particularly useful in breads, pizza doughs, pretzels, breadings, and other baked goods. They can also be used as a carrier for spices and smoked flavors.

5.4.1.1 *Manufacturing Brewer's Dry Yeast.* In a brewery fermentation, approximately 80 percent of the yeast slurry is surplus and provides the raw material for production of a range of food ingredients. Excess spent yeast that is delivered for dry yeast manufacture is subjected to strict tests and quality

control, and high standards of hygiene are maintained throughout the downstream processing. To achieve such standards, all vessels and equipment used throughout the plant are made of stainless steel and other Good Manufacturing Practice (GMP) guidelines are closely followed.

Typical Brewer's spent yeast is bitter from the presence of beer solids and hop components (humulones and isohumulones) adhering to the surface of the yeast cells. Prior to drying of fresh Brewer's spent yeast, it is washed and debittered to improve the palatability of the final product. Common practice is to debitter the yeast by means of an alkaline wash. The pH of the yeast slurry is first adjusted to approximately nine with a base such as sodium hydroxide to loosen the hop resins, which are firmly bound to the yeast cell wall at lower pH. A separation followed by a water wash helps to minimize the bitter flavor of the yeast considerably. The pH of the final yeast cream is eventually adjusted to approximately 6 [1]. Some food manufacturers do not object to the presence of some bitterness in the yeast, because the presence of these bitter components leads to generation of desirable reaction flavors during subsequent processing. The bitterness level in the yeast is generally controlled by the washing procedure used during the processing. A more recent development is the use of ion-exchange resins to trap the bitter hop components to achieve precise control in the debittering process.

The debittering process is followed by separation of the yeast from soluble bitter components present in the liquid medium by a centrifugation step. A water wash following this operation is not unusual. The yeast is then pasteurized and spray dried; the pasteurization step destroys the fermenting power of the yeast and the drying step reduces the moisture of the product to approximately 5 percent. The process does, however, leave the heat sensitive vitamins and other desirable food flavors unimpaired. Alternatively, if the product is drum dried, it is generally ground and screened to produce a fine, free-flowing powder of desirable particle size.

5.4.1.2   *Manufacturing Baker's and Torula Dry Yeast.*   A culture maintained on a nutrient agar slant for approximately 6 to 8 days and usually requires six seed stages to produce a batch of Baker's yeast suitable for distribution. In trade or commercial propagations, typically 100- to 200-kL fermenters are used, and in each batch 45,000 lb of cake yeast are produced from a few million cells present initially. The success of a good trade propagation depends on a variety of factors. For example, according to the Crabtree effect, which is the reverse of the Pasteur effect, respiration or oxygen utilization is hampered greatly by high glucose concentration even under highly aerobic conditions, leading to production of high levels of ethanol. To prevent this wasteful production of ethanol, the wort (molasses–sugar medium) is added incrementally into the highly aerated medium at a slow rate so that the sugar added is assimilated by the yeast as fast as it is added and there is no accumulation of residual sugar. This is generally referred to as *Zulauf* or *fed-batch* process. Currently, more modern equipment such as a mass spectrometer-type gas analyzer coupled

with a computer is beginning to provide more advanced feed-control systems to minimize sugar accumulation and maximize biomass production in the reactor. The yeast produced is then separated, washed, pasteurized, and dried to a moisture content of approximately 5 percent in a spray dryer.

A similar procedure can be used to produce Torula or *Kluyveromyces marxianus* dry yeast. Torula yeast can also be produced in the United States by continuous fermentation. The reader can find useful reviews on the production of yeast in Reed and Nagodawithana [1,12].

5.4.2 Yeast Extract

During the last 30 years, there has been a rapid growth in demand for savory flavors for use in formulated foods, and the use of meat extracts for such applications has failed to remain competitive because of its high cost. As mentioned above, in more recent years, HVP has been used to impart a savory, meaty flavor to various food products. However, although it served as an effective replacement for meat extracts initially, HVP began to lose its attractiveness as a food ingredient because of the high content of salt, MSG, MCP, and DCP. To overcome this problem, researchers began to investigate the use of yeast for the production of extracts as suitable replacements for HVP. These studies led to the development of a number of yeast extracts capable of imparting a rich, savory, meaty flavor when applied to foods at relatively low concentrations. Yeast extracts have a flavor strength substantially higher than that of dried yeast on an equal solids basis because of the concentration effect in extracts following the separation of the cell wall fraction during processing. Autolyzed yeast products also have a higher flavor intensity than dried yeast, but not as high as yeast extracts.

Production of quality flavors from yeast at minimal cost is a major prerequisite for the successful marketing of yeast-derived products. Achievement of the desired properties is largely determined by process conditions like temperature, pH, time, and the action of agents like ethyl acetate or enzymes like papain that might be necessary to enhance the autolysis process. With the understanding of flavor chemistry together with application of new techniques and improved downstream processing, it has been possible for manufacturers to develop a whole series of highly desirable products with savory richness at a cost attractive to the food industry.

The key factors in choosing the starting material for processing are the price and availability of the yeast itself, and the desired properties (flavor profile, color, etc.) of the final product. Spent Brewer's yeast as a by-product in the brewing industry costs less than Baker's or other types of primarily grown yeast. Although both types are used regularly for the production of extracts and autolysates for the food and fermentation industries, Brewer's yeasts tend to produce lower yields and darker colored extracts with characteristically different flavor profiles.

As mentioned earlier, the taste of spent Brewer's yeast without pretreatment can be overwhelmingly bitter. However, the bitterness level of the final extract can be controlled by the extent to which spent yeast is cleaned prior to processing. This may include a single wash, a series of washes, or in certain instances an alkaline wash to remove most of the bitterness. However, the latter treatment can influence the final extract yield. Pretreatment is also required for Baker's and other types of primary grown yeasts if grown on molasses. That is, an initial separation of the mother liquor and at least two water washes are required to remove the residual molasses adhering to the yeast cells. However, some processes require the presence of a limited level of molasses with the yeast to provide a characteristic flavor to the final product.

There are three manufacturing processes for the production of yeast extracts: autolysis, plasmolysis, and hydrolysis [1,13]. These three processes are described below.

5.4.2.1 Autolysis.   Autolysis, which is essentially a self-digestion of the yeast, requires the mediation of several endogenous degradative enzymes. The autodigestion process can be initiated by the application of carefully controlled conditions such as heat and pH, which kill the cells without inactivating the degradative enzymes. As the cells die under these conditions, disorderliness occurs within the cells, causing the internal membranes to disintegrate and release degradative enzymes into the cell matrix. These hydrolytic enzymes act on the corresponding macromolecular substrates, ultimately causing solubilization of the cell constituents. With the disintegration of the outer cell membrane, the soluble constituents leak out into the surrounding environment. Cell death under suboptimal conditions may result in minimal solubilization, because either the conditions are less favorable for the degradative enzymes, or the enzymes are destroyed totally as in the case of extreme heat.

A good-quality viable yeast is essential to produce an extract with the required properties. Starting yeast creams must be free of *Salmonella*, *Clostridium perfringens*, *Streptococcus faecalis* and other pathogenic microorganisms, but yeasts that are treated at high temperature or under any other condition lethal to the degradative enzymes are unsuitable. Brewer's yeast contains microbial contaminants, hop resins, trub (product derived from malt), and other particulate matter; a higher quality yeast extract is usually obtained from fresh Baker's and other primary grown yeasts because they are cleaner and have a higher protein content.

In a typical autolysis (see Fig. 5.3), the viable yeast cream at 15 to 20 percent solids is subjected to a temperature in the 45 to 60°C range for 24 to 36 hours at a pH of approximately 5.5. From the outset, the proteases and glucanases play a key role in the solubilization of the cell contents. More than 40 proteolytic enzymes have been identified in yeast, but only a few of these can be regarded as vital for the degradative process. Under autolytic conditions, these proteases are capable of degrading the yeast proteins into soluble amino acids and peptides. The action of $\beta(1$-$3)$- and $\beta(1$-$6)$-glucanases on the glucan-mannan matrix of the cell wall make it permeable even to large molecules like proteins and soluble carbohydrates. The product yield and the flavor charac-

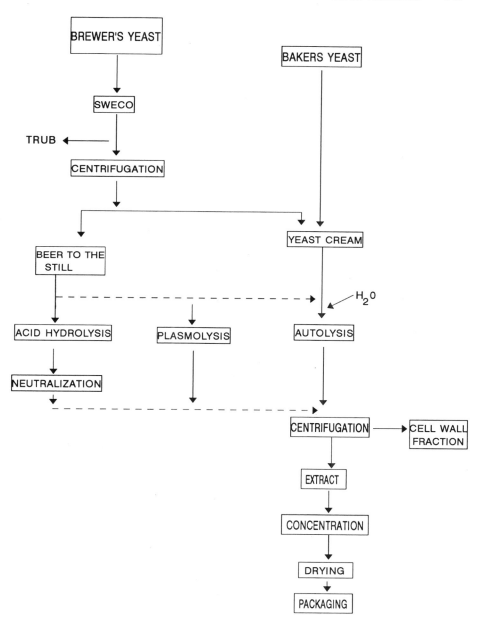

FIG. 5.3 Flowchart for the commercial production of Brewer's and Baker's yeast extracts.

teristics of the finished product are largely dictated by the processing conditions used during autolysis.

Often the conditions that are best for autolysis favor the growth of certain types of microorganisms, causing flavor problems in the final product. It is therefore mandatory to maintain good manufacturing practice throughout the process to suppress these microorganisms. A typical autolysis is generally brought to completion by elevating the temperature of the medium high enough to inactivate the degrading enzymes; this treatment also serves as a means of lowering the microbiological count prior to any further processing.

The solubilized yeast extract is then separated from the insoluble cell wall debris by centrifugation. The liquid fraction is filtered and adjusted to the required pH either with sodium hydroxide or hydrogen chloride as determined by the pH change needed in the medium. Use of these food grade reagents in the process results in the formation of salt (sodium chloride), which is perceived as undesirable by today's consumer. However, the NaCl formation can be minimized by the use of potassium hydroxide in place of sodium hydroxide for the adjustment of pH, without seriously affecting the salt perception. This is because components in yeast extract have the ability to reduce the bitter taste normally associated with potassium chloride.

Following the pH adjustment, the liquid is concentrated under vacuum in evaporators. The product that is made is a rich, savory, brown paste that may subsequently be spray dried to a light brown powder. Pastes containing 70 to 75 percent solids are commercially available and are popular in Europe. However, these products must be handled with care because of high susceptibility to microbial contamination. At every stage of the process, careful checks must be made to ensure high quality and acceptable microbiological content of the final extracts.

5.4.2.2 Plasmolysis.
The plasmolytic approach to yeast extract manufacturing enjoys wide acceptance in Europe, although it is less popular in the United States. The technique makes use of the phenomenon of plasmolysis to bring about the death of the yeast cells without the inactivation of the degradative enzymes. This method is known for its rapid initiation of the cell degradation process.

Concentrated solutions of organic and inorganic salts, sugars and other soluble ionic substances exert a plasmolyzing effect on living cells. This is manifested by the contraction or shrinking of the cytoplasm following the loss of water by exosmosis. That is, when the osmotic pressure of the medium surrounding the yeast cell exceeds that of the cell cytoplasm, a rapid loss of water from the cytoplasm leads to the immediate contraction of the cell. Eventually, the protoplasm pulls away from the inner surface of the cell wall from the release of the hydrostatic pressure of the cell contents against the wall. The condition in which the cell wall has contracted to its limit with the protoplasm not yet receded from the wall is termed *incipient plasmolysis*. Under these conditions, the protoplasm tends to occupy only a limited space within the cell cavity. Although the cell is recoverable if treated appropriately at the early stages, prolonged plasmolysis results in cell death and subsequent initiation of degradative processes within the cells. These degradative processes can be enhanced by elevated temperature (40 to 60°C) and other conditions unfavorable to yeast. The sign of death to the cell is the disruption of the outer cytoplasmic membrane. The protoplasmic constituents begin to break down from enzymatic activity (Fig. 5.4), and the degraded cell constituents finally appear in the soluble fraction due to the rapid leakage from the cells. A typical process at commercial scale is initiated with the addition of salt (sodium chloride) to yeast in the form of pressed cake at a moisture content of approximately 70 percent.

Plasmolysis has several advantages over autolysis. For instance, plasmolysis is a simple process requiring no special equipment. Common salt used as the plasmolyzing agent is readily available and is known for its flavor-enhancing properties as well as its bacteriocidal properties. Unlike a typical autolysis, chances for microbial contamination in this process are remote; indeed, final

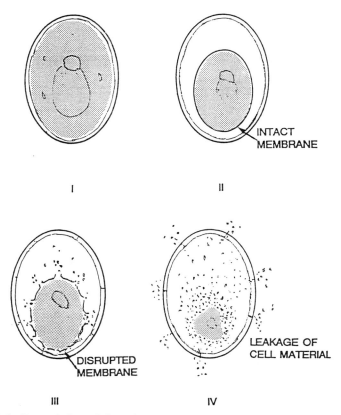

FIG. 5.4 Degradation of the cell matrix following plasmolysis of the yeast cell.

extracts produced by plasmolysis are generally known for their low microbial counts. However, despite these beneficial features, these extracts have limited use because of their high salt concentrations. Consumers worldwide are now demanding low sodium, low salt food products, and the high sodium yeast extracts made by the plasmolytic process do not offer this advantage to food processors in formulating food products for the health-conscious consumer. Also, plasmolysis may not be as effective as the standard autolysis (described above) because of the salting-out effect on the degrading enzymes.

5.4.2.3 Hydrolysis. During the past 60 years, chemists and food scientists have made significant progress in the development of unique flavors through the hydrolysis of vegetable proteins. Their manufacture has been outlined already in this chapter. Adoption of this technology for the manufacture of yeast hydrolysates has provided an alternate approach to the production of yeast extracts for the food industry.

The acid most commonly used is hydrochloric acid; it is a strong acid, and its sodium salt is acceptable in food products. The proportion of acid incorporated in the hydrolytic mixture is generally based on the amount of salt (sodium chloride) that is desired in the final extract. In a typical yeast hydrolytic procedure, the dried yeast is slurried to a solids concentration of 65 to 85 percent and heated to 100°C for 19 to 20 hours in the presence of the acid at very low pH; then it is cooled and neutralized with sodium hydroxide or sodium carbonate to a pH of 6.5. The slurry is centrifuged and the partially clear liquid is filtered and concentrated to approximately 80 percent solids to be sold as a paste, or concentrated to 35 to 40 percent solids and then spray dried to a moisture content of 5 to 6 percent to produce a free-flowing powder. Sometimes the liquid stream is treated with activated charcoal to eliminate color and odor compounds and with the amino acid phenylalanine, which is intensely bitter.

Unlike traditional yeast extracts, yeast hydrolysates generally have a stronger meaty, savory flavor than yeast extracts made by autolysis because of the high level of glutamic acid in these products and also the reaction flavor compounds generated during high-temperature processing. Hydrolysates are known also for their high process yields and higher productivity. However, there are several drawbacks that make acid hydrolysis unattractive. For example, due to the highly acidic nature of hydrolyzing agents, the reactors and other processing equipment must be made of corrosion resistant materials such as exotic alloys, or equipped with a glass or other inert lining, to prevent pitting and complete deterioration. Furthermore, when temperatures exceeding 100°C are required, it is necessary to construct special tanks that can withstand high pressure, and these tanks must meet the required safety standards. A plant that conforms to these strict standards can be very costly. Another consideration is the hazardous nature of hydrogen chloride, which can cause heavy casualties in the event of an accident.

There are other problems with hydrolyzed yeast proteins. For example, these extracts lack the familiar yeast extract flavor and are more akin to HVPs in flavor properties. Considering the similarities that exist with HVP processing, there is a high probability of finding unacceptable levels of chlorinated compounds (e.g., MCPs and DCPs), which are hazardous to human health if consumed excessively as mentioned earlier. Furthermore, the application of these extracts is restricted because of their high salt content. As an alternative, low-salt, low-sodium extract preparations can be produced by use of sulfuric acid and subsequent neutralization with calcium hydroxide. The insoluble calcium sulfate can easily be separated from the mother liquor by a simple decantation. However, this process is more costly.

5.4.3 Autolysates

Autolyzed yeast extract has previously been described as a concentrate of the soluble material produced by yeast autolysis. The soluble concentrate is obtained by the separation of the insoluble cell wall fraction from the solubles by centrifugation. However, some food processors prefer to use the total content of the cells, that is, without the separation of the insoluble cell walls. This product is referred to as *autolysate*. The flavor of an autolysate is inherently weaker than that of a yeast extract on an equal solids basis due to the presence of the inert cell wall material.

Although hop resins adhering to the surface of Brewer's yeast cells have minimal effect on extract manufacture, their presence can have an adverse influence on autolysate production. This is because of the presence of the cell wall (coated with hop resins) in the autolysates but not in extracts. For this reason, it is highly desirable to treat the spent Brewer's yeast to release bitter-tasting hop resins prior to initiation of autolysis. Several washes with water can minimize the bitter taste; even though autolysates treated this way are still partially bitter, they are useful in certain food applications. Further reduction in bitterness can only be achieved by an initial alkaline wash followed by several washes with water. This treatment is capable of loosening the hop components present on the yeast cell surfaces, thereby increasing their solubility in the aqueous medium. The yeast separated and washed after this treatment tends to be less bitter and more bland compared to the initial spent Brewer's yeast, or compared to Brewer's yeast washed only with water. However, the heavy yield losses associated with cleaning make the process less attractive to the manufacturer. Bitterness in autolysates can also be reduced by passing yeast through an ion-exchange column, but this increases the production cost of the product.

Yeast autolysates are not clear in solution like yeast extracts, but this is expected because the insoluble cell wall fraction is not removed from the autolysates. Autolysates are less expensive than extracts and are in demand for applications in which the clarity of the final food is of minor importance. Autolysates are used extensively in formulated meat products because of their high water-binding capacity and their ability to enhance meaty, savory flavor.

5.5 FLAVOR ENHANCERS

For general descriptive purposes, *taste* is defined as the sensation derived from all the chemical sensory receptor systems of the oral cavity. In our regular diet there is a wide variation in the tastes of natural foods, and almost without exception natural things that taste good are generally accepted as safe and make one feel satisfied. The types of sensations elicited are a function of the classes of compounds present in the food. Thus taste serves as a helpful guide in regulating the consumption of needed foods. However, there are certain compounds in the nutritional ecosystem of humans that have little or no flavor or aroma attributes themselves, yet are capable of enhancing the flavor of other foods. Those flavor enhancers that improve the savoriness, meatiness, mouthfeel, and continuity of certain food systems are discussed in this section.

Traditionally, Japanese cultures have used natural foods like sea tangle (*Laminaria japonica*), dried bonito, and black mushroom (*Lentinus edodus*) as condiments to improve the palatability of foods. However, the active ingredients responsible for flavor enhancement were not recognized until the early part of the 20th Century. The potentiating action of MSG [14], 5'-IMP [15], and 5'-GMP [16,17] were recognized at that time. These flavor potentiators are now commercially available either as crude extracts or as products of high purity. These products are particularly effective in enhancing the savory flavor of meat products, meat analogs, gravies, sauces, soups, and so on. The 5'-nucleotide, xanthosine 5'-monophosphate (5'-XMP), also has flavor-enhancing properties but to a lesser extent. Numerous other experimental 5'-nucleotides with many times the flavor-potentiating power of 5'-GMP and 5'-IMP have also been identified [18], but many of these derivatives also possess a bitter aftertaste, which severely limits their application in food systems.

Much of the current research in taste is based on the concept of four basic tastes: sweet, sour, salty, and bitter. However, the lines of evidence presented by more recent studies suggest that a few types of tastes cannot be completely analyzed and described in terms of these four basic tastes. For example, 5'-ribonucleotides (like 5'-GMP and 5'-IMP) and MSG have been reported to have a distinct savory and delicious taste in certain foods where the taste realm extends beyond the four basic tastes. This uniquely different taste is now referred to as *umami*, which is derived from the Japanese word meaning *savory* or *delicious*. Experimental data on receptor mechanisms of umami substances in animals have shown that receptor sites for umami are independent of those specific for the four basic tastes. These data suggest that the range of taste is broader and the taste of umami may therefore be considered another basic taste. However, the question of whether umami represents an independent and uniquely different entity of oral sensation is not resolved yet [19].

The flavor-enhancing effect of umami substances on the palatability of different types of foods has been extensively investigated [20]. Monosodium glutamate has been reported to have some sweet and salty properties with taste thresholds of 6.25×10^{-4} M (0.012 percent) in pure water [21]. Lilly [22] observed that MSG at 0.5 to 1.0 percent concentration exhibits a sweet-saline character together with a pronounced feeling of satisfaction that continues for nearly 30 minutes. The degree of taste response with MSG present is also known to increase more slowly with increasing MSG concentration. These characteristics are due to the solution properties of MSG and its interaction with the receptor sites. The MSG molecule, depending on its degree of hydration, is known to influence access to these receptor sites. However, the interaction of MSG with water shows a large variation with solute concentration. At higher concentrations of MSG, a marked change in solution properties occurs, resulting in a possible cyclization of the MSG molecule. This usually takes place as a result of the formation of a hydrogen bond between the amino group and the distal carboxyl group of the side chain of the glutamic acid molecule. The reduced increase in taste response in the presence of MSG as the MSG concentration is increased presumably can be attributed to the initial rapid cyclization followed by slow opening of the ring structure with the progress of time. The intensity and duration of the taste response is dictated largely by both the ability of the MSG molecule to migrate to the receptor region on the tongue and the duration of the binding to the receptor site.

5.5.1 Synergism

In the 1960s, Kuninaka [23–25] observed the remarkable synergism that exists between L-glutamate and the 5'-mononucleotides 5'-GMP and 5'-IMP. This type of synergism, not common among other taste stimuli, is one of the most remarkable properties of these flavor potentiators. Synergism is usually evaluated by examining whether or not the response to a mixture of two components is greater than the sum of the two individual responses. Although the receptor mechanism of the synergism between MSG and 5'-nucleotides is still unknown, several hypotheses have been proposed [26–29]. According to one hypothesis [28], the receptor for glutamate shows an allosteric transition due to the binding of 5'-nucleotides (5'-GMP or 5'-IMP) to its regulatory subunits. The conformational change that is elicited by this binding results in the exposure of additional glutamate binding sites of the receptor proteins, thereby enhancing the taste response. According to another possible mechanism for the synergism, 5'-IMP binds to an excitatory purinergic receptor that induces the formation of cyclic nucleotides within the receptor cell. These cyclic nucleotides in turn bind to an allosteric plasma protein receptor from the intracellular side, leading to an increase in available glutamate receptor sites [27]. Yet another mechanism was proposed by Torii and Cagan [29], who used biochemical binding studies in their work. They postulated that 5'-nucleotides act to increase the extent of L-glutamate binding by exposing the "hidden" or "buried" receptor sites for glutamate.

The enhancement of the binding of labeled L-glutamic acid ([³H]Glu) to receptor proteins of bovine circumvallate (CV) papillae by 5'-GMP was also examined by Torii et al. [30]. The CV tissue used was the side wall of papillae, rich in taste buds. The tongue epithelium, which is devoid of taste buds, was used as the control. The binding of the Glu was seen only on CV tissue, with its binding enhanced by the presence of 5'-GMP. The data strongly suggest that Glu might be the taste ligand responsible for umami character in foods and that 5'-GMP might play some role in increasing the number of Glu molecules bound at the taste bud receptor site. These findings are similar to results of sensory tests in humans and in nerve recordings from Chorda tympani of rats, strongly suggesting that mammals other than humans can also perceive umami taste quality.

Utilization of umami substances in numerous food systems is widely practiced in the food industry. Despite reports suggesting that large doses of MSG may have adverse neurophysiological effects on certain individuals who are sensitive to the chemical, MSG is currently the most widely used flavor potentiator in foods, and has been for many years [31]. Its application is most extensive in the countries bordering the Pacific rim. However, because of the negative publicity that MSG has received in the recent past, food manufacturers have now begun to move away from the use of MSG and seek other alternatives; some of these are possible by virtue of the synergism discussed above. For example, with the addition of other umami substances like 5'-GMP and 5'-IMP, it has been possible to significantly reduce the level of MSG in many foods. If appropriate levels of these two 5'-nucleotides are used, sodium intake can also be reduced by approximately 30 percent without decreasing palatability of foods or decreasing the degree of satisfaction of the meals.

5.5.2   Structure

The existing body of information suggests that commonly available naturally occurring flavor potentiators are either L-amino acids containing five carbon atoms or purine ribonucleoside 5'-monophosphates having six oxy groups. Monosodium glutamate can be considered typical of the former structure, whereas GMP, IMP, and XMP belong to the latter. Monosodium salts of the amino acids and disodium salts of the 5'-nucleotides offer the maximum effectiveness in flavor potentiation in food systems. Although L-glutamic acid has essentially the same basic structure as MSG, the former compound has reduced flavor potentiation properties. Furthermore, of the D and L isomers of MSG, only the naturally occurring L form offers flavor enhancing activity.

The portions of the 5'-nucleotide molecule that appears to be critical in the interaction with the regulatory site of the taste receptor are shown in Fig. 5.5. The keto function at position six of the aromatic ring of the purine moiety appears to be critical for the generation of flavor-enhancing properties. However, 5'-XMP was found to be less effective than 5'-GMP or 5'-IMP [31]. A hydrogen group at position 2 gives the structure of 5'-IMP. Furthermore, replacing the hydrogen atom at position two on the ring by other groups like hydroxyl (5'-XMP) or amino functions (5'-IMP) did not significantly change

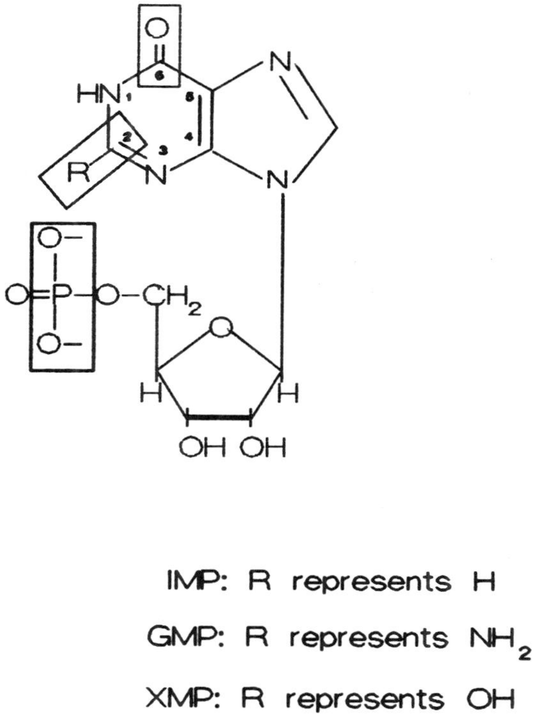

IMP: R represents H

GMP: R represents NH$_2$

XMP: R represents OH

FIG. 5.5 Structure of the 5'-nucleotide molecule, with the critical groups necessary for flavor potentiation boxed.

flavor-enhancing properties of the molecule. The flavor-potentiating activity of deoxyinosine 5'-monophosphate and deoxyguanosine 5'-monophosphate has also been demonstrated [32]. That is, the OH group at the 2' position of 5'-nucleotides could also be replaced by hydrogen without adversely affecting the flavor-enhancing property. In addition, Honjo et al. [33] demonstrated that the two OH groups on the phosphorus of the 5'-nucleotide molecule are also critical for efficient flavor potentiation. The order of effectiveness of 5'-nucleotides in humans is GMP > IMP > XMP and adenosine monophosphate is usually judged ineffective. The pyrimidine nucleotides are also generally known to be ineffective.

5.5.3 Industrial Production of Monosodium Glutamate

Monosodium glutamate is ubiquitous in the human diet. It was originally extracted from various natural products, but the quantities currently required by the food industry cannot be obtained economically in this manner. The estimated global demand for MSG is approximately 500,000 metric tons annually. The common method of commercial production is by fermentation using special strains of *Corynebacterium glutamicum* (formerly *Micrococcus glutamicus*). These strains contain a defective tricarboxylic acid (TCA) cycle with low α-ketoglutarate dehydrogenase activity (Fig. 5.6). This enzyme is responsible for the conversion of α-ketoglutarate to succinate; suppression of this biochemical reaction causes the accumulation of α-ketoglutarate, which is subsequently converted to glutamate by reductive amination. Most microorganisms have a high level of glutamate dehydrogenase activity, permitting the conversion of glutamate to α-ketoglutarate or vice versa. The specially selected strains of *C. glutamicum* have the equilibrium directed toward the formation of glutamic acid.

Biotin is a critical component of the glutamate fermentation medium. The concentration of biotin in the medium (typically less than 5 μg/L) has to be low enough to make the membranes of the microorganisms leaky. This allows the glutamic acid formed during the fermentation to leak out into the surround-

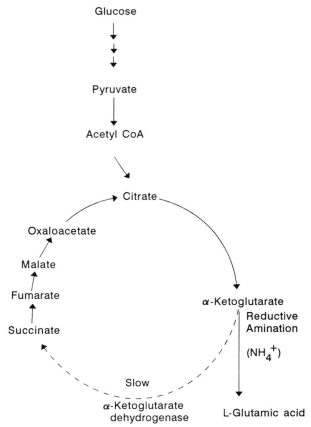

FIG. 5.6 Glutamic acid accumulation by *Corynebacterium glutamicum*, due to a pathway with a defective TCA cycle.

ing medium. Cane molasses is often used in glutamic acid fermentations, although its biotin content is high enough to inhibit glutamic acid production. For this reason the molasses has to be treated to reduce its biotin content.

Typically, urea or ammonia is used as a nitrogen source, and minerals like Fe^{+2} and K^+ play an important role in modulating the yield of glutamic acid. The pH at the optimum fermentation temperature of 30 to 37°C is regulated at 7 to 8, and aeration rate is controlled at 0.5 to 1.0 vvm (volume of air/volume of medium/minute). The fermentations are generally 35 to 45 hours long, with a final glutamic acid concentration of approximately 80 g/L.

The fermentation and downstream processing schemes are shown in Figure 5.7. After fermentation, the cells are removed by centrifugation or ultrafiltration, then addition of hydrogen chloride causes glutamic acid to precipitate. At this stage, the α-form crystallizes preferentially over the rod-shaped β form. However, the latter is easier to filter and wash, so the α form crystals are converted to the β form by heating as a suspension in water [34]; the β-form crystals are then filtered and washed. Glutamic acid is then redissolved by neutralization with sodium hydroxide, decolorized, filtered, concentrated, recrystallized, centrifuged, and dried.

5.5.4 Industrial Production of 5′-Nucleotides

Like MSG, flavor enhancing purine 5′-ribonucleotides were originally produced from natural sources. For example, fresh muscles of marine fish served as a satisfactory source for the production 5′-IMP. In mid-1950, production techniques involving fermentation and exploitation of microorganisms came into widespread use for 5′-inosinic and 5′-guanylic acids. Today, several processes for the production of these flavor enhancing 5′-ribonucleotides are being used. Only the most important methods are listed here:

1. Direct fermentation of sugars into 5′-GMP and 5′-IMP.
2. Direct fermentation of sugars into nucleosides, with subsequent phosphorylation into corresponding 5′-ribonucleotides.
3. Degradation of microbial RNA to 5′-nucleotides using 5′-phosphodiesterase. Subsequent conversion of 5′-AMP to 5′-IMP using adenylic deaminase enzyme is also an option.
4. Any combination of the above three procedures.

Almost all 5′-IMP and 5′-GMP of high purity available today are commercially produced by using the direct fermentation technique [31].

5.5.4.1 Fermentation Method. Microorganisms synthesize the metabolites necessary for their growth and sustenance but avoid their overproduction by a phenomenon termed metabolic regulation or feedback control. For efficient overproduction of the desired 5′-nucleotides, it is essential to remove these feedback controls without destroying the other metabolic machinery of the cells.

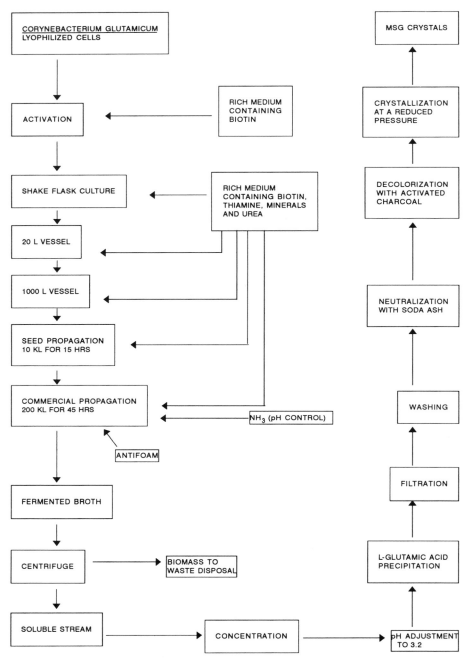

FIG. 5.7 Flowchart for the commercial production of monosodium glutamate by fermentation.

Organisms modified in this manner are called auxotropic mutants; in these mutants, specific control is disrupted by the removal of a particular protein or enzyme, or by a reduction in the concentration of the compound that causes the feedback control. Additionally, in some instances membrane permeability is enhanced by control of certain nutrients in the surrounding medium. Auxotrophic mutants of *Brevibacterium ammoniagenes* are the organisms of choice because of their characteristic overproduction of 5'-GMP and 5'-IMP. Limitation of Mn^{+2} has resulted in an increased production of 5'-nucleotides because of low feedback inhibition due to leakage of final product out of the cell [35].

5.5.4.2 Hydrolysis.

Crude yeast extracts rich in 5'-GMP and 5'-AMP are produced commercially by enzymatic degradation of crude RNA. The enzyme used for this conversion is 5'-phosphodiesterase. Yeast is used as the source of RNA because of its high RNA content (2.5 to 15 percent) [36]. Furthermore, certain strains of yeast are generally recognized as safe (GRAS), and can be grown economically to produce large quantities of biomass rich in ribonucleic acids.

The RNA content is strain dependent. Among yeasts, *Candida utilis* has been found to contain the highest RNA content (10 to 15 percent) on a dry solids basis. Baker's yeast strains (*S. cerevisiae*) are generally lower in RNA content (8 to 11 percent) but can also be used economically for the production of 5'-nucleotide-rich extracts. To maximize the RNA content, it is necessary to harvest the yeast during log phase, when the rate of protein synthesis is at its peak [37].

The hydrolysis process for the commercial production of 5'-nucleotide rich-extracts (shown in Fig. 5.8) consists of two important steps [38]. The critical first step consists of a treatment that releases RNA from the yeast cell into the surrounding medium. The second step is an enzymatic treatment that converts RNA to 5'-nucleotides. Most of the RNA in the cell is extracted by heating the yeast suspension to 90 to 100°C for 1 to 3 hours. By controlling the pH of the medium in the 6.0 to 6.6 range, it is possible to maintain the undesirable nucleases in the unreactive form prior to their complete destruction at high temperature. (These nucleases convert RNA to 2'- and 3'-nucleotides, which unlike the 5'-nucleotides have no effect on flavor in food systems.) Subsequent heat treatment causes the destruction of all enzymes and the release of soluble low-molecular-weight components from the cell. The second phase of the process starts with cooling of the RNA-rich medium to 50 to 60°C at pH 6.5; this is followed by the initiation of the RNA hydrolysis using the enzyme 5'-phosphodiesterase. This and other 5'-forming enzymes which are found in yeast at low concentrations are generally masked by the other more active nucleases that degrade nucleic acids to nucleotides other than 5'-nucleotides. For this reason, the 5'-phosphodiesterase generally must come from an external food-grade source (some exceptions are noted below). Some commercially available enzymes are derived from certain fungi such as *Penicillium citrinum* or certain *Actinomyces* such as *Streptomyces aureus*. However, these organisms are not considered GRAS in the United States, so that enzymes derived from them cannot be used. As an alternative, the presence of 5'-phosphodiesterase activity in cereal germs has been demonstrated by Schuster [39]. Also, malt rootlets (which are considered GRAS) are by-products in the malting industry and are now serving as a reliable source of 5'-phosphodiesterase activity.

Researchers have developed genetically modified edible yeast strains capable of producing 5'-nucleotide–rich yeast extracts during autolysis without the addition of 5'-phosphodiesterase [40]. In theses strains the endogenous nucleases mediate the degradation of RNA to preferentially form 5'-nucleotides instead of 3'-nucleotides at pH 8.5 to 9.5 and 40 to 45°C. These mutants are selected on the basis of their sensitivity to 5-fluorouracil (5-FU); however, they are generally found to be slow growers. Because of this serious deficiency, the approach has so far met with limited success.

Kitano et al. [41] also demonstrated the degradation of yeast RNA to 5'-nucleotides (and subsequent secretion into the medium) without the addition of 5'-phosphodiesterase. These investigators added certain buffers and surfactants to the yeast, leading to enhanced activity of the endogenous 5'-nucleases. Unfortunately, most of the surfactants that are shown to be effective are currently not suitable for food use.

The yeast extracts discussed in this section contain four types of 5'-nucleotides: 5'-GMP, 5'-uridine monophosphate (5'-UMP), 5'-cytidine monophosphate (5'-CMP), and 5'-adenine monophosphate (5'-AMP), corresponding to the four bases present in RNA (Fig. 5.9). Of these, only 5'-GMP provides the flavor-potentiating property to the extract. Although 5'-AMP offers no flavor enhancing property, it can serve as a precursor of the flavor enhancing compound 5'-IMP (Fig. 5.10). Crude extracts rich in 5'-IMP along with 5'-GMP are currently produced commercially by use of the two enzymes, 5'-phosphodiesterase and adenylic deaminase, acting in sequence [1]. These yeast-derived products have both flavor-enhancing properties and background flavor. The two remaining 5'-nucleotides (5'-CMP and 5'-UMP) do not possess flavor enhancing properties but have found use in the pharmaceutical industry for the production of antiviral drugs.

5.5.5 Amino Acids and Peptides

Many of the amino acids that exist in nature have a major influence on food flavor. For example, the taste of traditional Japanese foods like miso and soy sauce was found to be due to the release of amino acids from naturally occurring proteins during the fermentation process. The proper type and level of free amino acids that can significantly improve the taste of food products in naturally occurring or intentionally added flavor potentiators are also present. The sensations elicited by a large number of the naturally occurring L-forms of amino

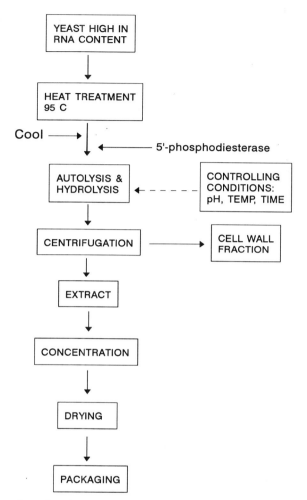

FIG. 5.8 Flowchart for the commercial production of 5'-nucleotide-rich yeast extracts.

acids are predominantly bitter. Most of the remaining amino acids are either sweet or salty (Table 5.2).

The influence of MSG and glutamic acid in enhancing the savory or umami characteristic of foods has been described earlier in this chapter. The contributions made by glycine, alanine, and arginine in exhibiting and enhancing the distinctly delicious umami flavor in crabs were reported by Konosu et al. [42] and Hayashi et al. [43]. Histidine and arginine have also shown enhancement of umami flavor in the presence of MSG, but only at a pH of 5 to 6.5 [21].

Probably, the most significant role that amino acids can play in food systems

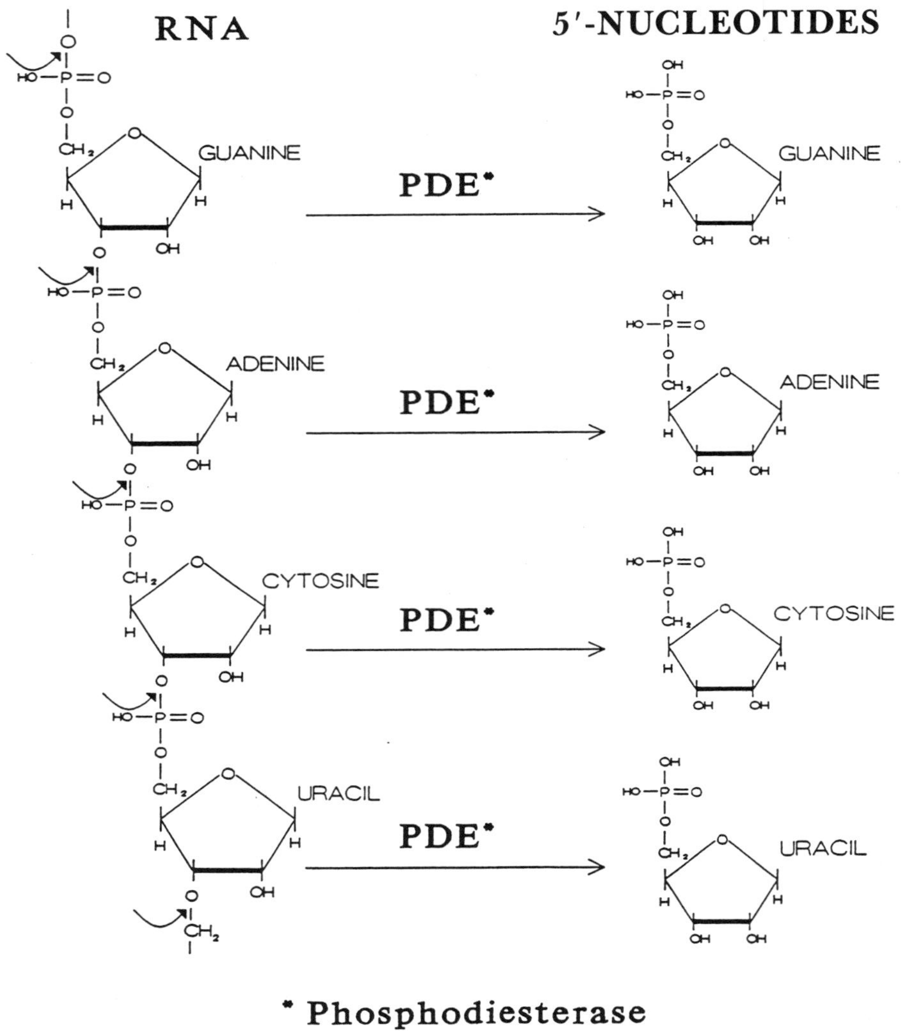

RNA   **5′-NUCLEOTIDES**

* **Phosphodiesterase**

FIG. 5.9 Degradation pattern of ribonucleic acid in the presence of 5′-phosphodiesterase.

is that of precursors for the production of flavor compounds during processing. Savory flavor and aroma development by the reaction of amino acids like cysteine with reducing sugars like glucose at different temperatures has been the subject of intensive study by a number of researchers [44–46]. The resulting flavor compounds are the products of the Maillard browning reaction as mentioned earlier. Studies have also been conducted to determine the components that improve the taste of meat during conditioning. In pork and chicken, differences in the levels of amino acids before and after conditioning (curing at

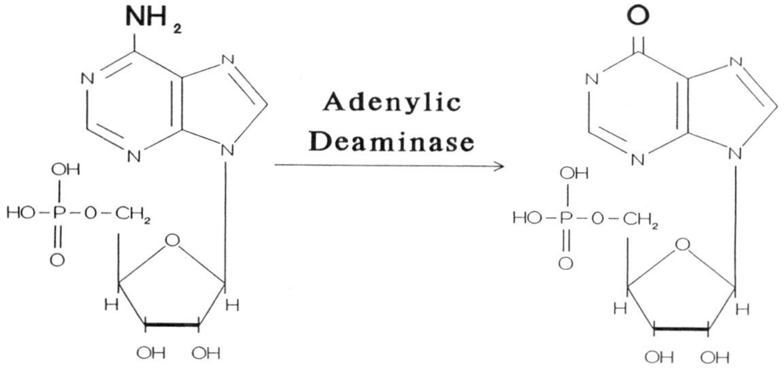

FIG. 5.10 Conversion of 5'-AMP to 5'-IMP by the action of adenylic deaminase.

ambient temperature) were very large, and these results corresponded to the brothy taste intensity after conditioning of meat as compared to before conditioning. These data suggest that the increase in free amino acids contributed to the improvement of meat flavor following the conditioning treatment [47].

In foods, peptides are also of special importance because of their taste characteristics. They can be sweet [48], bitter [49], salty [50], or act as flavor enhancers [51]. An octapeptide that may induce a delicious, savory taste has been isolated from papain-tenderized beef by Yamasaki and Maekawa [52]. Their studies revealed that the active region necessary for the savory taste has the following amino acid sequence:

Lys—Gly—Asp—Glu—Glu—Ser—Leu—Ala

Sensory studies with this octapeptide in beef gravy have shown that the predominant sensation is savory (60 percent) with lower levels of sweet (10 percent) and sour (30 percent) tastes [53]. Based on these findings, Yamasaki and Maekawa [53] synthesized an octapeptide similar in sequence to that iso-

TABLE 5.2 Flavor Characteristics of Amino Acids

Taste	Amino Acids
Bitter (intense)	Tryptophan, phenylalanine, arginine, isoleucine
Bitter (less intense)	Leucine, histidine,[a] valine,[a] ornithine, methionine, lysine[a]
Sweet (intense)	Hydroxyproline, proline, alanine[a]
Sweet (less intense)	Lysine,[a] valine,[a] alanine,[a] threonine, serine,[a] glycine, glutamic acid[a]
Sour (intense)	Aspartic acid, glutamic acid,[a] asparagine
Sour (less intense)	Histidine,[a] serine[a]

[a]Concentration dependent.

lated from the natural source. Both the synthetic peptide and the natural peptide isolated from extract of beef were identical in many respects.

Spanier and Miller [54] also reported on a peptide that enhances meat flavor and can make less flavorful cuts of meat taste like more expensive cuts. The peptide, designated *BMP* for beefy meaty peptide, may also have potential to serve as a substitute for MSG. According to the authors, work remaining to be done includes testing the peptide for heat stability, determining its shelf-life, and performing further sensory testing. Regrettably, the commercial production of such octapeptides may not be economically feasible because of the complexity that is associated with peptide synthesis.

5.5.6 Flavor Enhancement in Foods

Application of flavor potentiators is currently practiced widely in a number of food systems, yet the general observation that such potentiators do not provide the same degree of enhancement of the flavor in all food types has puzzled many flavor chemists and remains unresolved despite the extensive research carried out so far on the subject. For example, Maga [55] undertook an extensive organoleptic study of the dilute protein solutions derived from isolated and purified protein fractions from 17 food products selected from sources like meat, fish, vegetables, cereals, dairy, and eggs, and concluded that the three flavor potentiators (MSG, 5'-GMP, and 5'-IMP) influenced the flavor intensities in the different protein species to varying degrees. In all probability, the variability is at least partially due to the protein structure. According to this study, 5'-GMP and 5'-IMP exhibit their maximum influence on fish and vegetable proteins, with less impact on meats, cereals, milk, and eggs. Monosodium glutamate on the other hand has its most profound influence on meat, followed by fish, vegetables, cereals, eggs, and milk. Moreover, umami compounds exhibit the most pronounced effect on chicken, with a relatively weaker enhancement of beef, turkey, and pork, and the least impact on lamb. These results support the notion that two critical factors determine the resultant flavor intensity of different food types: (a) the nature of the proteins that comprise the food and (b) the specific type of the flavor potentiator used for flavor enhancement.

Results to date have shown the influence umami compounds exert in enhancing the savory flavor of foods. This effect in foods may be greatly influenced by the interactions of the meaty and the umami flavor components on the receptor sites of the oral cavity. However, food researchers need a clearer definition of meaty flavor in chemical terms to acquire a better understanding of savory flavor. More generally, much work must still be done in this area of flavor chemistry to elucidate the absolute nature of savory flavor.

5.5.7 Market Demand

The 1990 worldwide savory ingredient market was valued at a little less than $1 billion per year. According to the best estimates, the product that dominated this market was MSG, enjoying approximately 58 percent of the total market share. Other flavor ingredients like yeast extract, HVP, and IMP/GMP represented 20, 15, and 7 percent of the market share, respectively. Japan is by far the major producer of the three important flavor potentiators: MSG, 5'-GMP, and 5'-IMP.

Current annual global demand for MSG is approximately 500,000 metric tons, of which nearly 150,000 metric tons are produced in Japan. Other MSG-producing countries with lesser production capacities are Taiwan (55,000 tons), South Korea (48,000 tons), Thailand (30,000 tons), Italy (32,000 tons), France (23,000 tons), and the rest of Europe (68,000 tons), with China, Philippines, Indonesia and a few other countries together totaling approximately 100,000 tons. Per capita consumption of MSG is highest in Taiwan at 3 g/day. This is almost double that of Japan and ten times higher than that of the United States. In 1990, the total combined production capacity for IMP and GMP in Japan was estimated to be approximately 5,000 tons per year.

The U.S. extract market has been growing by approximately 20 percent per year. According to the 1989 Journal of Commerce, the U.S. extract and autolysate market was approximately $32 million that year, with Red Star Specialty Products, a division of Universal Foods Corporation, Milwaukee, accounting for almost 35 percent of all sales. The U.S. market is estimated to account for nearly 20 percent of the world market.

Today more than ever before there are market opportunity for yeast extracts and autolysates because MSG and HVPs are falling out of favor. Monosodium glutamate is perceived as causing "Chinese restaurant syndrome," and HVPs are experiencing problems following the detection of known carcinogens (i.e., dichlropropanol and monochloropropanol) in these products. Furthermore, both MSG and HVPs contain high levels of sodium, contrary to the public demand for low-sodium food products. For these reasons, many food companies have already severely restricted the use of these ingredients in new product development, and are aggressively searching for effective substitutes. These conditions have clearly created an excellent growth opportunity for yeast-derived flavor potentiators.

5.6 CONCLUSION

Over the centuries, red meat has occupied a prominent position in the diets of all affluent societies. This trend has now begun to change, especially during the last decade, for two critical reasons. In the developed nations, consumers have become exceedingly conscious of their health, and this has caused a dramatic reduction in the consumption of meat in their daily diet. But more importantly, meat has become unaffordable in developing nations. Even the Western World, which now produces and consumes vast quantities of animal meat, may in the future experience shortages that could result in dramatic changes in food habits. In response to these economic pressures and consumer demands, meat analogs will become increasingly popular, and as this occurs, production of meatlike savory flavors and meat-textured vegetable protein foods will be of ever increasing importance to a substantial segment of the world's population.

Many world organizations are actively involved in projects that related to problems in the Third World, with the major goal of overcoming world hunger. The world population is expected to double in the 21st century [56] and it is difficult to foresee an effective solution to this problem, especially in the developing countries. Even vegetable proteins that are currently in surplus may then be in short supply. It is the responsibility of food scientists and flavor chemists to develop savory richness in diets that are rich in vegetable proteins to make them palatable.

Several food companies, both in the United States and abroad, already have made significant advances to meet these future challenges. Some developments in the area of savory flavors are outlined in this chapter. Over the years a great deal has been learned about the chemistry of meat flavors, but many problems are still unsolved. For example, although a large number of flavor compounds have been isolated from heat-treated beef, it is yet uncertain how many of these compounds are vital for the typical meat flavor. The problem is more complex when the objective is to elucidate the chemistry of savory flavors. Only recently have food and flavor scientists begun to understand the probable mechanisms by which savory flavors are formed during processing. Currently, there is considerable interest in this area, and, with further improvement in analytical techniques coupled with greater understanding of reaction flavor technology, one can gain a better insight into savory flavor development and flavor chemistry. This should ultimately lead to the development of novel savory food products, which can become increasingly important throughout the world.

ACKNOWLEDGMENTS

My thanks are owed to Ed Schoenberg and Ben Cheng of the Red Star Specialty Products Division, Universal Foods Corporation, for their critical review of the manuscript. The invaluable help of Virginia Teat in the preparation of the manuscript is also gratefully acknowledged.

REFERENCES

1. Reed, G., and Nagodawithana, T.W., *Yeast Technology*. Van Nostrand-Reinhold, New York, 1991.
2. Pendergast, K., *Food Trade Rev.* **1,** 16 (1974).
3. Hurrell, R.F., in *Food Flavors* (I.D. Morton and A. J. Macleod, eds.), Elsevier, New York, 1982.
4. Heath, H.B., and Reineccius, G. *Flavor Chemistry and Technology*. Van Nostrand Reinhold, New York, 1986.

5. Shibamoto, T., in *Instrumental Analysis of Foods: Recent Progress* (G. Charalambous and G.E. Inglett, eds.). Academic Press, New York, 1983.

6. May, C.G., and Akroyd, P., German Patent 1,058,824 (1959).

7. May, C.G., and Akroyd, P., U.S. Patent 2,918,376 (1959).

8. May, C.G., and Morton, I.D., U.S. Patent 2,934,436 (1960).

9. Morton, I.D., Akroyd, P., and May, C.G., British Patent 836,694 (1960).

10. Bidmead, D.S., Giacino, C., Grossmann, J.D., and Kratz, P.D., U.S. Patent 3,394,016 (1968).

11. Giacino, C., U.S. Patent 3,394,017 (1968).

12. Nagodawithana, T.W., in *Advances in Cereal Science and Technology*, Y. Pomeranz, ed., Vol. 8. Am. Assoc. Cereal Chem., St. Paul, MN, 1986.

13. Nagodawithana, T.W., in *Foods and Feeds*, (D. K. Arora, K. G. Mukerji, and E. H. Martin, eds.), Vol. 3. Dekker, New York, 1991.

14. Ikeda, K., *J. Tokyo Chem. Soc.* **30,** 41 (1909).

15. Kodama, S., *J. Tokyo Chem. Soc.* **34,** 751 (1913).

16. Kuninaka, A., *J. Agric. Chem. Soc. Jpn.* **34,** 487 (1960).

17. Nakajima, N., Ichikawa, K., Kamada, M., and Fujita, E., *J. Agric. Chem. Soc. Jpn.* **35,** 797 (1961).

18. Yamaguchi, S., Yoshikawa, T., Ikeda, S., and Ninomiya, T., *J. Food Sci.* **36,** 814 (1971)

19. Yamaguchi, S., in *Umami: A Basic Taste* (Y. Kawamura and M. R. Kare, eds.). Dekker, New York, 1987.

20. Hanson, H.L., Brushway, M.J., and Lineweaver, H., *Food Technol.* **14,** 32 (1960).

21. Yamaguchi, S., and Kimizuka, A., in *Glutamic Acid: Advances in Biochemistry and Physiology* L.F. Filer, S. Garattini, M.R. Kare, A.R. Reynolds, and R.J. Wurtman, eds.). Raven Press, New York, 1979.

22. Lilly, A.E.V., *Chem. Prod.,* **17,** 333 (1954).

23. Kuninaka, A., *J. Agric. Chem. Soc. Jpn.* **34,** 487 (1960).

24. Kuninaka, A., in *Flavor Chemistry* (R.F. Gould, ed.). Am. Chem. Soc., Washington, DC, 1966.

25. Kuninaka, A., in *The Chemistry and Physiology of Flavors* (W.H. Schultz, E.A. Day, and L.M. Libbey, eds.), p. 515. AVI Publ. Co., Westport, CT, 1967.

26. Cagan, R.H., Torii, K., and Kare, M.R., in *Glutamic Acid: Advances in Biochemistry and Physiology* (L.F. Filer, S. Garattini, M.R. Kare, A.R. Reynolds, and W.J. Wurtman, eds.). Raven Press, New York, 1979.

27. Schiffman, S.S., and Gill, M., in *Umami: A Basic Taste* (Y. Kawamura, and M.R. Kare, eds.), Dekker, New York, 1987.

28. Cagan, R.H., *J. Neurobiol.* **10,** 207 (1979).

29. Torii, K., and Cagan, R.H., *Biochim. Biophys Acta* **627,** 313 (1980).

30. Torii, K., Mimura, T., and Yugari, Y., in *Umami: A Basic Taste* (Y. Kawamura, and M.R. Kare, eds.), p. 513. Dekker, New York, 1987.

31. Maga, J.A., *Crit. Rev. Food. Sci. Nutr.* **18,** 231 (1983).

32. Nakao, Y., and Ogata, K., *Annu. Rep. Takeda Res. Lab.* **22,** 47 (1963).

33. Honjo, M., Imal, K., Furukawa, Y., Moriyama, H., Yasumatsu, K., and Imada, A., *Annu. Rep. Takeda Res. Lab.* **22,** 42 (1963).

34. Ito, K., Mizoguchi, N., Dazai, M., Fujiwara, K., and Sakata, Y., U.S. Patent 3,565,950 (1971).

35. Kuninaka, A., in *Biotechnology* (H.-J. Rehm and G. Reed, eds.), Vol. 4, p. 71. Verlag Chem., Weinheim, Germany, 1986.

36. Nakao, Y., in *Microbial Technology, Microbial Processes*, (H.J. Peppler, and D. Perlman, eds.), Vol. 1. Academic Press, New York, 1979.

37. Katchman, B.J., and Fetty, W.O., *J. Bacteriol.* **69,** 607 (1955).

38. Tanekawa, T., Takashima, H., and Hachiya, T., U.S. Patent 4,303,680 (1981).

39. Schuster, L., *J. Biol. Chem.* **229,** 289 (1957).

40. Kanegae, Y., Sugiyama, Y., and Minami, K., European Patent Application 0,249,435 (1987).

41. Kitano, K., Akiyama, S., and Fududa, H., *J. Ferment. Technol.* **48**(1), 14 (1970).

42. Konosu, S., Yamaguchi, K., and Hayashi, T., in *Umami: A Basic Taste* (Y. Kawamura, and M.R. Kare, eds.). Dekker, New York, 1987.

43. Hayashi, T., Yamaguchi, K., and Konosu, S., *J. Food Sci.* **46,** 479 (1981).

44. Kiely, P.J., Nowlin, A.C., and Moriarty, J.H., *Cereal Sci. Today* **5,** 273 (1960).

45. Ledl, F., and Severin, T., *Chem., Mikrobiol., Technol. Lebensm.* **2,** 155 (1973).

46. Ledl, F., and Severin, T., *Z. Lebensm.-Unters.-Forsch.* **154,** 29 (1974).

47. Kato, H., and Nishimura, T., in *Umami: A Basic Taste.* (Y. Kawamura, and M.R. Kare, eds.). Dekker, New York 1987.

48. Brussel, L.B.P., Peer, H.G., and van der Heijden, A., *Z. Lebensm.-Unters.-Forsch.* **159,** 337 (1975).

49. Guigoz, Y., and Solms, J., *Chem. Senses Flavour* **2,** 71 (1976).

50. Tada, M., Shinoda, I., and Okai, H., *J. Agric. Food Chem.* **32,** 992 (1984).

51. Noguchi, M., Arai, S., Yamashita, M., Kata, H., and Fujimaki, M., *J. Agric. Food Chem.* **23,** 49 (1975).

52. Yamasaki, Y., and Maekawa, K., *Agric. Biol. Chem.* **42,** 1761 (1978).

53. Yamasaki, Y., and Maekawa, K., *Agric. Biol. Chem.* **44,** 93 (1980).

54. Spanier, A.M., and Miller, J.A., presentation to the American Chemical Society, San Francisco, April 6, 1992.

55. Maga, J.A., in *Umami: A Basic Taste* (Y. Kawamura, and M.R. Kare, eds.), p. 255. Dekker, New York, 1987.

56. Steinkraus, K.H., ed., *Handbook of Indigenous Fermented Foods.* Dekker, New York, 1983.

Dairy Flavors

DAVID C. EATON

Universal Flavors, Indianapolis, Indiana

Most flavors are mixtures of many component chemicals, but dairy and cheese flavors are especially complex. They are consequently difficult to decipher and successfully mimic. The current market trends are for increasing sales of naturally flavored products and of products with less fat, sodium, and cholesterol. This chapter discusses these issues, as well as the components of various dairy flavors and methods to produce them. Products of biotechnology are generally considered natural, and there are many important components of dairy flavors, such as diacetyl, fatty acids, esters, and 2-heptanone, where an enzymatic or fermentation route is practical.

6.1 HISTORY OF FERMENTED DAIRY PRODUCTS

Dairy flavors are almost all the result of one sort of fermentation or another; one notable exception being the flavor of fresh milk. The history of the production and use of these fermented dairy products goes back thousands of years, and various fermented products have been developed in most of the cultures that use dairy animals. Initially, the use of fermentation for dairy products was probably an accidental result of storing milk too long. Since milk is a very good source of nutrients for bacteria as well as man, it spoils quickly by souring due to the action of lactic acid bacteria. It was soon realized that this product was still edible, although sour in taste, and that a product of consistent quality could be made by storing milk in unwashed containers from a previous batch. The bacteria on the surface of the container form a crude starter culture, and once it was realized that the quality of the previous fermentation influenced the current fermentation, it was possible to improve the cultures by a simple selection process. By selecting for the best flavor, it was possible to quickly develop a yogurtlike product in form and perhaps in taste.

Bioprocess Production of Flavor, Fragrance, and Color Ingredients, Edited by Alan Gabelman, ISBN 0-471-03821-0 © 1994 John Wiley & Sons, Inc.

169

The fermentation processes for dairy products and flavors have been greatly refined over the years since the first early discoveries, but the general procedure of allowing a lactic fermentation followed by an aging period is still the same. Development of the microscope had a profound effect on the understanding of these fermentation processes, but for the most part the traditional procedures have changed little. However, the marketplace demands for more intense, lower cost, healthier foods have stimulated an increased level of research into the development of flavors, which is now beginning to have an impact on new products.

6.2 TRENDS

Dairy products and flavors are not immune from the recent trend toward more healthful foods. With recommendations to reduce daily cholesterol intake to less than 100 mg/1000 calories (maximum 300 mg/day) and to lower total dietary fat from the current average level of approximately 40 percent of the total caloric intake to less than 30 percent [1,2], there will be continuing pressure to lower the fat content of dairy products. Low-fat milk and yogurt products are already common, and the quality of lower fat cheese is improving. Fat is very important to taste, but it has an even greater contribution to mouthfeel. For this reason it is essential for many such products to contain fat property replacers such as food gums, microcrystalline cellulose, and microparticulated proteins (e.g., Simplesse®, a product of the NutraSweet Company). However, one problem with many of these products is that the consumer can still usually perceive differences in taste or mouthfeel. Development of improved fat flavors (without the fat) and of low-calorie fat replacers for mouthfeel and bulking is consequently an area of great interest to the food industry.

Another area of current health interest is the effect that salt (sodium) can have on increasing blood pressure. Unfortunately, cheese generally contains a substantial amount of salt—in the case of cheddar it is about 2 percent by weight. Cheese-containing snack foods, such as potato chips, are even more salty. Salt is traditionally used in the production of cheese, both to control and modify the microbial flora, and to enhance the cheesy taste expected by the consumer. For this reason the salt content of a cheese can only be reduced a small amount before the microbiological safety is jeopardized or the taste becomes so bland that it is unacceptable. Salts of minerals other than sodium can also taste salty, so one solution to reduce sodium ion content of food has been to use potassium chloride as a replacement. Unfortunately, although potassium chloride is salty in taste, it also causes bitterness. Although the addition of extra spices is sometimes helpful, no other acceptable salt substitutes have been identified. However, recently there was an interesting report that certain peptides can function as salt substitutes [3]. Although the effectiveness of peptides is still controversial [4–6], a salt substitute such as this could have tremendous impact if successfully used to replace even some or all of the salt in the diet.

Another trend of recent prominence is the consumer demand for natural food products. This trend has been very much in evidence recently from the number of product introductions or reformulations emphasizing all-natural ingredients (see Chapter 2 for a discussion of natural; also see Section 6.3). Like the trend toward lower salt and fat, the demand for natural ingredients results from the desire of the consumer for a more healthy diet. Many consumers consider food with artificial ingredients to be less wholesome than a totally natural product. Of course fermented dairy products are natural, but any added flavorings are often artificial and would have to be labeled accordingly. Despite this labeling problem, many dairy and snack foods contain artificial flavors for reasons of cost, stability, or lack of a strong enough natural flavoring. There is consequently a strong incentive to develop better natural flavors for applications in snack and other foods, especially for stronger and more characteristic cheese flavors.

Sweetness is one flavor area where several excellent artificial substitutes are available, and these substitutes are now being seen in dairy products. The perceived disadvantages of sugar are that sugar adds calories, is empty of nutrition, and contributes to tooth decay. Sugar is not added to the traditional fermented dairy products, since the carbohydrates normally present in milk are sufficient for the fermentation. However, sugar in fruit flavorings contributes to the pleasant taste in fermented yogurt, which has become very popular in the United States in the last few years. Lower calorie versions of these yogurt products using less sugar or artificial sweetening are now available, and this trend is likely to continue.

6.3 LABELING

Changes in labeling requiring ever greater disclosure are one of the greatest driving forces for ingredient changes such as those mentioned above. Although some types of information on a product label may be supplied voluntarily, other information is often required by specific regulation. The regulatory trend clearly has been for increasingly specific data about product nutrition and content. In the United States, this trend has recently been influenced the passage of the Nutrition Labeling and Education Act of 1990 [7]. This legislation directed the Secretary of Health and Human Services to issue regulations by November 1992 for, among other things, product label health claims based on the nutritional content of the food and for greater specificity in regarding the ingredient statement. This date was extended and the final labeling regulations were published January 1993, and are now scheduled to go into effect May 1994. There will undoubtedly be continuing pressure to provide additional information on labels, particularly levels of components with health implications, including saturated fat, cholesterol, sodium, complex carbohydrates, sugars, dietary fiber, and protein. New regulations induce manufacturers to reformulate to improve the perceived benefits of their products based on the label image. If a change in ingredients alters the traditional taste of the product, there will be a potential opportunity for flavor addition to offset the taste differences.

The techniques available today usually make it possible to identify and chemically synthesize the ingredients of many flavors of interest. The cost of these artificial flavors depends on the complexity of the formulation, but nevertheless even a fairly complex chemically synthesized flavor is likely to be less expensive than a similar product of biotechnology. In some cases, and especially where the flavor is not a simple formulation, extracted flavors are cost effective and of course natural. However, if the food product contains any artificial flavor, then the food product must be labeled as ''artificially flavored'' [8], and, as discussed above, this generally is not considered desirable. The desire for a natural product is thus a major incentive for the development of flavors by biotechnology.

The natural-flavor label is an important attribute in several countries, including the United States. Determining that a flavor is natural can get complicated in all but the simplest cases, and an expert determination is strongly recommended. Basically, a natural flavor can be produced in only certain restricted ways: (1) extracted directly from a plant or animal material by any of various physical means (e.g., by use of an organic solvent); (2) produced by microbiological fermentation; (3) produced by an enzyme reaction; or (4) produced by the use of any food preparation technique commonly used by a typical consumer such as cooking, acidification (vinegar), and so on. In addition, a natural flavor can only be made from materials that are themselves natural. In other words, a natural flavor cannot be made from any chemically modified ingredient (any treatment other than specified above), nor can it be made from a material derived from a petrochemical source.

The natural flavor labeling is not as important elsewhere (such as in many parts of Europe) where it is sometimes replaced with the nature-identical designation. Nature-identical flavors, as the name suggests, have the same chemical structure regardless of the method of production. In countries that use the nature identical designation, chemically synthesized ingredients are used whenever they are less expensive, so that there is less incentive to use biotechnology in the production of flavor.

Another labeling classification that the researcher developing a new flavor should consider is its kosher status. Kosher certification is used by people of the Jewish faith to assure that a food product is acceptable for consumption. A kosher food must meet certain standards of purity and cleanliness, so that kosher certification is often considered a desirable attribute in general, rather than just for religious purposes. A food is certified as kosher by a rabbi from a kosher certification organization. Not only must all ingredients of the product themselves be kosher, but the manufacturing methods and equipment must meet certain standards. The rabbi may need to visit the manufacturing facility to be assured that all requirements have been met. To avoid unpleasant surprises, the opinion of one of these organizations is best consulted during the early development phase of a project. The restrictions on food consumption during the Passover holiday are more strict, so that a kosher-for-Passover designation for a food is harder to get. In addition to other restrictions, the food usually cannot contain grain- or legume-derived ingredients. A detailed discussion of kosher and kosher-for-Passover is well beyond the scope of this chapter, but for those who would like a greater understanding, there are several excellent reviews available [9–11]. The main difficulty in developing a kosher dairy flavor is that it usually contains at least some dairy ingredients. According to Jewish custom, dairy- and animal-derived ingredients may not be mixed (fish is not considered animal). For this reason, all ingredients in a dairy flavor must be either dairy kosher or neither dairy nor animal derived (i.e., neutral or pareve) if the flavor is to be used in a dairy product (as is likely). However, it is much more desirable if only pareve ingredients are used. In this case, the flavoring may also, if desired, be used with meats or in products likely to be eaten with meats (such as bread).

The way a flavor is specified on the final product label is of tremendous importance in determining the applications in which a specific flavor can be used. A flavor will only be accepted for use in a specific food product if it can be labeled to the satisfaction of the manufacturer. The restricted availability of kosher and natural flavors means that these flavors will have a higher (usually much higher) value than a flavor that does not meet kosher requirements or one that must be labeled artificial.

6.4 CHEESE FERMENTATION PROCESS

To understand how a dairy flavor such as cheese flavor can be manufactured by biotechnology, it is important to first understand how that flavor naturally occurs. A discussion of cheese flavor is complicated because of the many varieties of cheese, but since the basic steps for production are similar for most cheeses, cheddar will be used as the example illustrating the production process.

Milk is made into cheese by coagulating and dewatering the milk protein (casein) into a material called the curd, which is aged to develop flavor. Cheese curd can be formed by the addition of chemical acidulants to milk, but for most cheeses including cheddar, the curd forms as a combined result of the action of an enzyme called rennin (see below) and the natural acidification that occurs as a result of bacterial fermentation. Bacterial starter cultures are used in modern cheesemaking to ensure product uniformity. These bacteria characteristically use the homolactic biochemical fermentation pathway in their metabolism, which means that only lactic acid is produced from the metabolism of the milk sugar, lactose. In particular, the use of a *Lactobacillus casei* strain (a homolactic bacterium) in the starter culture has been recommended for the fastest production of a strong cheddar flavor [12]. Although the lactic acid in cheese is nonvolatile and adds little flavor, it is important to the tanginess of cheese and the resultant lower pH creates the proper environment for chemical reactions during the long ripening process. During initial cheddar cheese-curd formation, only a small amount of the lactose in the milk is fermented. The remaining lactose is consumed during the first 2 to 3 weeks of aging by the developing secondary flora and by further action of the starter culture bacteria. The lactic acid remains relatively unmetabolized and is the major acid found in the final cheese. In contrast, if the heterolactic fermentation pathway were predominant, then the metabolic end products would also include acetic acid and/or ethanol and carbon dioxide. These end products are less desirable because formation of esters can sometimes be a problem in cheese, and ethyl butyrate and ethyl hexanoate in particular are a concern since they are associated with a fruity flavor defect in cheese [13]. In cheese with a fruity flavor defect, the ethanol concentration has sometimes been found to be as much as 16-fold greater than normal. The production of carbon dioxide could interfere with curd development; furthermore, in the special cases where gas formation is desirable, it is only desirable much later in the process, after the curd has formed (see Section 6.5.4).

The details of the cheddar cheese production process are well known and are shown in Figure 6.1. The starter culture is added to the milk as one of the first steps of cheese manufacture, and along with secondary microflora helps develop the cheddar flavor. After a short incubation, curd formation is induced by the addition of rennin (see below). The curds are then separated from the liquid portion, known as *whey*, and finally salted to the desired flavor. The cheese is placed into cold storage at approximately 4°C, or sometimes somewhat higher, to age. Within 24 hours after salting, the pH drops to its lowest point, approximately 5.1. After 6 months, the pH has moved back up to 5.6, partly because of the degradation of amino acids to amines and other more basic materials.

Over 80 percent of milk protein is casein; this protein forms a colloidal suspension in milk, with the remainder of the milk protein being soluble in the whey [14]. Casein is composed of α- and β-casein and a lesser amount of the

Cheddar Cheese-Making Procedure

- ■Add lactic acid bacterial starter culture to the milk
- ■Incubate at 32 C until pH decreases to 6.5 - 30 min.
- ■Add rennin to induce coagulation of casein to form the curd
- ■Cut curd into small cubes to help contract and expel whey
- ■Warm to 38 C to further contract curd
- ■Pile cubes of curd until they fuse (called Cheddaring)
- ■Slice fused curd into small pieces
- ■Salt to desired sodium concentration (600-700 mg/100 g)
- ■Press into blocks of cheese
- ■Age at 4 C (6-12 months) or force ripen at 10-16 C (until flavor develops)

FIG. 6.1 A common procedure for production of cheddar cheese from milk. Other cheeses generally follow a similar procedure.

carbohydrate-containing κ-casein. The α- and β-casein would normally be insoluble under the conditions occurring in milk, but they are rendered soluble by the κ-casein, which coats the surface of the protein globules to form micelles [15]. However, in most cheeses, the enzyme rennin (chymosin) is added to coagulate the protein and form the curd. Rennin acts by hydrolyzing the carbohydrate-containing portion of κ-casein, and, once κ-casein is hydrolyzed, the casein micelles are destabilized and coagulate. Although rennin also acts on other proteins, such action is slow enough that it contributes little to the overall proteolysis of cheese [16].

Traditionally rennin has been obtained from the stomach of calves. However, enzyme obtained from this source is expensive because the supply is scarce. Recently, rennin produced by fermentation using recombinant microorganisms has become available [17]. This rennin is identical in action to the enzyme obtained from the traditional source, it is less expensive, and it is not meat-derived so use in kosher products is not a problem. For these reasons, the recombinant renin probably will eventually almost totally replace the animal-derived product. There are several enzymes, including pepsin and fungal neutral proteases, which are available as cheaper substitutes for rennin, but their use is generally not considered to result in as desirable a flavor. Many proteases cannot be used because the peptides formed have bitter taste. Plant proteases such as papain and ficin generally produce the most bitterness.

Proteolysis occurs in cheese due to the presence of proteases in the original milk, proteases produced by the bacterial and fungal microflora, and proteolytic activity of added rennin [18]. For this reason, free amino acids and peptides make up an important part of the water-soluble fraction of aged cheese. The water-soluble fraction has an important impact on the general cheese background flavor, but the amino acids and peptides also contribute to the complex characteristic flavor of a cheese such as cheddar [19,20]. Therefore, as one would expect, the concentration of free amino acids correlates with the degree of ripening of the cheese [21–23]. Of the free amino acids that accumulate, methionine and leucine contribute the most to the total flavor [24], with leucine being present in the highest concentration [12, 16, 25]. Methionine is probably also a flavor precursor, because it is converted by *Lactobacillus helveticus* to various flavored compounds such as furanone, pyrrolidine, and pyrazines [26].

6.5 CHARACTERISTIC FLAVOR CHEMICALS OF CHEESE VARIETIES

6.5.1 Cheddar—A Bacterially Ripened Cheese

Cheddar cheese is manufactured on a large scale in more countries than any other cheese and is easily the most popular cheese worldwide [18]. However, despite the great economic importance of cheddar cheese, little is known for certain about the chemical composition of the characteristic cheddar flavor, although many compounds are known to play a role. It is likely that cheddar flavor is built up from a complex interaction of many component flavors with no one chemical solely responsible.

Despite the lack of specific knowledge of its composition, certain things can be done to enhance cheddar flavor. The impact of sharp aged cheddar may be at least partially due to phenol, which is found in concentrations as high as 225 ppb [27]. In this research, it was found that the addition of 150 ppb phenol to bland cheddar enhanced the clean sharp flavor. Although this is an interesting result, and phenol is found naturally in cheddar cheese, phenol is not a normal flavoring agent so that its safety as an added ingredient would have to be proven. There is a U.S. Patent for rapid (5 to 7 days) production of aged cheddar flavor by fermentation of milk [28]. The bacteria used in this patent were previously classified as species of *Micrococcus* which are actually sometimes found on other cheeses (e.g., brick) where they contribute to the aroma [29]. Unfortunately the bacteria recommended in the patent probably would not be considered food grade today, since they are now classified as *Bacillus cereus* (a cause of food poisoning!) and *Staphylococcus caseolyticus* (related staphylococci are human pathogens). *Micrococcus varians* is not mentioned in this patent but might be a likely substitute microorganism.

There is interest in the possibility that methanethiol (methyl mercaptan; Fig. 6.2) might be one of the key components of cheddar flavor. Methanethiol has a sauerkrautlike aroma at high concentrations, and by itself cannot produce a cheddar flavor. Nevertheless, some investigators have found that methanethiol

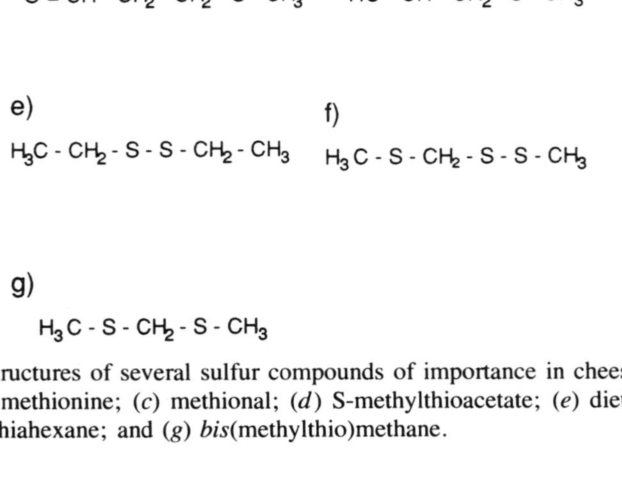

a) CH_3-SH

b) $\underset{\underset{\underset{OH}{|}}{\overset{\overset{NH_2}{|}}{CH}-CH_2-CH_2-S-CH_3}}{O=C}$

c) $O=CH-CH_2-CH_2-S-CH_3$

d) $HO-\overset{\overset{O}{\|}}{CH}-CH_2-S-CH_3$

e) $H_3C-CH_2-S-S-CH_2-CH_3$

f) $H_3C-S-CH_2-S-S-CH_3$

g) $H_3C-S-CH_2-S-CH_3$

FIG. 6.2 Structures of several sulfur compounds of importance in cheese: (*a*) methanethiol; (*b*) methionine; (*c*) methional; (*d*) S-methylthioacetate; (*e*) diethyldisulfide; (*f*) 2,4,5-trithiahexane; and (*g*) *bis*(methylthio)methane.

concentration correlates with the intensity, although not necessarily with the quality, of cheddar flavor [30–32]. Others contend that methanethiol and other sulfur compounds correlate only with the age of the cheese [33]. Lindsay and Rippe [34] investigated the relationship between methanethiol and cheddar cheese aroma by adding the enzyme methioninase to cheese to produce methanethiol from methionine (Fig. 6.2), thereby generating extra methanethiol as the cheese aged. The experimental conditions were (1) control cheese, (2) cheese plus methioninase, and (3) cheese plus encapsulated methioninase. Even after only 1 day of aging, measurable levels of methanethiol had accumulated in the enzyme-treated cheeses, but this did not correspond to the development of cheddar flavor.

After several months of aging, the cheeses containing the methioninase developed a more intense sulfury, toasted cheese flavor than the control, and, although this flavor was not unpleasant, it was not characteristic of cheddar cheese. As aging continued, the flavor improved and became more cheddar-like, but the improved flavor was associated with an increase in hydrogen sulfide and a decrease in methanethiol. These results apparently rule out methanethiol as playing a dominant role in cheddar flavor; it is perhaps more likely that methanethiol plays an essential, although supporting, role along with other compounds in a complex cheese aroma. Examples of these other compounds include diacetyl (2,3-butanedione; buttery flavor) and dimethyl sulfide, which are also found in good quality cheddar cheese [35]. In research that supports this idea, Manning and Robinson [36] found that only four compounds (hydrogen sulfide, methanethiol, dimethyl sulfide, and diacetyl) contributed most of the aroma of cheddar cheese.

It appears to be impossible to make a cheese with characteristic cheddar flavor using skim milk [37]. In fact, typical cheddar flavor does not develop when the fat content of the cheese is less than 50 percent of the total solids [29,38,39]. The source of fat is also important. In an experiment where various sources of fat were incorporated into the cheese by homogenization into skim milk, only a weak cheddar flavor was produced when vegetable oil was substituted for the milk fat [40]. In fact, the investigators found that mineral oil, which is inert and bland in flavor, produced a better cheese than vegetable oil. These results suggest that the lipid breakdown products of milk fat have flavors superior to those of vegetable oils. The homogenization process was also found to effect the results, because when milk fat was rehomogenized back into skim milk, the flavor was inferior, unless gum acacia was used as an emulsifying agent. Thus, the appropriate fat–water interface appears to be important in the flavor development process, perhaps because of the known preference of lipase enzymes for the lipid–water interface.

Volatile fatty acid concentrations increase during ripening of cheddar cheese because of the action of lipases, that is, enzymes that split fat molecules into free fatty acids. The primary source of these lipases is synthesis by the bacterial flora of the cheese, although some also come from the milk itself and additional lipase may be added to intensify the cheesy flavor. The source of the lipase is critical to good flavor. For example, a lipase from the mouth (pregastric esterase) acts on butter fat or vegetable oil to give a flavor free of rancidity and soapiness, but many lipases (including most derived from plants and animals) will cause these off-flavors.

During the aging of the cheese the fatty acids formed by the action of lipase can go on to form methyl ketones, lactones, and other flavor compounds. In this manner volatile fatty acids are thought to create the necessary aroma background, but they are not sufficient for the complete flavor [38]. In particular, they probably are responsible for the peppery taste of cheeses such as cheddar, although complete removal of fatty acids does not affect the characteristic cheddar aroma [32,36]. Cheddar cheese does not usually have any propionic, valeric, or isovaleric acids but has an average of 900 ppm acetic and 115 ppm of butyric acid [41] (more recently the complete absence of propionic acid has been disputed [20]). It has been suggested that acetic, butyric, and caproic acids (in a ratio of 8:1:0.3) are essential to the formation of quality cheddar flavor [38].

Lactones are formed both enzymatically and nonenzymatically from fatty acids, so that lipase-treated cheeses have higher lactone levels [42]. Lactones are also often added to synthetic cheese flavors. For example γ-dodecalactone, δ-decalactone, δ-dodecalactone, and δ-tetradecalactone (Fig. 6.3) are present in cheddar cheese, and the addition to simulated cheese was found to produce a smoother, mellower flavor [43–45]. Two other lactones, γ-decalactone and δ-undecalactone, have been recently detected in cheddar, but their significance to the flavor is unclear [46]. Lactones of several straight-chain hydroxy fatty acids also have dairy-type aromas, including δ-hydroxynonalactone and 4,4-dibutyl-γ-butalactone (Fig. 6.3) [47].

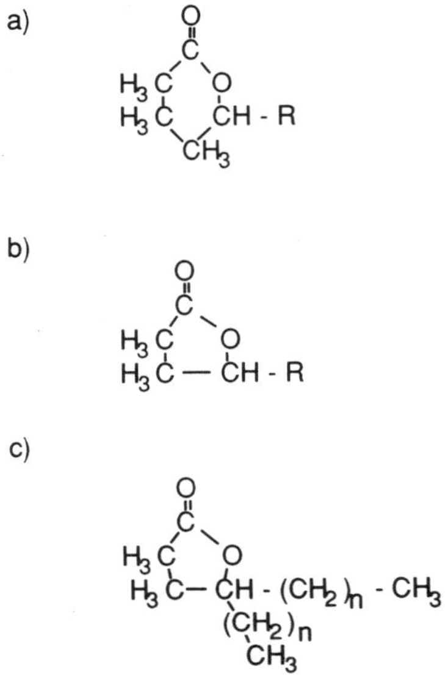

FIG. 6.3 Structures of lactones: (*a*) δ-lactones; (*b*) γ-lactone; (*c*) 4,4,-dibutyl-γ-butalactone.

Ketones and aldehydes are important in many dairy flavors. Aldehydes can be formed during production of cheese by a Strecker degradation. The Strecker degradation involves reaction of a dicarbonyl compound with an amino acid to yield an aldehyde of one less carbon. It has been found that heating enzymatic digests of casein or skim milk in the presence of pyruvic acid yields a strong odor, which when diluted has a pronounced aroma of cheese [48]. Adding a distillate of this material to cottage cheese imparts a cheddar flavor. Methional (Fig. 6.2) is probably produced in cheese from a Strecker degradation of methionine, and, in the presence of other aldehydes from cheese, it has a toasted cheese aroma. The concentration of 2-pentanone (along with methanethiol) has been found to correlate well with cheddar intensity, and was found to be a good indicator of maturity [49,50]. Butanone is present in cheddar, but its concentration decreases with age, and it is never found in concentrations high enough to effect the flavor [43]. Acetaldehyde is found in yogurt, where it gives a "green nutty" flavor, but it is found in smaller concentrations in cheese and probably contributes little to the cheese flavor.

Only low concentrations of pyrazines and related compounds are present in aged cheddar, but, because they have very low odor thresholds, they probably have an effect on the flavor. Although processed American cheese is manufactured from cheddar, it has a distinctive flavor of its own. It was found to contain 2,5-dimethyl- and 2,3,5-trimethyl pyrazine, which were suggested to be the main contributors to the weak nutty flavor characteristic of this cheese (Fig. 6.4) [51].

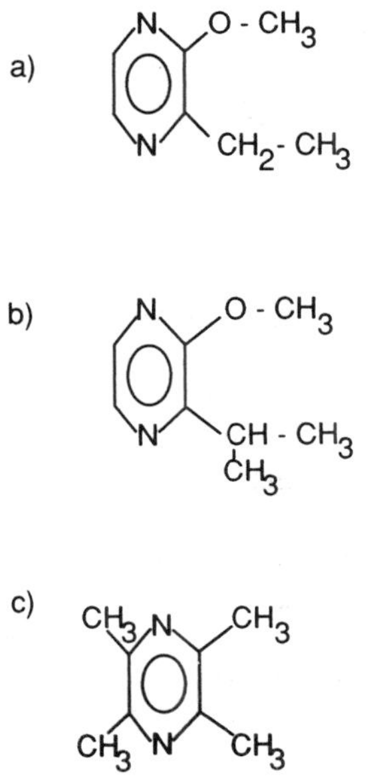

a)

b)

c)

FIG. 6.4 Structures of several pyrazines important in dairy flavors: (*a*) 2-methoxy-3-ethyl pyrazine; (*b*) 2-methoxy-3-isopropyl pryazine; and (*c*) tetramethyl pyrazine.

6.5.2 Bacterial Surface-Ripened Cheese

These cheeses include Limburger, brick, and Muenster. *Brevibacterium linens* is necessary for ripening of these cheeses, and produces the characteristic brownish-red surface coloration. However, the first microorganisms to appear on the surface of the cheese in significant numbers during ripening are salt- and acid-tolerant yeasts [52,53]. The surface of the cheese is initially acidic (pH 5.0), and yeasts grow well on the surface film under these conditions but disappear after approximately 2 weeks. The yeasts oxidize the surface lactic acid and degrade amino acids to amines, raising the pH to 5.9 or higher. At this point (6 days) *B. linens* begins to dramatically increase in numbers, but this organism is strongly aerobic and consequently is limited to the surface smear layer of the cheese. The pH of the smear layer slowly increases to as high as 7.2 through the proteolytic activities of *B. linens*, whereas the interior of the cheese increases to a lesser extent [54]. The size of the ripening cheese is kept small so that the surface-to-volume ratio is high and a uniform flavor develops throughout the cheese.

Although *B. linens* is not lipolytic, it is strongly proteolytic and also metabolizes several of the resultant amino acids. Leucine, phenylalanine, and tyrosine are converted to 3-methyl-1-butanol, phenylethanol, and phenol, respectively; these compounds are regarded as important flavor components of bacterial surface-ripened cheeses [55]. The importance of methanethiol, hydrogen sulfide, and other volatile sulfur compounds is still somewhat unclear in cheddar cheese (see above), but in bacterial surface-ripened cheeses methanethiol and hydrogen sulfide are essential for the characteristic putrid aroma [56]. Ammonia, indole, dimethyldisulfide, and butyric, isovaleric, hexanoic, octanoic, and decanoic acids also contribute significantly to the Limburger aroma [57–59]. The role of pyrazines in surface-ripened cheese is not clear, but numerous pyrazines have been detected in these cheeses, including 2,3-dimethyl, 2,5-dimethyl, 2,6-dimethyl, ethyl, ethylmethyl, trimethyl, tetramethyl, and ethyltrimethyl pryazine [59].

6.5.3 Fungal Surface-Ripened Cheese

These cheeses include Camembert, Roquefort, blue, and Brie. Brie can be considered to be both a bacterial and fungal surface-ripened cheese since it is ripened by both *Penicillium camemberti* and *B. linens*. Camembert, characterized by a white surface, is ripened by only *P. camemberti*. Roquefort and blue cheese are characterized by blue veins and are ripened by *Penicillium roqueforti* (or *P. glaucum*). With these cheeses a film of yeasts, including *Kluyveromyces lactis*, *K. fragilis*, *Saccharomyces cerevisiae*, and *Debaryomyces hansenii*, initially develops on the surface [60,61]. This microflora causes deamination of amino acids and oxidation of lactic acid, resulting in an increase in pH. Such an environment is suitable for *Penicillium*, which forms the surface growth that is characteristic of these fungal surface-ripened cheeses. In the case of Roquefort and blue cheese, passages are bored into the cheese to allow air to enter and thus allow *P. roqueforti* to develop in the interior of the cheese.

The fungi of these cheeses are strongly proteolytic and produce large quantities of peptides and amino acids. Proteolysis in blue cheese can be extensive, with more resultant protein degradation than is typical of other cheese varieties [62]. Free amino acids can represent 30 percent of the total nitrogen in some types of blue cheese [63]. Camembert cheese has a free amino acid content of only 9 to 12 percent, because of the extensive deamination of amino acids. As a result of this deamination, approximately 21 to 27 percent of the total nitrogen is present as ammonia [64]. This ammonia level is high enough to effect the flavor.

In addition to proteases, *P. roqueforti* also has strong lipolytic enzymes that release fatty acids that are important contributors to the final flavor [65]. The formation of fatty acids (in particular, octanoic acid) is also considered the rate-limiting step in the formation of 2-heptanone, which is necessary for characteristic blue cheese flavor. High concentrations of other methyl ketones are also found in blue cheese, including 2-pentanone and 2-nonanone (Table 6.1) [66]. δ-Lactones are more important in blue cheese than cheddar cheese flavor, with δ-tetradecalactone and δ-dodecalactone present at the highest levels [42]. These authors also found that the addition of microbial lipases results in the production of fourfold greater levels of these lactones from the accumulation

TABLE 6.1 Flavor Components of Blue Cheese Compared to Those Used in a Synthetic Blue Flavor

Component	Concentration (mg/kg cheese)	
	Synthetic Mix	Real Cheese
Acetic acid	550	826
Butyric acid	964	1448
Hexanoic acid	606	909
Octanoic acid	514	771
Acetone	6.2	3.1
2-Pentanone	30.3	15.2
2-Heptanone	69.5	34.8
2-Nonanone	66.3	33.1
2-Undecanone	17.0	8.5
2-Pentanol	0.9	0.4
2-Heptanol	12.1	6.1
2-Nonanol	7.0	3.5
2-Phenylethanol	2.0	—
Ethylbutyric acid	1.5	—
Methylhexanoic acid	6.0	—
Methyloctanoic acid	6.0	—

Source Reprinted with permission from D.F. Anderson and E.A. Day, *J. Agric. Food Chem.* **14**, 241 (1966). Copyright 1966 American Chemical Society.

of the hydroxy acid precursors, greatly enhancing the quality of the resultant blue cheese flavor.

Although methanethiol probably does not play a large role in cheddar flavor (see above), its concentration is much higher (in the μg/g range) in at least some types of blue cheese, and consequently it plays a more clear-cut and dominant flavor role there [50]. Sulfur compounds such as *bis*(methylthio)-methane (Fig. 6.2), which is formed by the reaction of methanethiol and formaldehyde, are important components of Camembert flavor [67]. *Bis*(methylthio)methane is also found in Gouda (a Swiss-type cheese) and is believed to be an important component of the flavor for this cheese as well [68]. Typical Camembert cheese also contains other sulfur compounds, including 3-methylthiopropanol, diethyldisulfide, 2,4,5-trithiahexane, and *bis*(methylthio)methane (Fig. 6.2). Another volatile compound important in Camembert cheese is 1-octen-3-ol [69], present at a concentration of 5 to 10 ppm, which is 5 to 10 times higher than in Roquefort. This compound makes a critical contribution to the distinctive flavor of Camembert, and it also adds to the pleasant moldy-type flavor of other fungal surface-ripened cheeses, although to a lesser extent.

6.5.4   Swiss—Bacterially Ripened Cheese

Swiss-type cheeses include Swiss, Emmental, and Gouda. The unique difference in production of this type of cheese is the use of a bacterial starter culture containing *Propionibacterium freudenreichii* subsp. *shermanii* [70]. As aging begins, *P. freudenreichii* ferments the lactic acid (produced by lactic acid starter culture bacteria) to propionic acid, acetic acid, and carbon dioxide. The carbon dioxide soon exceeds its solubility and forms the characteristic holes of traditional Swiss cheese. The concentrations of propionate and acetate in aged cheese reach as high as 0.59 percent and 0.37 percent respectively, and both acids contribute to the distinctive Swiss cheese flavor [71]. Although *P. freudenreichii* is in general nonproteolytic, it has peptidases specific for liberation of proline. Aged Swiss cheese contains as much as 0.7 percent proline, which contributes the sweet flavor characteristic of Swiss cheese [72]. A taste panel found that propionic acid and a minimum concentration of 0.25% proline are necessary for the best Swiss flavor. Butyric acid concentration is normally very low, and higher levels contribute only undesirable flavors to the cheese. A synthetic Swiss cheese flavor of "fair quality" can be prepared from a mixture of the above compounds plus various other volatiles (Table 6.2) [73].

Pyrazines probably play a role in creating the nutty flavor of Gouda [68] and Emmental cheese [74]. Many pyrazines have been detected, including tetramethyl, trimethyl, dimethyl, diethyl-3-methyl, ethylmethyl, 3-ethyldimethyl-, 2-ethylmethyl, and ethyltrimethyl pryazine. Although these pyrazines are not the main flavor components, they may add the nutty flavor which is desirable in these cheeses.

TABLE 6.2 Flavor Components of Swiss Cheese Compared to Those Used in a Synthetic Swiss Flavor

Component	Concentration (mg/kg cheese)	
	Synthetic Mix	Real Cheese
Fatty Acids		
Acetic Acid	1862	3724
Propionic acid	2960	5919
Butyric acid	165	329
2-Methylbutyric acid	50	100
3-Methylbutyric acid	6	13
Caproic acid	58	115
Caprylic acid	47	94
Amino Acids		
Proline	6000	—
Glycine	1600	—
Serine	1950	—
Threonine	1950	—
Aspartic acid	2500	—
Glutamic acid	3000	—
Cysteine	760	—
Tryptophan	2200	—
Histidine	3700	—
Lysine	2200	—
Selected Volatiles		
Dimethyl sulfide	0.11	0.11
Diacetyl	0.8	0.8
Acetaldehyde	1.4	1.4
Acetone	1.6	1.6
Butanone	0.3	0.3
2-Methylbutyraldehyde	0.42	0.42
2-Pentanone	0.98	0.98
2-Heptanone	0.45	0.45
Methylhexanoic acid	1.5	1.5
Ethylbutyric acid	0.6	0.6

Source Reprinted with permission from J.E. Langler, L.M. Libbey, and E.A. Day, *J. Agric. Food Chem.* **15**, 386 (1967). Copyright 1967 American Chemical Society. The fatty acid data in the original table were obtained from Langler and Day, and the amino acid data were obtained from Hintz et al. [72].

6.6   FERMENTATION AND ENZYMATIC METHODS FOR PRODUCTION OF DAIRY FLAVORS

6.6.1   Enzymatic Development of Cheese Flavors

Cheese is a relatively expensive product to manufacture because of the cost of the raw materials and the length of the time required for aging. Since enzymes (or the organisms containing the enzymes) are critical for the development of the final flavor, it is a logical step to add extra enzymes to accelerate the flavor development process or intensify the resultant flavors. For example, the addition of *Candida lipolytica* lipase to blue cheese leads to a more rapid formation of fatty acids that are essential to the flavor [75]. However, blue cheese is strongly flavored, and the flavor will tolerate relatively high levels of fatty acids. Increasing the fatty acid levels in most other cheeses tends to increase the cheesy character, but this must be done with great care or the flavor of the resultant cheese could be impaired. This is probably because the added lipases tend to encourage the release of high levels of C_{10}, C_{12} and C_{14} fatty acids, which give a rancid flavor. It is also important to be aware that lipases catalyze formation of esters in the presence of fatty acids and alcohols (see Section 6.6.4).

Animal-derived lipases tend to produce the more desirable shorter chain acids, but off-flavor notes can also be present. Italian cheeses such as Romano, Provolone, and Parmesan are dependent on animal lipases for their characteristic slightly rancid flavor [76–78]. Lipases from yeasts or fungi can also be quite useful in producing Italian cheese-type aroma from fat. For instance, Romano cheese aroma can be made by treating a 20 percent butterfat emulsion (1.5 percent Tween 80) with crude lipase from *Candida rugosa* [79]. The procedure calls for the emulsion to be held at 37°C for 3 hours and then at room temperature for 3 days. The authors reported that a good Romano cheese aroma was produced with only some ketonic and soapy off-notes, but unfortunately the flavor was only slightly cheesy and had a bitter and waxy taste. Another recipe calls for addition of *Aspergillus niger* lipase to butter fat with lecithin as an emulsifier, followed by stirring at 39°C [80]. A strong cheesy Parmesan-Romano aroma was easily produced after stirring for only 2 to 3 hours, but a good balanced flavor was not achieved [81]. This is because the soapy, rancid flavors that are also made during the reaction begin to strongly predominate after more than 3 hours.

Proteases have also been added to cheese in an attempt to cause accelerated ripening, since protein breakdown is an integral part of cheese aging. The lactic acid starter culture bacteria have cell wall proteases with various activities [82,83], but the addition of extra proteases can sometimes speed up ripening of the cheese. On the other hand, the effects of proteases are not totally beneficial, because, in some cases, they can release bitter-flavored peptides [84]. Alkaline proteases tend to cause the most bitterness, whereas neutral proteases may give acceptable results, producing a stronger, but still typical cheese flavor [85]. However, even neutral proteases need to be used with caution or perceivable bitterness in the final cheese can result. Usually proteases are not added alone, but in combination with a lipase and a pregastric esterase. The utility of these lipase–protease enzyme mixtures for cheese ripening has not escaped the dairy industry, and consequently enzyme mixtures for use in foods are available from several sources [86,87]. These enzyme mixtures are in most cases clearly superior to just protease or lipase alone.

6.6.2 Methyl Ketones

It has been known for some time that 2-heptanone and other methyl ketones are a dominant part of blue cheese flavor and that these methyl ketones are produced by various *Penicillium* species [88]. Because of the flavor dominance of 2-heptanone, the addition of 2-heptanone to a food product (such as salad dressing) containing only a small amount of authentic blue cheese will greatly enhance the flavor intensity. This potential economic benefit has inspired a considerable amount of research on the development of industrial processes to produce 2-heptanone from octanoic acid (caprylic acid) or dairy products high in octanoic acid. In general, the strategies employed have been to use spores and sometimes mycelia of the fungus *Penicillium roqueforti* in a medium based on milk, whey, or casein.

In 1963, Knight [89] was issued a patent for a process to make blue cheese flavors. In this process, milk is inoculated with *P. roqueforti* in a submerged aerobic fermentation. Interestingly, it was discovered that, if the vegetative cells were eliminated and only fungal spores were used in the inoculum, the fermentation was faster and the product had a better flavor. This might appear surprising since spores usually have a much lower level of metabolism than vegetative cells and do not even form lipases or proteases. However, spores can quickly convert fatty acids to ketones and will make 2-heptanone if supplied with octanoic acid [90]. On the other hand, the vegetative cells were found to have less ability to produce 2-heptanone. Most vegative cultures of *P. roqueforti* also have a substantial number of spores, and it appears that in most cases it is the presence of the spores that results in 2-heptanone production. Because lipases are necessary to release the fatty acid precursors of 2-heptanone from fat, the missing lipases should be added when a spore inoculum is used. When spores are allowed to germinate by providing a growth medium, the germinated spores quickly synthesize the missing lipases but lose the ability to make 2-heptanone. The source of fat is not absolutely critical, so skim milk can be used if vegetable oil is added to replace the milk fat.

It was later discovered that a more accurate blue cheese flavor can be generated by fermenting a medium containing sodium caseinate and butterfat, but without milk [91]. The resultant cheese flavor has a more favorable proportion of methyl ketones relative to protein-derived background flavors. Interestingly, it is claimed that neither casein nor any other caseinate salt such as calcium caseinate can be substituted for the sodium caseinate. This is somewhat surprising, because blue cheese itself contains substantial levels of calcium. Perhaps the explanation for this is that casein and calcium caseinate have a lower solubility than the sodium caseinate and thus are not used as easily. In the actual cheese production process, the incubation time is so much longer than in this process that the low solubility of the caseinate does not matter. Although animal or vegetable fats can be used in this process, use of butterfat will produce the greatest flavor fidelity. In the optimum configuration of the process, the flavor is made by incubating *P. roqueforti* for approximately 2 to 5 days at 25°C under agitated aerobic conditions. The medium contains preferably 10 percent sodium caseinate with 5 percent butterfat and is pasteurized both before and after the fermentation.

Watts and Nelson [92] offered an improved production method for the spore inoculum that is preferred in the Knight process [89] (see above). It uses a milk-based medium for production of the inoculum, which differs from the medium used by Knight in that the former uses lactic acid to reduce the pH and a substantial amount of added salt or sugar. The high ionic strength provided by salt or sugar, used at a concentration of up to 7 percent by weight (total), is known to enhance spore production. The process also has been modified to accommodate these changes in the growth medium. Although the use of milk is preferred, blends of milk and nonfat milk solids or whey solids can be substituted. After the milk is sterilized, 1 to 3 percent salt or sugar is added to increase the osmotic strength, then the mixture is inoculated with *P. roqueforti* and incubated aerobically for 3 to 4 days at approximately 25°C. An additional 4 to 6 percent salt or sugar is then added and the incubation is continued for 1 to 2 more days. During this time, the aeration rate is kept low to prevent loss of the volatile ketones, but this results in a decrease in the oxygen transfer rate; so to compensate, the vessel is operated at elevated pressure (15 to 25 psig). During the incubation, there should be two or three sequential additions of free fatty acids to simulate the slow release of fatty acids that occurs naturally in blue cheese. The free fatty acids used for this addition can be produced from milk fat by hydrolysis with calf pregastric esterase [76]. The ketone content of the final product is about ten times that of blue cheese, but the flavor intensity is only about four times greater. In particular, the product lacks the full nutty background flavor, probably because of the low level of protein-derived flavors (perhaps pyrazines; see Section 6.5.4).

The above processes use spore preparations to carry out the fermentation, and all produce a 2-heptanone rich material, but none completely reproduces the full blue cheese flavor. Usually, 2-heptanone is used as a supplement to blue cheese in a food product, therefore, the lack of full flavor does not usually matter, but, if a full flavor is important, then a mycelial process can be used. Dwivedi and Kinsella [93] found evidence that a full-flavor product can be produced by using fungal mycelia (probably containing spores) in a submerged continuous process. However, their process is fairly long and complicated. The process takes 5 days and uses three fermentation stages, including (1) a vessel for 3 to 4 days of growth, (2) a second vessel for 1 day of aging, and (3) a third vessel for contact of the mycelia with hydrolyzed fatty acids to produce the actual flavor. The fungal mycelia produced in this manner have only a low level of lipase activity, so fatty acids are produced from milk fat using added enzymes. In spite of the fact that animal-derived lipases typically create bitterness (from contaminating proteases) and soapiness (from the lipases) in a flavor, in this case the investigators did achieve satisfactory results using a crude pancreatic lipase to hydrolyze the milk fat. The octanoic and decanoic acids produced in this way were then preferentially metabolized, producing 2-heptanone and 2-nonanone [94]. Even though the fermented product had only a low concentration of short-chain fatty acids that are normally considered to be important for cheesy flavor, a taste panel found that there was no significant difference between the taste of the fermented product and that of blue cheese when used in a chip dip. This suggests that fatty acids are not critical to the blue cheese flavor, but although the fermented flavor chip dip did not need any blue cheese, it did contain cottage cheese, which was presumably used for background cheesy-type flavors.

More recently 2-heptanone itself, instead of the full blue cheese flavor, has been the target of investigations. Spores can be entrapped in calcium alginate, and then used as a biocatalyst for conversion of octanoic acid to 2-heptanone in a medium otherwise containing only a small amount of ethanol and calcium chloride [95]. In these experiments, octanoic acid was added slowly, with the actual rate of addition controlled by the pH of the fermenter. In other words, the pH of the system was controlled by the replacement of the acidic substrate as it was consumed. A slow continuous air stripping (0.06 vvm) was used to remove the 2-heptanone product from the reaction vessel (350 to 400 mL working volume fermenter agitated at 300 rpm). At pH 6.5 and a substrate concentration of 4 mM, the system was stable for 300 hours with a productivity of 0.975 mM/L/hour, with nearly 100 percent conversion of octanoic acid to product. Since no other carbon source was available, the energy for the endogenous metabolism came from a spore storage material [96–98]; after 300 hours, this energy source apparently became depleted, and the reaction rate began to decrease.

Another process for 2-heptanone production was developed by Jolly and Kosikowski [99]. This process uses spores and acid whey powder but also calls for *Aspergillus* lipase and either butter or coconut fat. The use of acid whey was found to be critical because substituting sweet whey resulted in some bitterness. The addition of lipase accelerated the flavor production because spores are deficient in lipase. Coconut fat was used because of its high content of octanoic acid (19 percent by weight as compared to 1.7 percent in butterfat), the precursor to 2-heptanone.

6.6.3 Lactones

Although γ-decalactone (γDL) has a fruity, peachy flavor, it also has a creamy flavor and consequently has some uses in dairy flavors [47]. δ-Decalactone (δDL) is also creamy and may be used with γDL. These lactones can be produced by several methods. Both γDL and δDL were produced during fermentation of glycerol by the fungus *Ceratocystis moniliformis*, but unfortunately the concentrations were not determined [100]. The yeast *Sporidiobolus salmonicolor* also produced and accumulated γDL when grown with glycerol [101]. High concentrations of glycerol were most effective at inducing the accumulation since γDL apparently acts as an osmotic stabilizer for the cells [102]. Interestingly, when compounds related to glycerol (such as sorbitol) were used instead, glycerol itself was produced by the cells and γDL was not accumulated [103]. The reason for this effect of glycerol may be that it also functions as an osmoregulator when in the cell, but high concentrations of glycerol in the medium inhibit its synthesis by the cells, thus requiring the synthesis of a secondary osmoprotector, γDL. Unfortunately, although osmotic regulation is an interesting use of γDL, this process is probably is not practical, because the fermentation time was 20 days and only 5 mg/L of γDL was produced.

Various γ-lactones are produced during fermentation of lipid precursors by yeasts of the genus *Pityrosporum* [104]. The lipids used include lecithin, oleic acid, and triolein. The lipid provides both glycerol (which may also induce γDL in this yeast, as explained above) and fatty acids, which are metabolized to a mixture that includes γ-hexalactone, γ-heptalactone, γ-octalactone, γ-nonalactone, γDL, γ-undecalactone, and γ-dodecalactone. Generally γDL is produced at the highest levels, but the practicality of the process cannot be easily judged, because only relative concentrations were determined.

A good method to produce γDL uses castor oil as a substrate for microorganisms capable of β-oxidation, and as much as 0.7 percent γDL is produced in a 1 week fermentation [105]. In addition to castor oil, the growth medium contains a carbon source such as glucose to improve cell growth. The exact composition of the medium depends on the choice of the microorganism; several different fungi and yeasts can be used, including *Aspergillus oryzae*, *Candida rugosa*, *Geotrichum klebahnii*, and *Yarrowia lipolytica*. The chosen microorganism is used in a fermentation to hydrolyze the oil to fatty acids, some of which are then β-oxidized to hydroxydecanoic acid. If the oil is first enzymatically hydrolyzed, then a wider variety of microorganisms are effective. The hydroxydecanoic acid can then be made to spontaneously lactonize to γDL in the culture medium under fairly mild conditions (pH 3 at 70°C for 10 min).

The δ-deca- and δ-dodecalactones can be produced using yeast or fungi to hydrogenate unsaturated lactones obtained from the bark of the Massoi tree, as shown by van der Schaft et al. [106]. Massoi bark oil is commercially available and, depending on the purity, may contain approximately 80 percent 2-decen-5-olide and 7 percent 2-dodecen-5-olide, the precursors for δ-DL and δ-dodecalactone, respectively. In most experiments, van der Schaft et al. [106] used a purified Massoi oil containing mainly 2-decen-5-olide to produce pure δ-DL. Baker's yeast (*Saccharomyces cerevisiae*) was preferred, but some Basidiomycetes were also found to be effective. The best conversion was obtained in 8 hours of aerobic incubation at 35°C and pH 5.5 in a medium containing 1.5 percent yeast (dry basis), 15 percent glucose, 1.7 g/L Massoi lactone, and 2 percent β-cyclodextrin. The β-cyclodextrin, used to limit the inhibitory effects of the lactones on the yeast, increased the rate of conversion but not the final yield.

6.6.4 Propionic and Acetic Acids

The task of finding bacteria that make propionic acid as a major end product of fermentation is relatively easy, but some of these bacteria are potentially pathogenic (such as *Arachnia propionica*), and consequently would not be acceptable for commercial production. Fortunately, there are at least six anaerobic bacterial genera that contain nonpathogenic representatives that make propionic acid as the major product of fermentation.

Viellonella are anaerobes that ferment carboxylic acids such as lactic acid to propionate, acetate, carbon dioxide, and hydrogen. Even though *Viellonella* are anaerobes, at least some members of the genus (including *Viellonella parvula*, formerly *alcalescens*) contain cytochromes a and b, which apparently function in an oxidative electron transport to fumarate and nitrate [107]. Although the authors hypothesized that nitrate can stimulate growth, they found that it inhibited propionate production. *Viellonella* do not ferment carbohydrates, so mixed cultures with other microorganisms are necessary if a carbohydrate such as glucose is to be considered as the growth substrate. For instance, *Actinomyces viscosus* or *Actinomyces israelii* (which metabolize glucose to lactic, succinic, acetic, and formic acids) have been combined with *V. parvula* to ferment the lactic and succinic acids to propionic acid [108]. However, *Actinomyces* are probably not the best choice for a propionate-producing mixed culture, because both these species are opportunistic pathogens, and of course they produce acetic and formic acids, which *Viellonella* cannot metabolize. Homolactic (see Section 6.4) fermentative lactobacilli make only lactic acid as the end product of carbohydrate degradation, have a rapid growth rate, and are nonpathogenic, and so would be a better co-culture choice. When lactobacilli are grown on whey (which is approximately 80 percent lactose) with *Viellonella criceti*, the mixed culture produces approximately 2.0 percent propionate and 1.5 percent acetate in 48 hours [109]. Because 1.5 percent acetic acid is produced, this also could be considered an acetic acid process, although acetic acid alone probably would not be valuable enough to justify the process.

Propionibacterium freudenreichii (*P. shermanii*) is used to make Swiss cheese, and ferments carbohydrates or lactic acid to propionic and acetic acids. When grown on carbohydrates, lactic acid is produced first, then transformed to the latter two acids. Processes to make propionic acid using one or more strains of *Propionibacterium*, sometimes along with various other microorganisms, are well known, with patents issued as long ago as 1923 [110–113]. Advantages of using *Propionibacterium* include some tolerance to oxygen and production of only small quantities of hydrogen (a safety advantage), but a disadvantage is relatively slow growth. The effects of slow growth can be avoided to a certain extent by simply using mixed cultures with lactic acid bacteria which greatly speed up the production of lactic acid. For example, in the process developed by Ahern et al. [114], a whey-based medium supplemented with yeast extract is first fermented with a lactic acid-producing organism, followed by fermentation with either *P. freudenreichii* or *P. acidipropionici* [114]. In this case a propionate concentration of approximately 2.6 percent is reached in about 60 hours with a medium containing 12 percent solids. In a similar process, *P. freudenreichii* is grown on whey as a mixed culture with lactobacilli to yield approximately 3 percent propionic acid [115]. This process was improved by first acclimating the mixed culture on the whey/ yeast extract medium and by increasing the whey solids to 18 percent (from 12 percent) [116]. The end result was a yield of 6.5 percent propionic acid after 100 hours growth, consumption of all the lactose, and no accumulation of lactic acid.

Selenomonas ruminantium is well studied, because it is an important part of the microflora of the rumen of the cow. It grows on various substrates including glucose and glycerol, makes lactate as an intermediate product, and usually produces a final product mix of propionic acid, acetic acid, and succinic acid, in a molar ratio of approximately 1.8:1.0:0.2. The biochemical pathway is shown in Figure 6.5. Propionate is more reduced and acetate is more oxidized so they are made in a 2:1 ratio to keep the oxidation/reduction state of the fermentation in balance. Carbon dioxide is a product of the fermentation made in a 1:1 ratio with acetate and is also a growth requirement for *S. ruminantium* grown on lactate (carbon dioxide is fixed during the formation of oxaloacetate; see Fig. 6.5). During early fermentation, or if an inert gas sparge is used, the carbon dioxide level can drop too low unless carbon dioxide is added. L-Aspartate, *p*-aminobenzoic acid, biotin, and valerate also are required for growth on lactate, whereas only the latter two are required for growth on glucose [117]. In continuous culture, *S. ruminantium* exhibits rapid growth (dilutions rates up to 0.5/hour) in a glucose–mineral salts-based medium [118], but does not metabolize the lactate formed to acetate and propionate at dilution rates above approximately 0.2 hour [119].

Unlike many other anaerobic bacteria, *S. ruminantium* produces only traces of hydrogen during fermentation (a safety advantage) [120]. *Selenomonas ruminantium* is sensitive to free oxygen in its environment rather than just the redox potential [121]; despite this sensitivity, it does have a limited ability to detoxify oxygen. This ability is based on an oxygen-inducible nicotinamide adenine dinucleotide (NADH) oxidase, which can be coupled to metabolism, and a superoxide dismutase [122]. Thus at high cell densities small amounts of oxygen usually can be eliminated easily [123], but this also means that the fermentation product mixture becomes more oxidized and more acetate is made at the expense of succinate and propionate. *Selenomonas ruminantium* grows well with whey but even better with corn steep liquor. When using corn steep liquor in a lactic acid-based medium, as much as 2.3 percent propionic acid,

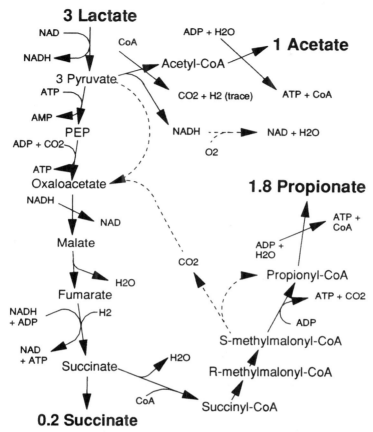

FIG. 6.5 Probable pathway of lactate metabolism by *Selenomonas ruminantium*. Typically, the overall reaction is that 3 lactates are metabolized to 1.8 propionates + 1 acetate + 0.2 succinate + 1 H_2O + 1 CO_2. The dotted arrow from CO_2 and pyruvate indicates that a transcarboxylase enzyme may play a role in transfer of the CO_2 group between S-methylmalonyl-CoA and oxaloacetate. The dotted arrow from NADH to NAD indicates that NADH is oxidized in the presence of oxygen. The CoA derivatives of acetate and propionate are both hydrolyzed by the unusual enzyme, acetate thiokinase [176]. PEP is phosphoenolpyruvate, CoA is coenzyme A, and NAD and NADH are respectively the oxidized and reduced forms of nicotinamide adenine dinucleotide, respectively.

1.1 percent acetic acid and 0.5 percent succinic acid are made in 72 hours [124]. Soy peptone can be substituted for the corn steep liquor, but in this case the propionate yield drops to 1.5 percent in 42 hours. Since this growth medium (unlike those for *P. freudenreichii* and *V. parvula*) does not contain any whey or other dairy products, the resultant propionic acid can be certified kosher pareve by an appropriate certifying organization (see Section 6.3).

Megasphaera elsdenii is the only species in the genus and ferments lactate to acetate, propionate, 4-carbon fatty acids, and valerate. When grown in complex media, *M. elsdenii* preferentially ferments lactate over simple sugars, with the end product being propionate [125]. The highest yield of propionic acid and the highest ratio of propionate to acetate were obtained when *M. elsdenii* was grown in a co-culture with *S. ruminantium*. *Clostridium propionicum* can ferment lactate to propionate, acetate, and sometimes acrylate, but unfortunately makes large quantities of hydrogen. *Megasphaera elsdenii* and *C. propionicum* use a different propionic acid pathway than the other bacteria and consequently sometimes can produce acrylate along with propionate.

Anaerovibrio lipolytica ferments glycerol or lactate to propionate and acetate but is fairly sensitive to oxygen, which could make commercial applications difficult. *Clostridium arcticum* ferments lactate to acetate and propionate and does not make large quantities of hydrogen, but grows slowly and requires temperatures under 25°C. *Coprococcus catus* produces propionate, butyrate, and acetate from lactic acid, and *Pectinatus cerevisiiphilus* produces acetate, propionate, and succinate from lactate, but in either case growth is probably slower than that of *S. ruminantium*.

6.6.5 Butyric Acid

Butyric acid is the major metabolic end product of many *Clostridium* species during fermentation of carbohydrates. Clostridia have so many advantages, including rapid growth, simple growth requirements, and high yield of butyric acid, that no other genus is competitive. In addition to butyric acid, clostridia also make acetic acid and substantial amounts of hydrogen (safety consideration) and carbon dioxide. When the pH is maintained near neutrality the clostridia generally only make acids, but if the pH is allowed to decrease as the acids are produced then two growth phases occur. In the initial acetogenic phase, only acids are produced, then at lower pH the acids, including butyric, are remetabolized to form butanol, ethanol, and acetone (solventogenic phase). Monot et al. found that the change to the solventogenic phase is triggered by a level of 1.5 g/L of undissociated butyric acid in the medium [126], but others found that solvent production can occur even at pH 7.0 and believe that total levels of both undissociated and dissociated fatty acids are the trigger for solventogenesis [127,128].

Mutation could be used to increase the fermentation yields of butyric acid by *Clostridium* species. One selection strategy is based on the use of so-called suicide substrates to select mutants unable to enter the solventogenic growth phase [129]. For example, several butyric and acetic acid analogs (such as chlorobutyrate and chloroacetate) are toxic to (normal) clostridia because they inhibit key metabolic pathways. These analogs have been used to select for mutants of *C. acetobutylicum* unable to remetabolize butyric acid to butanol.

Although some clostridial species are considered pathogenic, several are suitable for commercial use, including *C. acetobutylicum*, *C. beijerinckii*, *C.*

butyricum, and *C. tyrobutyricum*. *Clostridium tyrobutyricum* appears particularly suitable for production of butyric acid, producing 4.8 percent butyric acid and 0.5 percent acetic acid in 24 hours when grown on glucose supplemented with yeast extract [130]. The yield increases to as high as 6.4 percent butyric acid (and no acetic) in 43 hours if an enzymatic wheat flour hydrolysate is used. *Clostridium tyrobutyricum* also converts lactic acid to butyric acid, hydrogen and carbon dioxide [131], so it is interesting to speculate that this bacterium might be suitable for use as a mixed culture with lactic acid bacteria.

Clostridium botulinum produces butryic acid, but this organism would not be considered for commercial use because it also makes an extremely potent protein neurotoxin that causes botulism. There are actually six related but immunologically distinct variants of the toxin made by strains of *C. botulinum* and designated by the letter A to F [132]. Most are produced in a fully active form and all can be destroyed by heat. Surprisingly, other clostridial species with a long history of safe use for various purposes, including *C. butyricum*, a good butyric acid producer, can sometimes acquire the ability to make botulism toxin [133,134]. These strains, otherwise indistinguishable from the normal bacteria, were isolated from several cases of infant botulism. It appears that the toxin gene can be carried on a bacteriophage and can naturally be lost or regained by *C. botulinum* and other clostridial strains [135]. Transfer of the toxin gene by bacteriophage infection appears to be a rare event, but it would be wise to keep this possibility in mind when growing any *Clostridia*.

6.6.6 Esters

Esters are discussed here because of their relationship to cheese flavor, but they are also very important in other flavors, especially fruit flavors, and these aspects are discussed in detail in Chapters 1 and 3. Esters are produced in many fermentation processes, but are not usually produced in high enough yields to make recovery attractive. Often it is more economical to make the ester from the alcohol and the carboxylic acid using lipase as a natural catalyst. It is important to be aware of this since esters can be inadvertently made during lipase treatment of a fat where the intended purpose may only be to create volatile fatty acids (see Section 6.6.1). Often these reactions can be conducted in organic solvents because lipases (and many other enzymes as well) are stable and retain partial to full activity in such solvents [136]. Lipases are particularly well suited to this hydrophobic environment because they are often associated with the lipid membranes of cells. One advantage of using an organic solvent is that the reactants and products often have a much higher solubility in the solvent than in water. *Mucor miehei* lipase is especially active relative to other lipases in the production of various esters under anhydrous conditions [137]. For example, this lipase can be used to catalyze the reaction of butyric acid and butanol to form butylbutyrate; in this work *n*-heptane was found to be the best solvent [138]. Of particular relevance is a paper by Welsh et al. [139], who screened 27 commercial lipases for promotion of synthesis of low-molecular-weight esters using nonaqueous systems. Hexane, octane, and decane were used as the solvents and the highest activities were found with lipases from *Candida cylindracea, Pseudomonas fluorescens*, and *M. miehei*. Terpene alcohols such as geraniol, citronellol, and menthol, are also substrates of some lipases and can be used for ester production [140]. Lipases with this activity can be obtained from *Aspergillus niger, Rhizopus delemar, Geotricum candidum*, and *Penicillium cyclopoium*.

Lipases are relatively stable but expensive, so there can be a greater incentive to use immobilization than with other enzymes. *Candida cylindracea* makes a lipase that can be immobilized with glutaraldehyde for stable ester production in a continuous organic solvent process [141]. Using a purified lipase and water-saturated hexane, ethyl propionate is formed in this system at a rate of 0.007 mol/hour per gram of immobilized protein at 22°C. In a batch mode, equilibrium is reached in approximately 10 hours. In addition to ethyl propionate, ethyl butyrate, isoamyl propionate, and isoamyl butyrate can also be made at similar reaction rates. The immobilization results in a greater than 1/3 increase in enzyme activity, but also reduces enzyme heat resistance [142]. Crude enzyme preparations are much less expensive, and are thus preferable in cases where the presence of side reactions catalyzed by contaminating enzymes can be tolerated. For example, the fungus *Rhizopus chinensis* was cultivated so that it grew naturally inside cubes of polyurethane [143]. These cubes were given a simple water rinse, followed by extraction and drying with acetone, and finally vacuum drying. They were then placed in a column and used in a continuous process for interesterification of olive oil and methyl stearate in hexane. The half-life of the lipase in this system was 1200 hours at the optimal water content of the reaction, which was 100 ppm. Immobilization did not reduce enzyme activity, and in fact these cells had as much as four to seven times *higher* lipase activity than free cells [144].

6.6.7 Methanethiol

The presence of methanethiol (methyl mercaptan; see Fig. 6.2) is vital to the flavor of Limburger cheese and also probably necessary for good quality cheddar flavor, as discussed above. Furthermore, bacteria facilitate esterification of methanethiol with short-chain fatty acids, producing esters such as S-methylthioacetate (see Fig. 6.2), which also contribute to cheese aroma [145]. Unfortunately, methanethiol is difficult to use as a flavor, not only because it is very volatile (b.p. 6°C), but also because it reacts with itself in the presence of oxygen to form dimethyl sulfide. Thus it would be difficult to use successfully in flavor applications where air contact is inevitable, such as in cheese powder.

As discussed above (Section 6.5.1), the enzyme methioninase can be added to wet products such as cheese to slowly generate methanethiol from methionine, which would presumably avoid these problems of instability and loss. However, natural methionine is currently very expensive, so that the use of added methionine as a precursor to methanethiol is usually not cost effective.

Methanethiol can also be made by fermentation. Several bacteria, including *Brevibacterium linens* (an obligate aerobe) and *P. freudenreichii* (an anaerobe), are known to produce methanethiol, probably from degradation of methionine [146–149]. One plausible scenario would be to conduct the fermentation using a protein hydrolysate high in methionine as the substrate, followed by steam stripping of the highly volatile methanethiol. With this approach, the cost of the methanethiol would depend on the price of the protein hydrolysate and the concentration of methionine. Unfortunately, the methionine concentration is typically very low in most potential protein substrates. An alternate process could take advantage of the fact that various species of *Penicillium* fungi, including *Penicillium caseicolum*, are also able to produce methanethiol [150].

6.6.8 Diacetyl, Acetoin, and Acetaldehyde

Diacetyl (2,3-butanedione) is found in milk and many other dairy products, where it gives a buttery flavor and also enhances dairy flavors. Its level in good-quality cheddar cheese can vary, but usually is approximately 0.05 mg/100 g [151]. Higher levels do not appear to benefit cheddar flavor. On the other hand, diacetyl is found at much higher levels in butter (0.1 to 0.2 mg/100 g), where it has been proven in taste tests to be very desirable [152]. Although many different lactic acid bacteria can produce diacetyl during fermentation, certain strains are known to produce the largest quantities, and are often used as one component of mixed bacterial starter cultures.

The most-studied bacteria that form diacetyl are certain strains of *Lactococcus lactis*. These strains formerly were classified as *Streptococcus lactis* subsp. *diacetylactis*, but have been found to be significantly different from the other species in the *Streptococcus* genus. Consequently, all the lactic streptococci have been recently moved into their own genus, *Lactococcus* [153,154]. The subspecies status was eliminated [155] because the only major characteristic differentiating these bacteria from the others in their species is the production of diacetyl, and this is regulated by an easily lost plasmid [156]. Since diacetyl production could potentially be unstable, diacetyl levels should be closely monitored.

The lactic acid bacteria form diacetyl by degradation of citric acid, which is naturally present in milk [147,157,158]. Intermediates in the formation of diacetyl include oxaloacetate, pyruvate, and acetolactate [159]. The final step in the diacetyl pathway is the spontaneous oxidation of the acetolactate to form diacetyl. As a consequence of this, diacetyl is only formed under aerobic conditions, with acetoin (3-hydroxy-2-butanone) being formed instead in the absence of air [147,152,154]. The formation of acetoin has not been considered to be desirable, because it was not believed that acetoin has an effect on dairy flavor [58]. However, acetoin does have flavor properties that can be beneficial to butter flavors, so perhaps its flavor effect should be reconsidered. Even though milk contains some citric acid, the addition of extra citrate often stimulates production of diacetyl (or acetoin) and has the added benefit of stimulating the growth of several of the lactic acid bacteria [160–162]. The optimum pH for diacetyl production is less than 5.0, corresponding to the pH optimum of the citric acid transport system [163].

Lactococcus lactis has been treated with ultraviolet radiation to produce stable mutants that are able to produce greater amounts of diacetyl. For example, Kuila and Ranganathan [164] were able to achieve an increase in diacetyl production from 17 ppm to as high as 72 ppm. In addition, they occasionally found that acetaldehyde production also increased from 3 ppm to as much as 34 ppm in these mutants. Acetaldehyde is an important component of the characteristic flavor of yogurt, where it is formed by lactic acid fermentation. Higher acetaldehyde levels could be important for improving starter cultures intended for yogurt production. Other methods for acetaldehyde production are discussed in Chapter 3.

6.6.9 Pyrazines

Pyrazines are discussed briefly here because of their relationship to cheese flavor, but are covered in more detail in Chapter 4. As discussed above (Sections 6.5.1 and 6.5.4), pyrazines are an important component of Swiss and American cheese flavor, but some pyrazines (especially 2-methoxy-3-isopropyl pyrazine; see Fig. 6.4) have also been associated with flavor defects in cheese such as musty and potato-like odors [165,166]. Although 2-methoxy-3-isopropyl pyrazine is associated with dairy flavor defects, a discussion of its synthesis pathway is relevant because of the similarities to other pyrazines.

Bacteria in the *Pseudomonas* genus are well known for their high metabolic activity, and it is no great surprise that they can synthesize pyrazines. *Pseudomonas perolens* and *Pseudomonas taetrolens* can produce several 2-methoxy-3-alkyl pyrazines and tetramethyl pyrazine during stationary growth phase [167]. These authors found that pyrazine production was regulated by the level of phosphate in the medium, and that lactate and pyruvate stimulated production, but the maximum yield was only 0.04 mg/L. Mutagenesis with N-methyl-N-nitro-N-nitrosoguanidine and the leucine analog, 4-azaleucine, improved the yield to approximately 16 mg/L, a 400-fold increase. The presence of L-valine or L-leucine further increased the amount of 2-methoxy-3-isopropyl pyrazine formed. Leucine was converted to valine, which was probably one of the immediate precursors [168]. Using labeling experiments, the other pyrazine precursors were shown to be glycine and methionine, and a biosynthetic pathway was proposed with the nitrogen content of the medium playing a role in regulation [169]. When the ammonium ion content of the fermentation medium was increased from 15 mM to 50 mM, the synthesis of 2-methoxy-3-isopropyl pyrazine was shut off and only tetramethyl pyrazine was produced [170]. The biosynthetic route to pyrazines in bacilli may be similar, because amino acids also are stimulants for production of pyrazines by *Bacillus natto*, which produces approximately 5 to 37 mg/L of pyrazines [171]. With this organism L-threonine stimulates production of 2,5-dimethyl pyrazine, and L-serine stimulates production of tetramethyl pyrazine and trimethyl pyrazine.

As with *Pseudomonas*, the use of mutation to effect the biochemical pathway can be very successful in increasing yields of the pyrazines in other bacteria. The mutagen N-methyl-N-nitro-N-nitrosoguanidine was used to increase the production of tetramethylpyrazine by *Corynebacterium glutamicum*, which normally makes only trace levels [172]. A mutant was selected that had requirements for isoleucine, valine, leucine, and pantothenate. These requirements resulted from the loss of an enzyme of the isoleucine–valine pathway, which blocked the formation of these amino acids and allowed acetoin to accumulate; acetoin was in turn transformed to tetramethyl pyrazine. As much as 3 g/L of tetramethyl pyrazine was produced in a 5-day fermentation, which is a sufficient concentration to lead to crystallization upon cooling. Other microorganisms that might be of potential use in production of pyrazines are *Serratia rubidaea*, *Serratia odorifera*, *Serratia ficaria*, *Cedecea davisae*, and *Aspergillus oryzae* [173,174]. *Aspergillus oryzae* is capable of synthesizing at least 19 pyrazines, but most are made in only low concentrations.

It has been claimed that the final stages of pyrazine formation in cheese are dominated by chemical rather than enzymatic or bacterial reactions [175]. The reaction of dihydroxyacetone with lysine occurs rapidly at room temperature and generates a number of compounds, including pyrazines with a nutty aroma such as 2,5-dimethyl- and ethyldimethylpyrazine. The products of this type of reaction might still be termed natural if natural reactants are used.

6.7 SUMMARY

Dairy and especially cheese flavors are generally very complex. They are consequently difficult to decipher, and, even if their exact compositions were known, they would be difficult and costly to successfully mimic. On the other hand, there are many important components of dairy flavors, such as diacetyl, fatty acids, esters, and 2-heptanone, where an enzymatic or fermentation route is practical, and development of such technology is being driven by increasing sales of all-natural products. In the next few years the number of dairy flavor components that can be synthesized using biotechnology is expected to increase substantially.

REFERENCES

1. American Cancer Society Special Report, *Ca-A Cancer J. Clin.* **34,** 121 (1984).
2. American Heart Association Position Statement, *Circulation* **74,** 1465A (1986).
3. Tada, M., Shinoda, I., and Okai, H., *J. Agric. Food Chem.* **32,** 992 (1984).
4. Huynh-Ba, T., and Philippossian, G., *J. Agric. Food Chem.* **35,** 165 (1987).
5. Seki, T., Kawasaki, Y., Tamura, M., Tada, M., and Okai, H, *J. Agric. Food Chem.* **38,** 25 (1990).
6. Tamura, M., and Okai, H., *J. Agric. Food Chem.* **38,** 1994 (1990).

7. The Nutrition Labeling and Education Act of 1990, Public Law 101-535 (1990).
8. Code of Federal Regulations, 21 CFR 101.22.
9. Regenstein, J.M., and Regenstein, C.E., *Food Technol.* **33**, 89 (1979).
10. Regenstein, J.M., and Regenstein, C.E., *Food Technol.* **42**, 86 (1988).
11. Regenstein, J.M., and Regenstein, C.E., *Food Technol.* **44**, 90 (1990).
12. Puchades, R., Lemieux, L., and Simard, R. E., *J. Food Sci.* **54**, 885 (1989).
13. Bills, D.D., Morgan, M.E., Libbey, L.M., and Day, E.A., *J. Dairy Sci.* **48**, 765 (1965).
14. Scott, R., *Cheesemaking Practice*, 2nd ed. Elsevier Applied Science, Essex, England, 1986.
15. Farrell, H.M., *J. Dairy Sci.* **56**, 1195 (1973).
16. Dulley, J.R., *Aust. J. Dairy Technol.* **29**, 65 (1974).
17. Pfizer Information Sheet No. 2154, *ChyMax*, and Technical Data Sheet *Q&A ChyMax*. Pfizer, Inc., Dairy Products Division, Milwaukee, WI.
18. Kosikowski, F.V., *Cheese and Fermented Milk Foods*, 2nd ed. Kosikowski and Associates, Brooktondale, NY, 1982.
19. Sood, V.K., and Kosikowski, F.V., *J. Dairy Sci.* **62**, 1865 (1979).
20. Marsili, R., *J. Dairy Sci.* **68**, 3155 (1985).
21. Amantea, G.F., Skura, B.J., and Nakai, S., *J. Food Sci.* **51**, 912 (1986).
22. Puchades, R., Lemieux, L., and Simard, R.E., *J. Food Sci.*, **55**, 1555 (1989).
23. Lee, B.H., Laleye, L.C., Simard, R.E., Munsch, M.-H., and Holley, R.A., *J. Food Sci.* **55**, 391 (1990).
24. Aston, J.W., and Creamer, L.K., *N. Z. J. Dairy Sci. Technol.* **21**, 229 (1986).
25. Dilanian, Z.C., *Milchwissenschaft* **35**, 614 (1980).
26. Kowalewska, J., Zelazowska, H., Babuchowski, A., Hammond, E.G., Glatz, B.A., and Ross, F., *J. Dairy Sci.* **68**, 2165 (1985).
27. Dunn, H.C., and Lindsay, R.C., *J. Dairy Sci.* **68**, 2859 (1985).
28. Luksas, A.J., U.S. Patent 3,689,286 (1972).
29. Lubert, D.J., and Frazier, W.C., *J. Dairy Sci.*, **38**, 981 (1955).
30. Manning, D.J., *J. Dairy Res.* **41**, 81 (1974).
31. Manning, D.J., Chapman, H.R., and Hosking, Z.D., *J. Dairy Res.* **43**, 313 (1976).
32. Manning, D.J., and Price, J.C., *J. Dairy Res.* **44**, 357 (1977).
33. Aston, J.W., and Douglas, K., *Aust. J. Dairy Technol.* **38**, 66 (1983).
34. Lindsay, R.C., and Rippe, J.K., *ACS Symp. Ser.* **317**, 286 (1986).
35. Patton, S., Wong, N.P., and Forss, D.A., *J. Dairy Sci.* **41**, 857 (1958).
36. Manning, D.J., and Robinson, H.M., *J. Dairy Res.* **40**, 63 (1973).
37. Mabbitt, L.A., and Zielinska, M., *Int. Dairy Congr.* **2**, 323 (1956).
38. Patton, S., *J. Dairy Sci.* **46**, 856 (1963).
39. Ohren, J.A., and Tuckey, S.L., *J. Dairy Sci.* **52**, 598 (1969).
40. Foda, E.A., Hammond, E.G., Reinbold, G.W., and Hotchkiss, D.K., *J. Dairy Sci.* **57**, 1137 (1974).
41. Bills, D.D., and Day, E.A., *J. Dairy Sci.* **47**, 733 (1964).

42. Jolly, R.C., and Kosikowski, F.V., *J. Agric. Food Chem.* **23,** 1175 (1975).

43. Day, E.A., and Libbey, L.M., *J. Food Sci.* **29,** 583 (1964).

44. Wong, N.P., Ellis, R., LaCroix, D.E., and Alford, J.A., *J. Dairy Sci.* **56,** 636 (1973).

45. Wong, N.P., Ellis, R., and LaCroix, D.E., *J. Dairy Sci.* **58,** 1437 (1975).

46. Vandeweghe, P., and Reineccius, G.A., *J. Agric. Food Chem.* **38,** 1549 (1990).

47. Arctander, S., *Perfume and Flavor Chemicals.* Steffen Arctander, Montclair, NJ 1969.

48. Keeney, M., and Day, E.A., *J. Dairy Sci.* **40,** 874 (1957).

49. Manning, D.J., *J. Dairy Res.* **46,** 523 (1979).

50. Manning, D.J., and Moore, C., *J. Dairy Res.* **46,** 539 (1979).

51. Lin, S.S., *J. Agric. Food Chem.* **24,** 1252 (1976).

52. Kelly, C.D., *J. Dairy Sci.* **20,** 239 (1937).

53. Kelly, C.D., and Marquardt, J.C., *J. Dairy Sci.* **22,** 309 (1939).

54. Tuckey, S.L., and Sahasrabudhe, M.R., *J. Dairy Sci.* **40,** 1329 (1957).

55. Law, B.A., in *Advances in the Microbiology and Biochemistry of Cheese and Fermented Milk* (F.L. Davies, and B.A. Law, eds.). Elsevier, New York, 1984.

56. Grill, H., Patton, S., and Cone, J.F., *J. Dairy Sci.* **49,** 409 (1966).

57. Parliment, T.H., Kolor, M.G., and Rizzo, D.J., *J. Agric. Food Chem.* **30,** 1006 (1982).

58. Margalith, P.Z., *Flavor Microbiology.* Thomas, Springfield, IL, 1981.

59. Liardon, R., Bosset, J.O., and Blanc, B., *Lebensm.-Wiss. Technol.* **15,** 143 (1982).

60. Schmidt, J.L., and Lenoir, J., *Lait* **58,** 355 (1978).

61. Schmidt, J.L., and Lenoir, J., *Lait* **60,** 272 (1980).

62. Marcos, A., Esteban, M.A., Leon, F., and Fernandez-Salguero, J., *J. Dairy Sci.* **62,** 892 (1979).

63. Ismail, A.A., and Hansen, K., *Milchwissenschaft* **27,** 556 (1972).

64. Lenoir, J., *Ann. Technol. Agric.* **12,** 51 (1963).

65. Farahat, S.M., Rabie, A.M., and Farag, A.A., *Food Chem.* **36,** 169 (1990).

66. Anderson, D.F., and Day, E.A., *J. Agric. Food Chem.* **14,** 241 (1966).

67. Dumont, J.P., Roger, S., and Adda, J., *Lait* **56,** 595 (1976).

68. Sloot, D., and Harkes, P.D., *J. Agric. Food Chem.* **23,** 356 (1975).

69. Moinas, M., Groux, M., and Horman, I., *Lait* **53,** 601 (1973).

70. Kurtz, F.E., Hupfer, J.A., Corbin, E.A., Hargrove, R.E., and Walter, H.E., *J. Dairy Sci.* **42,** 1008 (1959).

71. Langler, J.E., and Day, E.A., *J. Dairy Sci.* **49,** 91 (1966).

72. Hintz, P.C., Slatter, W.L., and Harper, W.J., *J. Dairy Sci.* **39,** 235 (1956).

73. Langler, J.E., Libbey, L.M., and Day, E.A., *J. Agric. Food Chem.* **15,** 386 (1967).

74. Sloot, D., and Hofman, H.J., *J. Agric. Food Chem.* **23,** 358 (1975).

75. Peters, I., and Nelson, F.E., *Int. Dairy Congr.*, [*Proc.*], *12th, 1949*, vol. 2, p. 567 (1949).

76. Farnham, M.G., U.S. Patent 2,531,329 (1950).
77. Moskowitz, G.J., in *The Analysis and Control of Less Desirable Flavours in Foods and Beverages* G. Charalambous, (ed.). Academic Press, New York, 1980.
78. Harper, W.J., and Gould, I.A., *Butter, Cheese, Milk Prod. J.* **43**, 22 (1952).
79. Lee, K.-C.M., Shi, H., Huang, A.-S., Carlin, J.T., Ho, C.-T., and Chang, S.S., *ACS Symp. Ser.* **317**, 370 (1986).
80. *Lipase UL-8*, Tech. Bull. Amano International Enzymes Co., Inc., Troy, VA.
81. Eaton, D.C., unpublished research (1990).
82. Mills, O.E., and Thomas, T.D., *N. Z. J. Dairy Sci. Technol.* **15**, 131 (1980).
83. Monnet, V., Bockelmann, W., Gripon J.C., and Teuber, M., *Appl. Microbiol. Biotechnol.* **31**, 112 (1989).
84. Gordon, D.F., and Speck, M.L., *Appl. Microbiol.* **13**, 537 (1965).
85. Law, B.A., and Wigmore, A.S., *J. Soc. Dairy Technol.* **35**, 75 (1982).
86. Chr. Hansen's Laboratory, Milwaukee, WI.
87. Amano International Enzyme Co., Inc., Troy, VA.
88. Starkle, M., *Biochem. Z.* **151**, 371 (1924).
89. Knight, S.G., U.S. Patent 3,100,153 (1963).
90. Gehrig, R.F., and Knight, S.G., *Appl. Microbiol.* **11**, 166 (1963).
91. Luksas, A.J., U.S. Patent 3,720,520 (1973).
92. Watts, J.C., and Nelson, J.H., U.S. Patent 3,072,488 (1963).
93. Dwivedi, B.K., and Kinsella, J.E., *J. Food Sci.* **39**, 620 (1974).
94. Dwivedi, B.K., and Kinsella, J.E., *J. Food Sci.* **39**, 83 (1974).
95. Creuly, C., Larroche, C., and Gros, J.-B., *Appl. Microbiol. Biotechnol.* **34**, 20 (1990).
96. Lawrence, R.C., and Bailey, R.W., *Biochim. Biophys. Acta* **208**, 77 (1970).
97. Larroche, C., Tallu, B., and Gros, J.B., *J. Ind. Microbiol.* **3**, 1 (1988).
98. Larroche, C., Arpah, M., and Gros, J.B., *Enzyme Microb. Technol.* **11**, 106 (1989).
99. Jolly, R., and Kosikowski, F.V., *J. Food Sci.* **40**, 285 (1975).
100. Lanza, E., Ko, K.H., and Palmer, J.K., *J. Agric. Food Chem.* **24**, 1247 (1976).
101. Tahara, S., Fujiwara, K., Ishizaka, H., Mizutani, J., and Obata, Y., *Agric. Biol. Chem.* **36**, 2585 (1972).
102. Gervais, P., and Battut, G., *Appl. Environ. Microbiol.* **55**, 2939 (1989).
103. Gervais, P., and Pecot, I., *J. Biotechnol.* **19**, 211 (1991).
104. Labows, J.N., Webster, G., and McGinley, K.J., U.S. Patent 4,542,097 (1985).
105. Farbood, M.I., and Willis, B.J., U.S. Patent 4,560,656 (1985).
106. Van der Schaft, P.H., ter Burg, N., van den Bosch, S., Cohen, A.M., *Appl. Microbiol. Biotechnol.* **36**, 712 (1992).
107. De Vries, W., van Wijck-Kapteyn, W.M.C., and Oosterhuis, S.K.H., *J. Gen. Microbiol.* **81**, 69 (1974).
108. Distler, W., and Kroncke, A., *Arch. Oral Biol.* **26**, 123 (1981).
109. Mays, T.D., and Fornili, P.N., U.S. Patent 4,794,080 (1988).
110. Sherman, J.M., and Shaw, R.H., U.S. Patent 1,459,959 (1923).

111. Sherman, J.M., and Shaw, R.H., U.S. Patent 1,470,885 (1923).

112. Sherman, J.M., U.S. Patent 1,865,146 (1932).

113. Stiles, H.R., U.S. Patent 1,913,346 (1933).

114. Ahern, W.P., Andrist, D.F., and Skogerson, L.E., U.S. Patent 4,743,453 (1988).

115. Bodie, E.A., Goodman, N., and Schwartz, R.D., *J. Ind. Microbiol.* **44,** 913 (1982).

116. Bodie, E.A., Anderson, T.M., Goodman, N., and Schwartz, R.D., *Appl. Microbiol. Biotechnol.* **25,** 434 (1987).

117. Linehan, B., Scheifinger, C.C., and Wolin, M.J., *Appl. Environ. Microbiol.* **35,** 317 (1978).

118. Mink, R.W., Patterson, J.A., and Hespell, R.B., *Appl. Environ. Microbiol.* **44,** 913 (1982).

119. Wallace, R.J., *J. Gen. Microbiol.* **107,** 45 (1978).

120. Scheifinger, C.C., Linehan, B., and Wolin, M.J., *Appl. Microbiol.* **29,** 480 (1975).

121. Marounek, M., and Wallace, R.J., *J. Gen. Microbiol.* **130,** 223 (1984).

122. Wimpenny, J.W.T., and Samah, O.A., *J. Gen. Microbiol.* **108,** 329 (1978).

123. Samah, O.A., and Wimpenny, J.W.T., *J. Gen. Microbiol.* **128,** 355 (1982).

124. Eaton, D.C., and Gabelman, A., U.S. Patent 5,127,736 (1992).

125. Das, N.K., U.S. Patent 4,138,498 (1979).

126. Monot, F., Engasser, J-M., and Petitdemange, H., *Appl. Microbiol. Biotechnol.* **19,** 422 (1984).

127. Holt, R.A., Stephens, G.M., and Morris, J.G., *Appl. Environ. Microbiol.* **48,** 1166 (1984).

128. Huang, L., Gibbins, L.N., and Forsberg, C.W., *Appl. Environ. Microbiol.* **50,** 1043 (1985).

129. Junelles, A-M., Janati-Idrissi, R., El Kanouni, A., Petitdemange, H., and Gay, R., *Biotechnol. Lett.* **9,** 175 (1987).

130. Fayolle, F., Michel, D., Marchel, and Ballerini, D., European Patent 350,355 (1990).

131. Bryant, M.P., and Burkey, L.A., *J. Bacteriol.* **71,** 43 (1956).

132. Smith, L.D.S., Sugiyama, H., *Botulism: the Organism, Its Toxins, the Disease,* 2nd Ed., Charles Thomas, Springfield, IL, 1988.

133. McCroskey, L.M., Hatheway, C.L., Fenicia, L., Pasolini, B., Aureli, P., *J. Clin. Microbiol.* **23,** 201 (1986).

134. Hall, J.D., McCroskey, L.M., Pincomb, B.J., Hatheway, C.L., *J. Clin. Microbiol.* **21,** 654 (1985).

135. Zhou, Y., Sugiyama, H., Johnson, E.A., *Appl. Environ. Microbiol.* **59,** 3825 (1993).

136. Carrea, G., *Trends Biotechnol.* **2,** 102 (1984).

137. Gatfield, I., and Sand, T., German Patent Application 3,108,927 (1982).

138. Fayolle, F., Marchal, R., Monot, F., Blanches, D., and Ballerini, D., *Enzyme Microbiol. Technol.* **13,** 215 (1991).

139. Welsh, F.W., Williams, R.E., and Dawson, K.H., *J. Food Sci.* **55,** 1679 (1990).

140. Tsujisaka, Y., Japanese Patent 81,000,038 (1981).

141. Carta, G., Gainer, J.L., and Benton, A.H., *Biotechnol. Bioeng.* **37**, 1004 (1991).

142. Blair, C., M.Sci. Thesis, University of Virginia, Charlottesville (1989).

143. Kyotani, S., Nakashima, T., Izumoto, E., and Fukuda, H., *J. Ferment. Bioeng.* **71**, 286 (1991).

144. Nakashima, T., Fukuda, H., Kyotani, S., and Morikawa, H., *J. Ferment. Technol.* **66**, 441 (1988).

145. Cuer, A., Dauphin, G., Kergomard, A., Demount, J.P., and Adda, *J., Agric. Bioi. Chem.* **43**, 1783 (1979).

146. Tokita, F., and Hosono, A., *Jpn. J. Zootech. Sci.* **39**, 127 (1968).

147. Sharpe, M.E., Law, B.A., and Phillips, B.A., *J. Gen. Microbiol.* **94**, 430 (1976).

148. Sharpe, M.E., Law, B.A., Phillips, B.A., and Pitcher, D.G., *J. Gen. Microbiol.* **101**, 345 (1977).

149. Cuer, A., Dauphin, G., Kergomard, A., Dumont, J.P., and Adda, J., *Appl. Environ. Microbiol.* **38**, 332 (1979).

150. Tsugo, T., and Matsuoka, H., *Proc. Int. Dairy Congr., 16th, 1962*, vol. IV, p. 385 (1962).

151. Calbert, H.E., and Price, W.V., *J. Dairy Sci.* **32**, 515 (1949).

152. Pette, J.W., *Int. Dairy Congr. Proc. 12th, 1949*, vol. 2, p. 572 (1949).

153. Schleifer, K.H., in *Bergey's Manual of Systematic Bacteriology*, (P.H.A. Sneath, ed.), vol. 2. Williams & Wilkins, Baltimore, MD. 1986.

154. Schleifer, K.H., Kraus, J., Dvorak, C., Kilpper-Bali, R., Collins, M.O., and Fischer, W., *Syst. Appl. Microbiol.* **6**, 183 (1985).

155. Hardie, J.M., in *Bergey's Manual of Systematic Bacteriology*, (P.H.A. Sneath ed.), vol. 2. Williams & Wilkins, Baltimore, MD. 1986.

156. Moller-Madsen, A.A., and Jensen, H., *Contrib. Int. Dairy Congr., 16th, 1962* p. 255 (1962).

157. Pette, J.W., *Neth. Milk Dairy J.* **2**, 12 (1948).

158. Federov, M.V., and Kruglova, L.V., *Dokl. Acad. Nauk SSSR* **103**, 161 (1955); *Dairy Sci. Abstr.* **17**, 1036 (1955).

159. Seitz, E.W., Sandine, W.E., Elliker, P.R., and Day, E.A., *Can. J. Microbiol.* **9**, 431 (1963).

160. Harvey, R.J., and Collins, E.B., *J. Bacteriol.* **86**, 1301 (1963).

161. Cogan, T.M., *J. Appl. Bacteriol.* **63**, 551 (1987).

162. Schmitt, P., and Divies, C., *J. Ferment. Bioeng.* **71**, 72 (1991).

163. Harvey, R.J., and Collins, E.B., *J. Bacteriol.* **83**, 1005 (1962).

164. Kuila, R.K., and Ranganathan, B., *J. Dairy Sci.* **61**, 379 (1978).

165. Morgan, M.E., Libbey, L.M., and Scanlan, R.A., *J. Dairy Sci.* **55**, 666 (1972).

166. Dumont, J.P., Roger, S., and Adda, J., *Lait* **55**, 479 (1975).

167. McIver, R.C., and Reineccius, G.A., in *Biogeneration of Aromas* (T.H. Parliment and R. Croteau, eds.). Am. Chem. Soc., Washington, DC, 1986.

168. Gallois, A., Kergomard, A., and Adda, J., *Food Chem.* **28**, 299 (1988).

169. Cheng T.-B., Reineccius G.A., Bjorklund, J.A., and Leete, E., *J. Agric. Food Chem.* **39**, 1009 (1991).

170. McIver, R.C., Ph.D. Thesis, University of Minnesota, Minneapolis, (1990).

171. Ito, T., Sugawara, E., Miyanohara, J.-I., Sakurai, Y., and Odagiri, S., *Nippon Shokuhin Kogyo Gakkaishi* **36,** 762 (1989).

172. Demain, A.L., Jackson, M., and Trenner, N.R., *J. Bacteriol.* **94,** 323 (1967).

173. Gallois, A., and Grimont, P.A.D., *Appl. Environ. Microbiol.* **50,** 1048 (1985).

174. Liardon, R., and Ledermann, S., *Z. Lebensm. Unters. Forsch.* **170,** 208 (1980).

175. Griffith, R., and Hammond, E.G., *J. Dairy Sci.* **72,** 604 (1989).

176. Michel, T.A., and Macy, J.M., *Abst. Annu. Mtg. Am. Soc. Microbiol. 1988,* Vol. 88, p. 215 (1988).

Production of Food Colorants by Fermentation

GUNNARD JACOBSON and JOHN WASILESKI
Universal Foods Corporation
Milwaukee, Wisconsin

In an age of increased food processing, color is an extremely important determinant in the palatability and appeal of food. Synthetic pigments traditionally used by food processors continue to be used with success, but there is increasing consumer preference for natural food additives. Despite the obvious opportunity, penetration into the food industry by fermentation-derived pigments is limited at present to segments of the β-carotene and riboflavin markets. This largely is due to the higher production cost of fermentation pigments than synthetic pigments or pigments extracted from natural sources. Biological innovations will improve the economics of pigment production by creating new or better microorganisms that will lend themselves to traditional high-productivity fermentations. Technical innovations will improve the productivity of novel fermentation systems such as plant cell culture or microalgal systems to market-competitive levels. Additional opportunities for fermentation-derived pigments exist in rare or difficult-to-synthesize pigments such as phycocyanins, xanthophylls, or Monascus *pigments, where biological improvements will also play a role.*

7.1 INTRODUCTION

People visually perceive an object as a function of the wavelengths of light that it reflects or emits. These wavelengths of light are detected by appropriate photoreceptors in the eye's retina. Color, then, is defined by that very small portion of the electromagnetic spectrum that the human eye can detect. The significance of this is immense. The importance of color to the human percep-

Bioprocess Production of Flavor, Fragrance, and Color Ingredients, Edited by Alan Gabelman,
ISBN 0-471-03821-0 © 1994 John Wiley & Sons, Inc.

tion of self and the surrounding world extends from the use of red ochre at Cro-Magnon man's burial sites; to the decree of Augustus Caesar that only the Emperor of Rome and his family could wear Tyrian purple; to the color some might choose to paint the nurseries of their infants; and, to the colors we associate with death, purity, and anger. Many of these associations are cultural, but research has shown that color does have the real ability to modify our moods and perception of pain.

With respect to foods, color can influence our taste thresholds and ability to identify flavors. Man has always used color to make preliminary judgments on his food's ripeness, taste, palatability, and safety. An orange or banana tinged with green is thought of as unripe, a bright red cherry as sweet, and a greenish or brown piece of raw beefsteak as possibly unfit to eat [1,2]. Also there is subconscious association of the requisite color for a given flavor. Orange flavors must be bright reddish orange; mint, yellow-green; peach, light dull gold; cherry, bright-bluish red; plum, reddish-navy; chocolate, dark reddish-brown; and so on. [3]. These associations must be maintained in an age of industrialization that provides consumers with a wide variety of prepackaged processed foods.

The color/flavor link rooted in the consumption of natural foods is now reinforced artificially by food processors who manufacture artificially flavored foods and beverages. In addition, today's food processors find it necessary to use added food colorants for other reasons. These include

1. Replacement of color lost during processing, storage or cooking by the consumer
2. Intensification of natural colors
3. Correction of seasonal variations of natural color in a product so that a consistent product is presented to the customer

The concept of food color enhancement for the enjoyment and festivity of eating a well-prepared meal has been exploited by cooks, food processors, and food purveyors for centuries. The ancient Egyptians made colored candies and used carmine (probably derived from the Madder plant or the scale insect, Kermes) to color spices. The Roman historian Pliny the Elder described the coloration of wine and bread with berries. In medieval England, the wealthy imported from Alexandria sugar colored with madder and kermes, both anthraquinone dyes, and with Tyrian purple, an indigoid dye [1,3,4].

The onset of the Industrial Revolution and the resultant movement away from an agrarian lifestyle led from foods being grown and produced at home to foods being acquired, processed, and distributed by traders, with the resultant loss of color and flavor from processing, storage, and distribution. The resulting need to preserve the color/flavor link, unfortunately, led to the widespread use of colorants to adulterate foods and mislead the consumer. In England, the problems generated by unscrupulous processors were so bad that they resulted in the earliest legislation concerning food ingredients, the Adulteration of Food and Drink Act of 1860. Examples of these excesses include cheese colored with red lead, pickles colored with copper sulfate, and candy colored with lead and copper salts. Coloring allowed some substances to imitate others (e.g., thorn leaves were tinted with copper oxide and sold as tea).

The development of aniline dye chemistry in the 1850s resulted in the rapid application of these dyes in coloring foodstuffs. Their range of colors was far

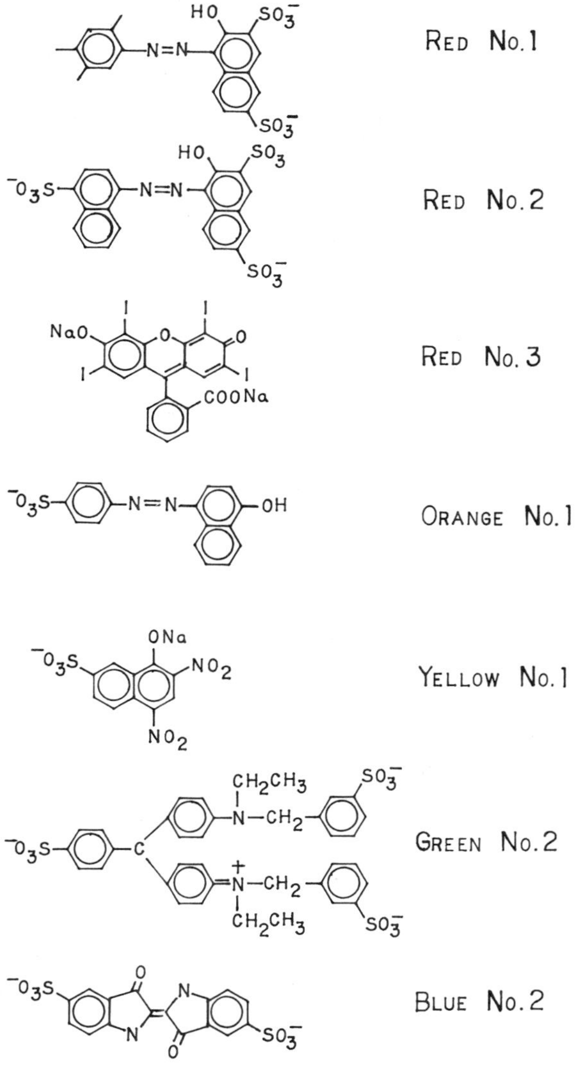

FIG. 7.1 Food dyes allowed by the 1906 Pure Food and Drug Act: Ponceau 3R (Red No. 1); Amaranth (Red No. 2); Erythrosine (Red No. 3); Orange 1; Naphthol Yellow S (Yellow No. 1); Light Green SF Yellowish (Green No. 2); Indigo Carmine (Blue No. 2).

greater than natural pigments, they were more stable to light, heat, and oxidation, and their tinctorial power was far greater. In the United States, the Pure Food and Drug Act of 1906 legalized the food use of the seven dyes shown in Figure 7.1. Of these dyes, only Indigo Carmine (FD&C Blue No. 2) is chemically similar to dyes found in nature; it is the sodium disulfonic acid salt of indigo, a plant pigment used since Egyptian times as a dye for fabrics and cosmetics. Today, of these original seven dyes, only FD&C Red No. 3 and FD&C Blue No. 2 are currently in use [3,5]. Since 1906, many other synthetic dyes have been developed successfully for use as food colorants. Singly or in combination they provide the food processor with a wide range of safe, economical colors that increase the appeal of a wide variety of foods and beverages.

In addition to FD&C colors, pigments identical to those found in nature have also been synthesized for use as food dyes. For the purpose of this chapter, we refer to pigments that are normal components of foods but are chemically synthesized as *nature identical* and to those food pigments that are extracted from food sources or other natural sources such as flowers, algae, fungi, or bacteria as "natural."

The advent of the biotechnology revolution is leading the food industry to explore new approaches to food handling, processing, and packaging. Although there has always been an interest in nature identical colors, biotechnology now makes it possible to produce natural colors by fermentation or tissue culture cheaply enough to be competitive with FD&C pigments.

7.2 NATURAL COLORS IN FOODS

MacKinney and Little [6] classify the majority of the colors occurring naturally in foods into five basic groups (Table 7.1). Four of these groups are pigments associated with the plant or animal tissues from which the food was derived. The fifth is considered a derived food colorant, that is, the result of heat or other processing, on food components such as carbohydrates. Examples are the formation of Maillard reaction products between amines and carbohydrates and pigmented compounds arising from aldehyde formation (melanoids and caramels). This latter group will not be discussed in this chapter.

Figure 7.2 shows basic structures of the tetrapyrrole-based food pigments. These are responsible for the greens in vegetable-derived foods and the pink-to-red and brown colors in animal flesh foods. The presence of and the oxidation state of the metal ion in the tetrapyrrole complex is important to the color imparted by the chromophore. In the case of chlorophyll, the replacement of the magnesium ion (Mg^+) by two hydrogen atoms results in the formation of pheophytine. This can occur during canning or in the presence of weak acids. Aesthetically, this results in a decrease in the intensity of the green color and a shift in color to an olive green. Chlorophyll pigments are also subject to photooxidation; dehydrated spices like parsley, oregano, can lose color if not protected from light. The iron moiety of heme is subject to oxidation from the

TABLE 7.1 Naturally Occurring Food Pigments

Component	Occurrence
TETRAPYRROLES	
Chlorophylls	Green leafy vegetables
Hemes	Meats
ISOPRENOIDS (CAROTENOIDS)	
Carotenes	Carrots, tomatoes
Xanthophylls	Capsicum peppers, salmon
BENZOPYRANS	
Anthocyanins	Grapes, berries, apples
Flavones and flavanones	Nuts, onion skins, tea
BETALAINES	
Betacyanin	Beets
Betaxanthines	Beets
PROCESS DERIVED PIGMENTS	
Caramels	Honey, syrups
Melanoids	Syrups

ferrous to the ferric state. This shift in iron oxidation state is accompanied by a shift from a red to a brownish color. Color degradation can also occur when the porphyrin ring is opened, for example, by oxidation resulting in the formation of bilins. This can give some meats a greenish cast [6].

The hemes in blood have no use as added food colors because they are not permitted in most countries. However, as blood they do find use as ingredients in such foods as blood puddings and soups. Chlorophyll has some use as a food colorant in confectionaries but also finds its main application as a food ingredient and flavorant, such as the use of spinach juice to pigment pasta.

There has been interest in the bilin pigments associated with algae and blue-green algae (cyanobacteria) for use in confectioneries and frozen sherbets. Work in Japan suggests that the production of these green-to-blue pigments may be amenable to large scale production by fermentation.

Figure 7.3 depicts the structures of several of the isoprenoid-based pigments known as carotenoids. These carotenoids and xanthophylls are probably more widely distributed in nature than any other group of pigments. Carotenoids are found in bacteria, fungi, yeasts, and plants. Some of these pigments are incorporated readily into the external pigmentation patterns of many animal species, including insects, crustaceans, fish, and birds. In some cases (like salmonid fishes), carotenoids are deposited in the flesh and ova. Several carotenoids have pro-vitamin A activity and consequently are important to the nutritional as well as the aesthetic value of the foods containing them [7–9]. Traditionally,

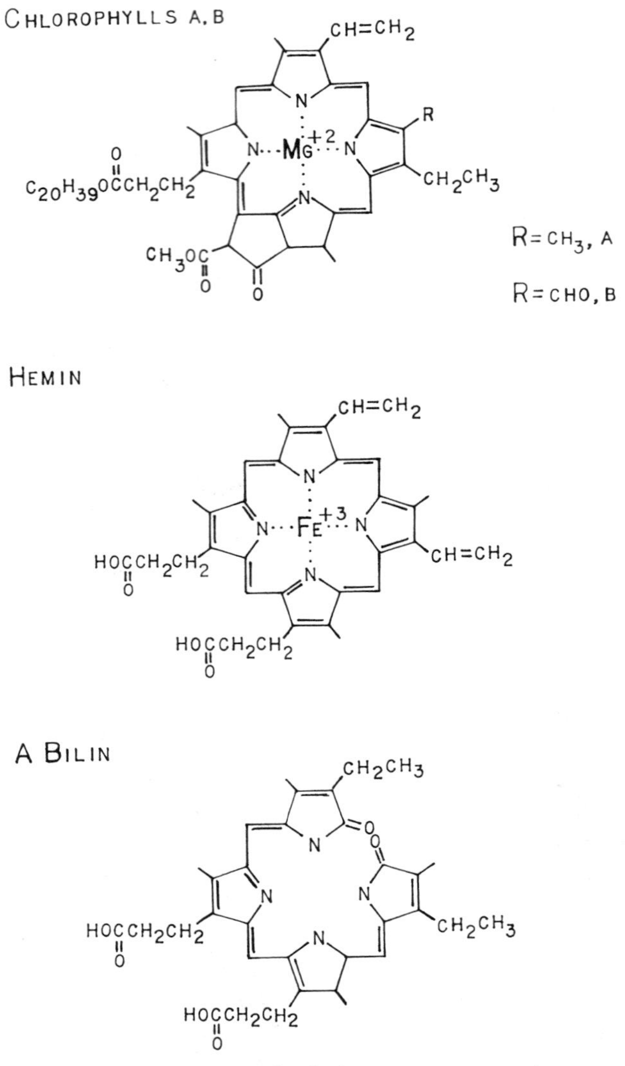

CHLOROPHYLLS A,B

$R = CH_3$, A

$R = CHO$, B

HEMIN

A BILIN

FIG. 7.2 Natural food pigments: tetrapyrroles.

carotenoids have been used as food colorants in many cultures. These naturally occurring food colorants include bixin obtained from annatto (achiote) seeds and crocetin obtained from the spice saffron. Carotenoids are relatively economical to synthesize, and synthesized carotenoids, such as β-carotene, β-apo-8'-carotenal, canthaxanthin, and astaxanthin, have been marketed. They have found application in a wide range of products, including dairy products, frostings, salad dressings, egg products, and beverages. The advantages of carotenoids as colorants are that they have good stability at the pH of most foods,

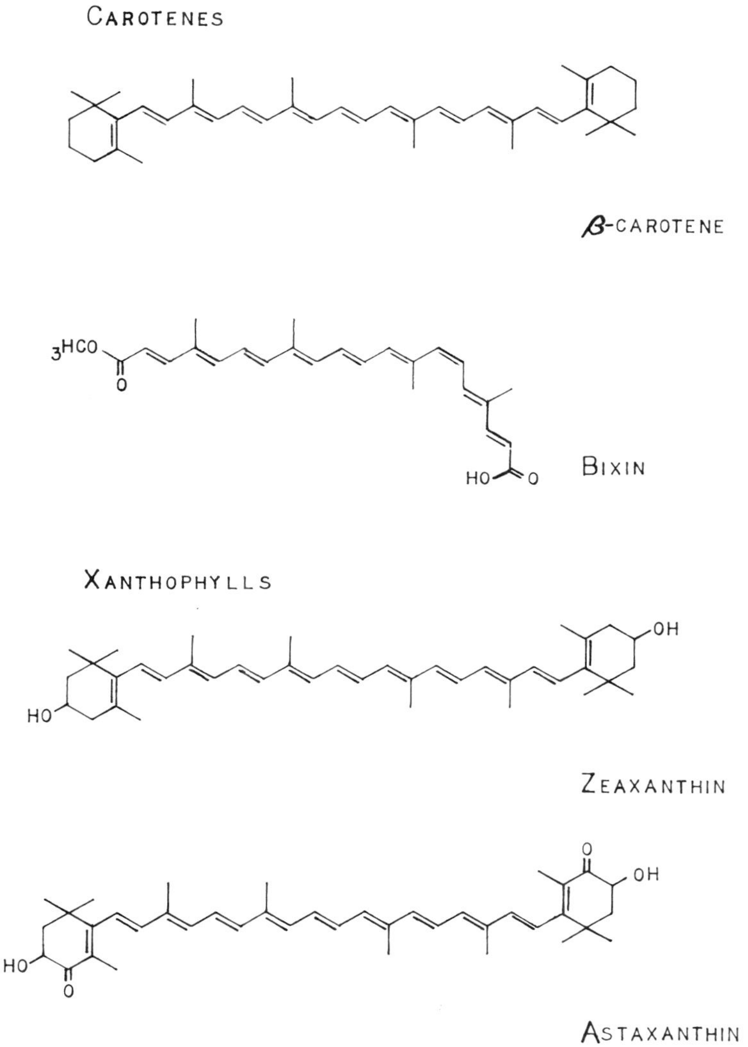
CAROTENES
β-CAROTENE
3HCO
O
HO
O
BIXIN
XANTHOPHYLLS
OH
HO
ZEAXANTHIN
O
OH
HO
O
ASTAXANTHIN

FIG. 7.3 Natural food pigments: carotenoids.

they are unaffected by reducing agents such as ascorbic acid, and their yellow-to-red color. Their disadvantages are that they are more expensive than the azo food dyes, they are more easily oxidized, and they don't have the range of colors available in synthetic dyes [10–12].

Benzopyran pigments include the flavonoids and anthocyanins. Representative structures of these compounds are shown in Figure 7.4. The flavonones and flavones have low tinctorial power and do not contribute greatly to food color. Some are substrates for polyphenolase enzymes and can contribute to enzymatic browning in some foods. In the presence of such metals as iron,

ANTHOCYANINS

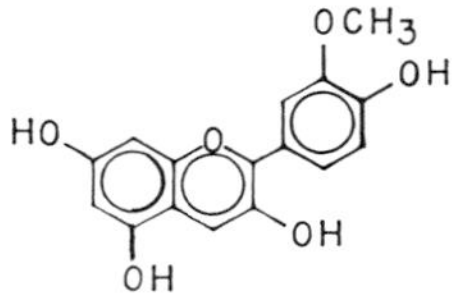

PEONIDIN

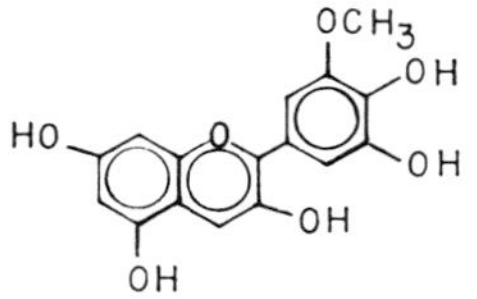

ENOCIONIN

FLAVONOIDS

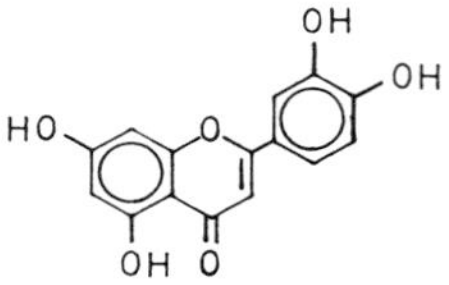

LUTEOLIN

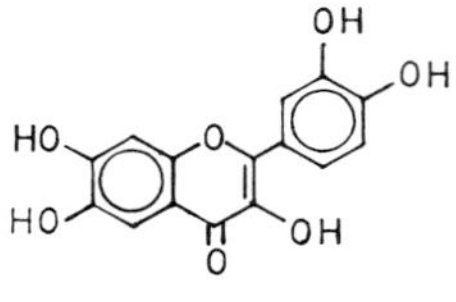

QUERCITIN

FIG. 7.4 Natural food pigments: benzopyrans.

some flavonoids give rise to green-to-blue discolorations [6]. The anthocyanins, however, are responsible for the orange, red, and blue-red colors of such fruits and vegetables as mangos, grapes, cranberries, strawberries, and red cabbage [13]. Anthocyanins extracted from grape skins have been used as a source of food color since the 1880s and cranberry juice concentrates have been used more recently to color beverages that are compatible with the flavor, color, and pH of the concentrate. Other applications include confectioneries and jams and jellies. The color of anthocyanins is pH dependent, with more intense colors at lower pH. These pigments are heat sensitive and subject to photooxidation. Consequently, adequate quantities of pigment must be added to account for losses during processing and storage. Some effort has been put into the production of anthocyanin pigments by plant cell culture [13]. Chapter 6 deals with the growing interest in plant cell culture.

The betalaines are a group of yellow, red, and purple betacyanin and betaxanthine pigments found in the plant family *Centrospermae*. This family includes the table beet (*Beta vulgaris*) and the American pokeberry (*Phytolacca americana*) [14,15]. Figure 7.5 shows structures of several betalaines. These pigments are also heat, light, and oxygen sensitive, and their use as food pigments has been limited to foods with short shelf lives such as dairy products and meats or in low pH products such as soft drinks. They are also sensitive to oxidizing agents such as sulfur dioxide. The production of these pigments by plant cell culture is also under study [16,17].

The use of natural pigments as color additives depends primarily on their

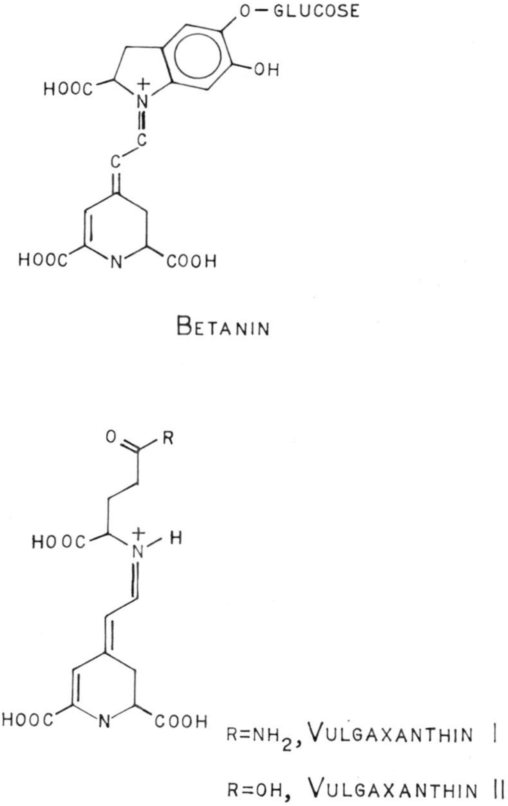

FIG. 7.5 Natural food pigments: betalaines.

economical extraction from natural sources. During extraction, they are subject to the same destructive forces as they are in application: oxidation, photooxidation, formation or loss of metal complexes, isomerization, and so forth. In addition, in concentrated form they are more subject to precipitation and polymerization.

Impacting the economics of natural pigments is the fact that, although many are good pigments, their low concentrations as extracted from natural sources result in lower tinctorial strengths. Also, they are less stable than synthetic azo dyes. Consequently, higher levels of colorant must be used to give the desired color at the time of purchase and consumption.

Some natural pigments have desirable colors but have other problems associated with their use, examples include riboflavin (vitamin B_2) and curcumin. Riboflavin has an attractive yellow color, but it has a green fluorescence and is bitter to the taste. Curcumin is the bright yellow pigment in the spice turmeric, but as extracted from *Curcuma* roots, it must be debittered if its odor and sharp taste are not desired [14,18–20].

Nevertheless, with the enhanced worldwide interest in developing food colorants from natural sources, these and many alternative or nontraditional (by Western standards) pigments are being evaluated. Improved methods to extract and stabilize these pigments are part of these efforts. Novel organisms from which food colorants can be extracted include the Cape Jasmine (*Gardenia jasminoides*), the fungi *Monascus purpureus* and *Monascus anka*, and various algae. These organisms produce carotenoids, iridoids, flavonoids, phycobiliproteins, and other pigments [21,22]. These pigments can range in color from yellow to orange and red to purple. The phycobiliproteins contain both red phycoerythrins and blue phycocyanins. Methods to improve extraction include such techniques as the addition of pectinases, cellulases, and proteases to crude extracts to dissociate pigment complexes; fermentation of extracts with yeast such as *Candida utilis* or *Saccharomyces cerevisiae* to remove sugars and nitrogenous compounds; supercritical carbon dioxide extraction; and a variety of membrane technologies, such as micro- and ultrafiltration and reverse osmosis, to purify and concentrate pigment extracts. Means to stabilize pigments include the use of additives such as ascorbic acid, metal ions, and various organic acids; encapsulating pigments into or adsorbing them onto starch, gelatin, alginate, silicate, or other carriers; the use of cyclodextrins to form inclusion complexes; condensation of anthocyanin or betalaine pigments with natural polyphenols derived from tea, for example, to yield more stable complexes; and chemical modification of pigments such as the acetylation of anthocyanins [18,21–24].

For many reasons natural pigments are more expensive than synthetic dyes by six- to eightfold [18]. Exceptions to this are some of the nature-identical pigments such as β-carotene, β-apo-8'-carotenal, and canthaxanthin, whose synthesis and commercialization took place in the 1950s and 1960s [7]. Nevertheless, naturally occurring pigments finding commercial application include carotenoids extracted from the flowers of the Aztec marigold (*Tagetes erecta*), annatto seed (*Bixa orellana*), and paprika (*Capsicum annuum*); anthocyanins from grapes (*Vitis vinifera* and *Vitis lambrusca*) and from the American cranberry (*Vaccinium macrocarpon*); and betalaines from the table beet (*B. vulgaris*).

Table 7.2 lists the FD&C dyes currently approved for food use in the United States and suggested natural alternatives to their use [25,26]. Some of these alternatives are current topics of research in that the production of betalaines and anthocyanins by plant cell culture (PCC) is under study [13,16, 17,27]. Other pigment-producing plants may also lend themselves to PCC, for example, *Indigofera tinctoria*, the plant from which indigo has been traditionally extracted. However, tryptophan-based pigments, like indigo, are well known in the microbial community and microbial routes may be more cost-effective than PCC. Other pigments may also be found from microbial sources. It has been known for some time that many carotenes, including β-carotene, and xanthophylls are present in algae, fungi, and bacteria. In addition, fungi such as the *Trichoderma polysporum* and *Trichoderma viride* [28] have been demonstrated to produce the anthraquinone pigment emodin.

It is these bacteria, fungi, and algae that lend themselves to industrial fermentations. To the best of our knowledge, the only food-approved colors with a history of being produced commercially by traditional bacterial, fungal, or algal fermentation are β-carotene and riboflavin. Some authors [22] have expressed doubts about the feasibility of pigment production by fermentation due to "production, extraction and legislative costs"; however, it is likely that fermentation processes will be competitive in some areas. Initially, these will be specialty areas involving compounds whose presence in nature is limited, expensive to extract, concentrate, and purify, or whose synthesis is expensive. With legislative approval such compounds will find wider application in foods/feeds, drugs, and cosmetics. To be competitive with synthetic nature-identical or extracted pigments these fermentation processes will necessarily have to be highly efficient. The old allied sciences of genetics, microbiology, and fermentation engineering have made enormous strides in the last 50 years. The products of this alliance include a diversity of high-value products such as antibiotics, amino acids, and vitamins and low-value commodity chemicals such as ethanol and citric acid. The new science of molecular biology may also play an important role in pigment production. Under the appropriate market conditions, it would permit, for example, the transfer of the genes for canthaxanthin synthesis from *Brevibacterium* KY-4313 into a food-grade microorganism such as Baker's yeast, *S. cerevisiae* [29]. However, it is not likely that this approach will be exploited in the near future.

The following section examines those efforts being made by fermentation processes to produce pigments that are competitive with extracted natural pigments or nature-identical pigments. The production of alternative natural pigments targeted as substitutes for synthetic dyes is also considered.

TABLE 7.2 FD&C Dyes Currently Approved for Food Use in the United States and their Natural Alternatives

| Current U.S. Approved Synthetic Dyes | | | Possible Natural Substitutes | | |
Dye	Type	Color	Current U.S. Approval	Not approved in U.S.	Pigment Types
FD&C Red 3[a]	Xanthine	Watermelon red	Beets, carmine, grapes	*Monascus* pigments	Betalaines, anthraquinone, anthocyanins
FD&C Red 40	Monazo	Orange red	Beets, carmine, grapes	*Monascus* pigments	Betalaines, anthraquinone, anthocyanins
FD&C Yellow 6	Monoazo	Orange	Annatto, β-carotene, paprika		Carotenes, xanthophylls
FD&C Yellow 5	Pyrazolone	Lemon yellow	Turmeric, riboflavin	*Monascus* pigments	Curcumin, alloxan
FD&C Green 3	Triphenyl methane	Sea green	Chlorophylls	Algal bilins	Chlorophylls, bilins
FD&C Blue 1	Triphenyl methane	Bright blue	—	Algal bilins	Chlorophylls, bilins
FD&C Blue 2	Indigoid	Royal blue	—	Algal bilins	Chlorophylls, bilins

[a]The FDA is currently considering delisting this dye. Adapted from Freund et al. [25,26].

7.3 NATURAL COLORS CURRENTLY PRODUCED BY FERMENTATION

7.3.1 Riboflavin

Riboflavin (vitamin B_2, Fig. 7.6), has a variety of applications as a yellow food colorant. Its use is permitted in most countries. For some applications riboflavin phosphate is preferred because of its higher solubility in water; both have vitamin B_2 activity. Applications include dressings, sherbet, beverages, instant desserts, ice creams, and tablets and other coated products [8,14,30]. Riboflavin has a special affinity for cereal-based products, but its use in these applications is somewhat limited due to its slight odor and natural bitter taste. Usage levels in foods must be below the taste threshold or be compatible with the food formulation (e.g., highly sugared beverages). In aqueous solution, riboflavin has a bright yellow-green fluorescence that can be quenched by using mildly acidic or alkaline conditions. Riboflavin is produced both fermentatively and semisynthetically; in the latter, ribose produced by bacterial fermentation is a reagent in the chemical synthesis of this vitamin. Approximately half of the world's supply is produced in this manner [31]. The vitamin used in human therapeutics has traditionally been synthetic, whereas that produced by fermentation has been used in poultry and livestock feeds [32].

There are numerous microorganisms that produce riboflavin fermentatively. Demain [33] has classified riboflavin fermentations into three categories: weak overproducers (100 mg/L or less, e.g., *Clostridium acetobutylicum*), moderate overproducers (up to 600 mg/L, e.g., yeasts such as *Candida flareri*, *Candida guilliermundii*, and *Debaryomyces subglobosus*), and strong overproducers (over 1 g/L, e.g., the molds *Eremothecium ashbyii* and *Ashbya gossypi*). Bacterial and yeast riboflavin production is sensitive to high iron concentration, whereas fungal fermentation is not. Fermentation with *A. gossypi* is preferred because of its higher yields and greater genetic stability; riboflavin levels of over 11 g/L have been reported. Fermentations using glucose, molasses, corn steep liquor, and corn, soybean, and cod-liver oils have been reported [32–34]. The presence of glycine and purines in the medium stimulates riboflavin synthesis. Processes involving many organisms have been patented; these in-

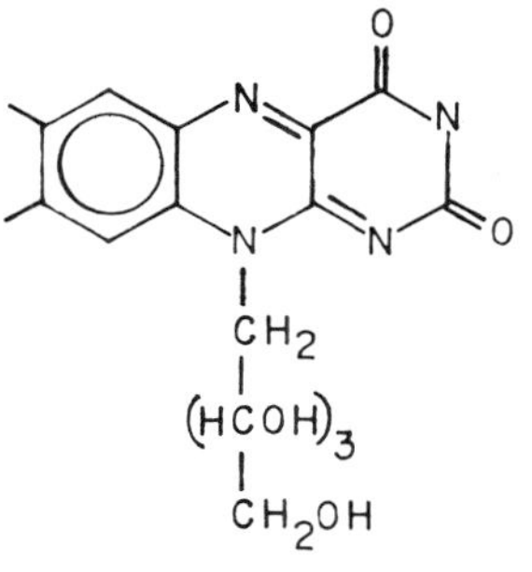

FIG. 7.6 Riboflavin.

clude the yeasts *Torulopsis xylinus*, *Hansenula polymorpha*, and *Pichia guilliermondii* and the bacteria *Achromobacter butrii*, *Micrococcus lactis* and *Streptomyces testaceus*.

Heefner et al. [35] claim a patent process by which *Candida flareri* produces 21 g/L riboflavin in a 200-hour fermentation (0.2 g/L per hour). Their patent claims selection techniques, purine analog resistance, techniques to obtain strains resistant to growth inhibition in spent or depleted medium, and increased riboflavin production at high iron concentrations. Kawai et al. [36] claim a process by which *Saccharomyces* yeasts produce riboflavin, even in the presence of iron, with productivities of 5 to 10 mg/L per hour. These workers also used purine mutants and mutants resistant to β-amino-1,2,4-triazole. A Russian patent describes a recombinant strain of *Bacillus subtilis*; an erythromycin-resistant gene was introduced into a plasmid containing the bacterial riboflavin operon and then transformed into an erythromycin-sensitive strain. Whereas the original strain produced 0.8 g/L riboflavin in 48 hours (a productivity of 0.017 g/L per hour), the recombinant strain produced 4.5 g/L in 24 hours (a productivity of 0.18 g/L per hour) [37]. Clearly such organisms can be further improved using similar recombinant techniques. A large body of literature has accumulated on the regulation of riboflavin biosynthetic pathways. The pathways appear similar in yeasts and bacteria such as *B. subtilis*. In yeast, six unlinked genes are involved [38]. Superior commercial strains that will dominate fermentation processes are likely to be bacterial, because the linked genes of the riboflavin operon in bacterial systems are more readily amenable to manipulation than are the scattered genes of the yeast and fungal riboflavin regulon.

7.3.2 *Monascus*

These fungi have been used since antiquity in the Orient to pigment foodstuffs such as rice, rice wine, and soybean curd cheese. *Monascus* pigments are traditionally produced on solid substrates such as rice or corn, which can be dried, ground, and incorporated into foods directly. Alternatively, the pigments can be extracted from the mycelium and modified chemically at high pH with other natural compounds, such as nucleotides, amino acids, and proteins, to yield red pigments that can be purified and used as colorants. Currently, *Monascus* pigments are being used in Japan to pigment meats [39]. Species involved are primarily *M. anka* and *M. purpureus*. The red, yellow, and purple pigments, as produced by the fungus, are not water soluble, but they readily react with amino groups. This leads to an opening of the pyran ring, a Schiff rearrangement, and resultant water solubility (Fig. 7.7).

These pigments are stable in the pH range of 2 to 10 and are heat stable. Their long history of use in the Orient could shorten the regulatory acceptance process in the United States [21,22,25]. However, difficulties that must be addressed in the commercialization of *Monascus* fermentations are complex. First, other secondary metabolites are produced by species such as *M. ruber*,

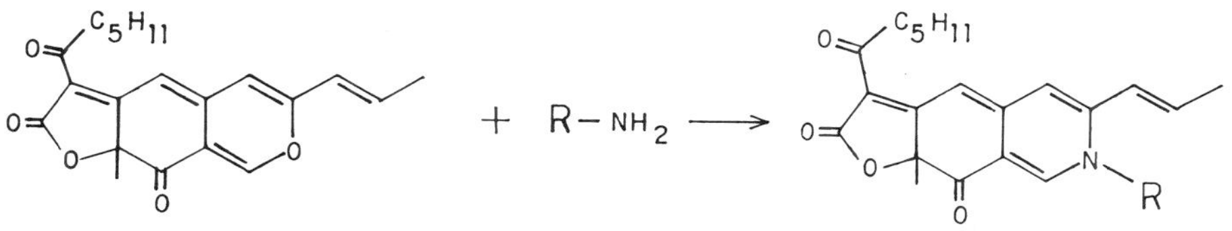

FIG. 7.7 Reaction of the *Monascus* pigment monascin with primary amines.

M. pilosus, and *M. purpureus*. These include antibiotics and hypocholesterolemic agents [39]. Secondly, pigment quantity is higher in solid-state koji-type fermentations than in submerged fermentations. Finally, pigment type is influenced by substrate and is retained intracellularly, especially under acidic conditions [21,22,39–41].

The reaction of *Monascus* pigments with amino groups that leads to water solubility can be exploited to modify the color of the resulting pigment, increase its stability, and limit absorption in humans. Moll and Farr [42] suggested the use of chitosan, a nondigestible polysaccharide widely distributed in insects and crustaceans, for this latter purpose. Wong and Koehler [43] reacted orange *Monascus* pigments with amino-acetic and amino-benzoic acids to generate red pigments. These compounds conferred heat and ultraviolet stability levels similar to more expensive *Monascus*–gelatin pigments. Broder and Koehler [44] reported that *M. purpureus* NRRL 2897 produced the most red pigments on 10 percent maltose, whereas corn and potato starches stimulated orange to red-orange pigments. Other workers have investigated the role of zinc and carbon/nitrogen ratios on growth and pigment production [41,45].

Lin and Hzuka [46] have generated strains of *M. baoliang* that produce pigment extracellularly. Wong and Koehler [47] have also reported on mutants that have increased pigment content, release more pigment into the medium, and have decreased antibiotic synthesis associated with them. Thus it seems entirely probable that the genus *Monascus* will be manipulated into highly productive industrial fungi like those commercial strains in the genera *Penicillium*, *Aspergillus*, and *Acremonium*.

The hyperproducing pigment strain studied by Kiyohara et. al. [39] grew slowly. These workers have used protoplast fusion technology to generate heterokaryons between *M. anka* and *Aspergillus oryzae*. Heterokaryotic colonies segregated white (*A. oryzae*) and red (*M. anka*) sectors. Stable diploids could be generated by treatment with ultraviolet light. These fusants grew faster than parental *Monascus* strains, but produced considerably less pigment—only 3.4 percent of that produced by the parents.

The production of pigment on large trays of steamed rice, which is the traditional method of the Orient, is extremely expensive [22]. However, submerged cultures produce only 10 percent of the pigment produced by solid state fermentations [48]. Evans and Wang [48] found that imbedding *Monascus* cells in calcium alginate beads reduced pigment production relative to non-embedded cells in shake flask cultures. That is, the presence of a solid support as in traditional koji fermentations could not stimulate pigment production. However, when the nonionic polymeric adsorbent resin XAD-7 was added to the medium, the pigment production rate increased significantly. It was suggested that the resin enhanced pigment removal from the mycelia, without which end-product inhibition or repression results. Nevertheless, the pigment concentration in this immobilized cell resin system was only 56 u (A_{500}) compared to 240 u for a solid-state fermentation.

Mak et al. [49] have compared agar surface, submerged, and roller bottle fermentations of *Monascus*. On PGY medium (peptone, glucose, yeast extract) roller bottle fermentations produced roughly 10-fold more pigment than submerged or surface agar fermentations. Roller bottle fermentations resulted in a higher final pH and suppressed fungal conidia. Availability of solid support and better gas exchange may have contributed to better pigment production. However, scale-up of such a system may be difficult.

7.3.3 Carotenoids

Carotenes and xanthophylls have a long history of use in the coloring of foods and beverages. They have application in fat-based foods such as butter, margarine, cheese, and oils. In beadlet or emulsion form, carotenoids have application in coloring water-based juices, beverages, soups, gravies, pasta, frozen dairy desserts, and so on. They are not affected by the presence of reducing agents such as tin, zinc, and aluminum, and, unlike some azo dyes, they can be used in vitamin C-fortified products without color loss. Major natural sources of carotenoids include paprika oleoresins, Aztec marigold extracts, annatto seed, and carrot oils. Pigments like β-carotene, β-apo-8'-carotenal, and canthaxanthin are currently approved for use in the United States. Unfortunately, recent findings show that canthaxanthin can form crystals in the retina when taken in therapeutic doses (prescribed to individuals abnormally sensitive to sunlight) [50]. This will probably have some influence on the future use of this pigment. However, carotenoids like canthaxanthin and astaxanthin do have desirable properties in that they are better natural antioxidants than β-carotene [51].

Carotenoid pigments are available from extracted vegetable sources and more recently from fermentation processes. These production processes compete with the nature-identical synthetic pigments β-carotene, β-apo-8'-carotenal, canthaxanthin, and astaxanthin. Borowitzka and Borowitzka [52] estimate the annual worldwide market for β-carotene is $50 to $100 million, most of which is produced synthetically. This is undoubtedly due to the fact that the current production cost of β-carotene by a microalgal fermentation, for example, exceeds that of the industrial synthesis by a factor of ten [53]. According to Benemann [53] and Borowitzka and Borowitzka [52], the selling price of synthetic β-carotene is about $500/kg, compared to up to $1000/kg, for microalgal β-carotene. Synthesized carotenoids, however, are not quite nature identical. Synthetic β-carotene is >99 percent the *trans* isomer whereas the β-carotenes from microalgae, fruits, and vegetables are a mixture of *cis* and *trans* isomers. It may be significant to the marketing of microalgal β-carotenes that the anticancer activity of carotenoids has been attributed to *cis–trans* mixtures [54].

When one examines the microbial community, one finds literally scores of carotenoid pigments that have been identified. They are produced by photosynthetic microbes such as algae, cyanobacteria (blue-green algae), and bacteria. They are also produced by nonphotosynthetic bacteria, yeasts, and fungi. Several of these organisms such as the yeast *Rhodotorula*, the fungi *Neurospora*, *Phycomyces*, and *Blakeslea*, and the bacteria *Rhodobacter capsulatus* (formerly *Rhodopseudomonas capsulata*) and *Erwinia* have been used as model systems to study the regulation, photoregulation, and genetics of carotenoid synthesis [55–62].

Blakeslea trispora and *Phycomyces blakeslea* have been grown at the 120 L pilot level [63,64]. These fermentations were expensive in that maximal β-carotene production of 2.5 to 3.2 g/L in 7 to 8 days was obtained only in the presence of expensive added inducers and stimulators such as β-ionone and iproniazide. Without such additives, pigment production was only approximately 0.85 g/L in 7 to 8 days. In addition, a fermentation broth of high viscosity worked best; such broths have high energy input requirements. These fungal fermentations, where only approximately 2 percent of the fungal biomass is β-carotene, are not that efficient.

Traditional fermentation technology, that is, submerged culture in batch, batch-fed, or continuous fermentation systems, has yet to be effectively applied to commercial β-carotene production. Yeasts and bacteria are more suited to these systems, and yields with these organisms are not sufficiently high to be commercially attractive. However, recent work on the carotenoid gene clusters from *Rhodobacter* and *Erwinia* now suggest that it would be possible to construct recombinant microorganisms for carotenoid production [56,65,66]. Marrs [67] showed that all of the carotenoid genes in *R. capsulata* are present in a 46-kb photosynthetic gene cluster and carried on the plasmid, pRS4021. At least nine genes involved in carotenoid synthesis have been mapped, and mutations within this cluster have been used to define gene function. *Rhodobacter* shares the common carotenoid biosynthetic pathway of other microorganisms up to the noncyclic C_{40} pigment, neurosporene. The major end products of the pathway in this organism, however, are spheroidene under anaerobic and spheroidenone under aerobic conditions.

Seven carotenoid genes occupying a 12.4-kb fragment have been cloned from *Erwinia herbicola* Eho10 [56]. These genes have been cloned into the common organism of biotechnology, *Escherichia coli*. Unlike *Rhodobacter*, the carotenoids identified in *Erwinia* include carotenoids commonly associated with foods. They include β-carotene (carrots), β-cryptoxanthin (citrus and corn), and zeaxanthin (corn). When transformed into *E. coli*, the *Erwinia* carotenoid gene cassette produces twice as much pigment (290 μg/g dry wt versus 140

μg/g dry wt). Inactivation of the gene kat F reduced pigment production by 85 percent [65]. Work by Misawa et al. [57] has revealed the following sequence of gene activity in carotenoid synthesis: lycopene-CrtY $\rightarrow$ β-carotene-CrtZ $\rightarrow$ zeaxanthin-CrtX $\rightarrow$ zeaxanthin-β-diglucoside. *Escherichia coli* transformants prepared by these investigators produced five- to sixfold more pigment than *Erwinia uredovora* and *E. herbicola*. Thus, with the proper manipulation of such gene cassettes, organisms can be designed to produce specifically lycopene (the predominant pigment in tomatoes), β-carotene, or zeaxanthin (the pigment from corn). The biotechnology industry has developed a wealth of information on the growth of *E. coli* to high densities in industrial fermentations producing pharmaceutical proteins. As data accumulate on the genetics of carotenoid pigment production in *Erwinia*, other bacteria such as *Brevibacterium* strain KY-4313 (which produces canthaxanthin [29]), and fungi such as *Phycomyces*, it will no doubt be possible to produce selected pigments by fermentation that are competitive with synthetic or extracted carotenoids.

In addition to the major light-harvesting pigment chlorophyll, microalgae synthesize a variety of accessory pigments including carotenoids. Under certain environmental conditions, the accessory pigments may be accumulated to a greater extent than chlorophyll [68]. All microalgae produce carotenoids that seem to function primarily as protective agents against photooxidation [54]. In addition, carotenoids may serve as accessory light-harvesting pigments and play a role in phototaxis.

The primary carotenoid pigment that is produced by microalgal fermentation today is β-carotene. Large-scale production of β-carotene by microalgal fermentation occurs today in Israel, Australia, and the United States [69]. Commercial quantities have been available for several years. A microalga in widespread use for production of β-carotene is *Dunaliella*. Two species *D. salina* and *D. bardawil* have been identified as being capable of producing 10 to 14 percent β-carotene on a dry weight basis [54,69–72]. *Dunaliella* is a unicellular, motile, halotolerant microalga. This microalga is an obligate photoautotroph. The halotolerant nature of this microorganism is significant, because it allows it to be cultivated in open saline mass culture relatively free of competing microorganisims and predators. This factor and a relatively high β-carotene content are the major reasons this microalga is commercialized for β-carotene. *Dunaliella* is also used in Israel to produce glycerol in mass culture [69].

Carotenogenesis in *Dunaliella* is affected by a number of factors. Three conditions that seem to be required for high β-carotene levels are nutrient limitation, intense light, and low water activity. β-Carotene accumulation in *Dunaliella* is increased by limiting nitrogen, phosphate, sulfate, or iron; nitrogen limitation is the most effective. Borowitzka and Borowitzka [69] suggested that any factor that leads to growth cessation induces higher levels of carotenoids. They postulated that it is not the nutrient limitation *per se*, but rather the effects the limitation has on the metabolism of proteins and nitrogen regulation. The cells must still be able to carry out photosynthesis to accumulate β-carotene, however.

High light intensity is a prerequisite for high levels of β-carotene synthesis in *Dunaliella*; this necessitates cultivation in very shallow ponds. Lers et al. [71] investigated the effects of photoinduction in *Dunaliella*. By growing the cells in low light and then switching to a high-intensity-light environment, they demonstrated two types of photoinduction of β-carotene: one that occurs immediately in growing cells and a late induction in stationary phase cells. They also demonstrated a gene activation role in these processes. A better understanding of the genes involved in the relationship between light intensity and β-carotene accumulation could lead to the manipulation of these genes to cause overproduction of β-carotene in suboptimal light.

High salinity is also required for high β-carotene levels in *Dunaliella*. As in most biochemical systems, there is a tradeoff, however, in that salinity levels that induce high levels of β-carotene (25 percent NaCl) are not optimal for high cell growth rates. Accordingly, two production strategies are employed [69]. One method uses a two-stage system where the microalga is first grown rapidly under low salinity conditions and then shifted to an induction medium of low nutrient levels and high salinity for greater β-carotene accumulation. The system can potentially become contaminated at the lower salinity stage. The other strategy uses a system of intermediate salinity in batch or continuous mode that is not optimal for growth or β-carotene accumulation but gives the best yield over time. The best approach must be determined by factors such as strain, site location, and available capital. Although β-carotene produced by microalgal fermentation is a viable business today, it is a struggling industry due to the high cost of production. Some of these issues are addressed in Section 7.5.

Astaxanthin, a xanthophyll (depicted in Fig. 7.3), is of current industrial interest. This pigment is present in a diversity of organisms such as dinoflagellates, algae, lichens, fungi, and yeasts [73–76]. This pigment accumulates in the food chain and is deposited in a variety of tissues in such animals as insects, birds, crustaceans, and fish [77–81]. The yeast *Phaffia rhodozyma* and the microalga *Haematococcus pluvialis* each produce astaxanthin as their primary carotenoid pigment. Natural isolates of *Phaffia rhodozyma* synthesize 200 to 400 $\mu g/g$ of total carotenoid pigments [82] and are capable of growing on a variety of industrial fermentation substrates such as alfalfa residual juice [83], molasses [84], and whey hydrolysates [85].

The microalga *Haematococcus pluvialis* is a unicellular motile alga capable of photoautotrophic and/or heterotrophic growth. Resting cysts of this microalga have been shown to accumulate up to 2 percent astaxanthin [86]. The encystment of *Haematococcus* occurs in response to nutrient deprivation or an increase in culture salinity. As with carotenogenesis in *Dunaliella*, a means to generate energy either by respiration or photosynthesis must be present for encystment and carotenogenesis to occur with *Haematococcus* [87]. The economics of astaxanthin production are favored by the relatively high concentration of astaxanthin in this microalga and the ease of separation of the dense cysts.

Gudin [88] notes that astaxanthin may be the most specific useful carotenoid of microalgae because it is more difficult to make by synthetic chemistry than β-carotene. As in the β-carotene market, astaxanthin produced by microalgae must compete with astaxanthin produced synthetically. Several companies in Norway, England, and the United States are engaged in research into the production of astaxanthin by microalgae but commercial production is limited at present.

7.3.4 Phycobiliproteins

Phycobiliproteins are chromoproteins that occur in a wide variety of microalgae. Phycobiliproteins of industrial interest are the red pigment phycoerythrin, used as a diagnostic reagent, and the blue pigment phycocyanin, used as a coloring agent in foods and cosmetics. The existing market for phycoerythrin is much smaller than for phycocyanin [53].

The phycobiliproteins c-phycocyanin and allophycocyanin normally comprise approximately 20 percent of the cellular protein of *Spirulina* and are quantitatively the dominant pigments in the cell [89]. In contrast to β-carotene systems where nitrogen limitation induces high pigment concentrations, high medium nitrogen levels induce high phycocyanin content in *Spirulina*, which is consistent with the role of phycocyanin as a nitrogen storage compound in *Spirulina*. Phycocyanin content is also affected by light intensity and quality. It is important to maintain a proper alternating light–dark cycle for good productivity because *Spirulina* can be photoinhibited by sustained high light intensity [90]. Cell density, turbulence, and pond depth are also important factors affecting productivity.

Spirulina is one of the most widely cultivated microalga in the world today because it grows relatively rapidly to a high cell density, prefers brackish environments of high alkalinity, and the filamentous nature of this microalga lends itself to cell harvesting. The robust growth of *Spirulina* is facilitated by the increased solubility of carbon dioxide, a growth promoter, at alkaline pH. The ability to thrive at a pH as high as 10 allows *Spirulina* to be cultivated continuously for extended periods as a near-axenic culture in relatively simple inexpensive systems. Because cell harvesting is a significant cost component of microalgal fermentation systems, the ease of cell separation by filtration or screening is a clear advantage to using this cyanobacterium, although care must be taken not to damage the relatively delicate cell walls [91].

The blue pigments of *Spirulina* can be extracted into aqueous solution. The use of alkaline carbonates or proteases facilitates dissociation of the protein color complex. Ascorbate has been used to aid color stability. These pigments have found application in confections, chewing gums, and frozen desserts [21].

The future of phycobiliprotein production by microalgal fermentation systems is promising. As new applications are found for pigments like phycocyanin, particularly in light of the consumer food preference for natural products, the markets for these pigments should increase. It is interesting to note that the phycobiliproteins produced for these markets could be by-products of microalgal processes used for other products [89]. This approach could be very cost effective.

7.4 PIGMENTS WITH POTENTIAL FERMENTATION ROUTES

7.4.1 Indigoids

Indigoid dyes derived from shellfish and plants have been used since antiquity. Their traditional use has been in the dyeing of textiles, but indigo has also been used in cosmetics [2]. Indigo itself is not approved for food use in the United States, but its disulfonic acid derivative, FD&C Blue 2, is approved.

Numerous synthetic routes to indigo have been developed and synthetic indigo began supplanting the natural dye over 80 years ago. More recently, while studying the conversion of naphthalene to salicylic acid, workers at Amgen discovered that a subset of the genes involved in naphthalene oxidation can lead to the microbial production of indigo [92,93]. Other workers have shown that xylene oxidase genes can function in the same manner [94]. However, indigo is not yet produced commercially by fermentation because productivities are still too low. For example, Ensley, et al. [92] reported that a recombinant strain of *E. coli* bearing the plasmid pE317 could produce only 25 μg/L indigo in 24 hours.

7.4.2 Anthraquinones and Naphthaquinones

The insect-derived anthraquinone glycoside, carminic acid, has been called "the best of all natural red pigments" [20]. This dye and other insect-derived anthraquinones such as kermes and laccaic acid have been used for centuries. The dyes, however, remain very expensive. Anthraquinone pigments are common in plants and fungi, and recently have been reported in the bacterium *Xenorhabdus* [95]. Although the anthraquinone pigments associated with those insect pigments used since antiquity may not be found in the microbial community, extensive screening may result in the discovery of organisms that produce these pigments or pigments that are related closely enough to facilitate legislative approval if economical fermentations are developed. This screening of the microbial community extends to the naphthaquinone pigments as well. The pigment shikonin is a naphthaquinone pigment produced commercially in Japan since 1983 via plant cell culture of *Lithospermum* sp. [96,97]. Naphthaquinone pigments are also synthesized by fungi in the genus *Fusarium* [98]. If extensive screening of the pigments produced by this large form genus identifies naphthaquinones similar to shikonin, the technology to produce them by fermentation is in place because organisms have been the subject of extensive work for the production of gibberellic acid [99–102].

7.5   FERMENTATION SYSTEMS

A key element in the economic feasibility of the production of natural pigments by fermentation is the design of the fermentation system itself; that is, the relationship between microorganism and bioreactor. Fermentation processes by nature are slow and dilute, especially when compared with chemical processes. However, they generally are less demanding in terms of temperature, pressure, and corrosion. It is beyond the scope of this chapter to discuss the many types of bioreactors in use by the biotechnology industry today. The reader is referred to excellent books on this topic [103,104].

Process economics of pigment production by fermentation can be improved by producing a higher concentration of product in less time. These goals can be summarized using the term *productivity*, which is equal to product concentration divided by fermentation time. Thus productivity has units of grams of product per liter of reactor volume per hour [105]. Alternatively, in algal shallow pond fermentations, the productivity can be described as grams of product per square meter of surface area per day [106]. A number of strategies have been employed to improve fermentation productivity: developing or creating a new microorganism; adopting a mode of operation that will maximize the productivity of a microorganism; developing new bioreactor designs to accommodate unique microorganisms; and improving the technology of existing fermentation systems. In practice, it is likely that more than one of these strategies will be employed for a given process.

There are three general modes of bioreactor operation: batch, fed-batch, and continuous [107,108]. Although most commercial processes are batch processes, fed-batch processes are becoming more common. With this mode of operation, productivity is improved by feeding one or more key substrates as nutrients to the fermentation at a controlled rate. Fed-batch technology has been used successfully to increase productivity of growth-associated products such as Baker's yeast and in the production of secondary metabolites such as penicillin [109–111]. Many secondary metabolites, including pigments, are subject to catabolite repression by glucose, other carbohydrates, or nitrogen compounds. By feeding these substrates at a controlled rate, their concentration in the medium is kept low and the repression is alleviated, improving system productivity.

A continuous fermentation process is one in which fresh nutrients are continually fed into the bioreactor at a controlled rate, and a product stream flows out continually, maintaining the system volume at a constant level. The microbial culture is poised in a single metabolic state rather than the constantly changing environment found in batch systems [105]. System productivity is enhanced by the minimizing of unproductive downtime in continuous versus batch processes. However, continuous process technology does not lend itself readily to the production of secondary metabolites.

An example of a fermentation system where new bioreactor designs are needed to improve process economics is the supported growth fermentations used in the surface culture of molds and filamentous fungi. Penicillin was manufactured originally by this technique and itaconic acid still is produced industrially by this method [109]. Also, as noted in this chapter, *Monascus* pigments are produced in the Orient by means of surface culture on stacked trays of steamed rice. However, the difficulty in economically improving pigment production by surface culture is the high surface-to-volume ratio, which makes it difficult to take advantage of the economies of scale. That is, because of the thin-layer design, a large surface area equates to a relatively small volume. Volumetric productivity can be high, but plant production is low. Penicillin manufacturing did not become a successful industry until it was possible to grow the mold in submerged culture. *Monascus* pigment production is feasible in surface culture because the solid support can become part of the dried pigment product for food use, little downstream processing is required, and the equipment for culturing the mold already exists for other fermentation processes.

One of the interesting aspects of biotechnology is the transfer of biosynthetic capability from one microorganism to another. As discussed earlier, the genes for pigment production now can be isolated and transferred to another microorganism along with an appropriate regulatory system. The use of bacterial strains as recipient cells is advantageous because of the ease of genetic manipulation in bacteria as well as their high biosynthetic rates. Bacteria can be propagated routinely in high cell density bioreactor systems with high oxygen transfer rates to give high productivities.

At present, the most significant production of pigments by fermentation is in microalgal mass culture. Microalgae have been used as a food source for centuries, but only in the last few decades have they been considered for the production of fuels and chemicals and for use in wastewater treatment [112]. In the last 5 years, microalgae have been used to produce specialty chemicals such as the natural pigment β-carotene. The success achieved in such a young industry indicates that as the technology matures, there is great potential for commercial production of pigments by microalgal fermentation. As evidenced by the number of failures, however, the profit margin of producing intermediate value chemicals by existing technology in competitive markets is so small as to entail considerable risks [53].

Process economics in microalgal fermentations are dominated by the requirement to achieve and maintain high productivities while keeping capital and operating costs to a minimum [113,114]. Unfortunately, simple systems that are relatively inexpensive to construct and operate have demonstrated low productivities and unstable performance. Better sustained performance can be obtained with complex systems, but these systems are more expensive and can only be justified for chemicals that command a market price greater than $100/kg. Since microalgal pigments are in the intermediate price range ($200 to $1000/kg) and market size of chemicals, use of some improved technologies for their production might be feasible [115]. This section addresses some of these technologies.

Microalgae grow photosynthetically by converting carbon dioxide and water to cell biomass and oxygen using light energy. The most efficient way to utilize this elegant biochemistry to mass cultivate microalgae and their associated products is in fermentation systems that use solar energy. Since pigment biosynthesis in microalgae is often related to light intensity, attempts to efficiently produce pigments, such as carotenoids, in conventional agitated tank-type reactors are not yet feasible. On the other hand, photobioreactor systems that use artificial light supplied internally or externally have been operated successfully at the bench and small pilot scales. However, these systems are difficult to scale up and are extremely expensive to operate, on the order of $10/kg of biomass for electricity alone. (At present, algal biomass costs approximately $1 to $20/kg to produce by conventional means [52].) Thus the use of photobioreactor systems can only be justified for the production of high value chemicals. At present, none of the known pigments produced by microalgae have applications that would justify this type of system [53].

Mass cultivation systems for microalgae can be divided into two basic types: horizontal ponds with induced turbulent flow and inclined baffled cascade-type systems where the microalgal culture is pumped from the bottom to the top of the unit [106,112,113]. Horizontal pond systems can be further subdivided into oblong-form raceway systems or large circular ponds. Most modern microalgal cultivation systems are of the raceway design, which incorporates long channels arranged in single or multiple loops [52].

Probably the most important design aspect for commercial microalgal cultivation systems is the need for adequate mixing to maintain high system productivity. Good mixing is important to maintain the proper supply of nutrients (especially carbon dioxide) to the cell, to help diffuse toxins (especially oxygen) from the cell, to maintain a relatively even temperature throughout the reactor, and to provide optimal light contact. Microalgal cultivation systems are relatively shallow, frequently less than 30 cm in depth, to use available sunlight efficiently. In shallow systems, there can be a temperature variation of several degrees without mixing. In high-density microalgal fermentations, the photic zone is very narrow, so adequate turbulence is required to distribute light to the cells evenly and allow the alternating light and dark conditions that are conducive to high productivity fermentations. Mixing must be vigorous enough to keep cells from settling to the bottom where they can decay and anaerobic digestion can occur. Mixing is achieved by a variety of methods including paddle wheels, pump/gravity systems, propellers, fluid injection systems, and airlift designs. Paddle wheel designs are currently the most popular because of efficiency, low energy usage, and durability [106]. It is interesting to note that the Australian β-carotene production facilities are of the open pond design, which rely on wind and convection currents for mixing, whereas the Israeli and U.S. facilities are of the paddle wheel raceway design. With the motile *Dunaliella* strain that is employed, the lower productivity of the pond system is offset by the lower capital and operating cost. However, Benemann [53] notes that the operating cost of mixing is only $0.01/kg biomass and predicts the Australians will convert their system.

Another important aspect of microalgal fermentation systems is the issue of a system that is open or closed to the atmosphere. Systems can be enclosed by a transparent covering or by the use of a tubular bioreactor. Closed systems prevent evaporation and moderate the temperature of the system by limiting evaporation and radiant cooling. Evaporative water loss with the resulting concentration of salts is a problem, particularly in dry areas where rainfall does not balance evaporation. Most commercial microalgal production facilities are located in subtropical or dry temperate regions to take advantage of the intense sunlight and moderate temperatures that occur there. The use of a closed system would provide distinct advantages in these regions, as well as expand the geographic and seasonal ranges over which microalgal systems can be effectively operated. Covering the system also prevents problems with dirt and insects and limits contamination. The disadvantages of covered systems are that less light reaches the algal biosystem, capital construction costs are higher, and some form of cooling is frequently required [113].

Closed photobioreactors are an area of active research in microalgal biotechnology. Chaumont et al. [116] concluded that the cost of producing algal biomass by means of an enclosed tubular system is comparable to that achieved in well-mixed open raceway systems. They suggest that enclosed systems that are inherently less prone to contamination are more appropriate for "sensitive" strains growing in nonextreme environments. The two major pigments produced by microalgal fermentations, β-carotene from *Dunaliella* and phycobiliproteins from *Spirulina*, each rely on extreme conditions in open systems to achieve near axenic conditions. Closed photobioreactor systems might allow the consideration of other microalgal strains that produce high-market-value pigments.

With microalgal as in all fermentation systems, a proper supply of nutrients in the appropriate form is essential to a high productivity process. Media cost comprises a significant portion of final product price. However, the relatively high market price for pigments probably justifies the use of purified chemicals in mass production systems [106]. The use of purified chemicals results in higher productivity with more consistent performance and lower incidence of contamination than ill-defined waste nutrient sources. Photoautotrophic algae can produce most of their organic requirements from simple media containing only the basic oxidized substrates: carbonate, nitrate, phosphate, and sulfate along with certain minerals [117].

Carbon dioxide is added to microalgal fermentations as a nutrient and for pH control. The compound is usually supplied as gaseous carbon dioxide and a variety of injection and diffusional techniques have been developed [113]. In general, the higher the medium pH, the better the efficiency of carbon dioxide application. In open systems, carbon dioxide addition is not necessary but improves productivity, whereas it is required in closed bioreactors. Carbon dioxide addition is expensive, up to 33 percent of the total cost in large scale systems, so this cost must be weighed against the gain in algae productivity and yield [52]. If one could use a "waste" or geological source of carbon dioxide, this would be very beneficial.

Microalgal cell harvesting is an important cost component of producing pigments by microalgal fermentations. The difficulties involved in separating microalgae from large dilute (<0.5 g/L) cell suspensions is one of the major challenges in production of fine chemicals by mass microalgal fermentations. A variety of mechanisms and combinations have been employed including centrifugation, screening/filtration, and auto- and chemically mediated flocculation [91]. In general, the more expensive a cell harvesting technique is, the better it functions. Harvesting is likely to continue to be an active area of research since for each microalgal process an optimum cost-effective method must be developed. There is, however, no "universal" method of harvesting microalgae.

One method for improving the productivity of a microalgal fermentation system for pigment production is by genotypic improvement of existing strains or selection and development of new species with desirable qualities. An obvious improvement would be an increase in the rate of biomass production or product formation. Pathway manipulation or alternative promoters might be employed to cause overproduction of the pigment of interest. Since microalgae are grown in open systems, resistance to environmental extremes or stresses could play an important role in determining process feasibility by allowing the fermentation to proceed under nearly axenic conditions. For example, a shift in temperature or pH optima may allow new fermentations to be considered. For photoautotrophs, a lower respiration rate in the dark would be advantageous. Finally, any factor that improves the separability of the microalgal cell, such as a higher sedimentation velocity, improves autoflocculation characteristics, or increases the tendency toward colonial or filamentous growth would be a distinct benefit.

In most cases, it would appear that the time-tested techniques of mutagenesis and selection are the most effective methods for improvement of the characteristics of a given microalgal strain. This is because genetic systems controlling these characteristics tend to be too ill-defined and the genetic techniques for microalgal (particularly eucaryotic) species too poorly developed to permit direct genetic engineering [118]. One interesting approach is to work with strains that are relatively poor producers of the product but have other desirable characteristics, such as high growth rate or ease of separation, and to genetically improve their productivity. *Chlorella* microalgal strains have been identified which are capable of growing rapidly under heterotrophic conditions. Endo and Shirota [119] have grown *Chlorella* heterotrophically at a doubling rate of approximately 2 hours. They note carotenoid levels of 0.5 and 0.3 percent under light and dark conditions, respectively. Farrow and Tahenkin [120] have patented a heterotrophic process with *Chlorella* to produce lutein, a xanthophyll, at up to 1 g/L of culture. If gene-manipulation techniques could be employed to increase pigment concentrations to significant levels, then conventional high-density cell culture techniques could be employed to develop high-productivity processes. However, it is probably more straightforward to use genetic engineering techniques to produce pigments in fungi, yeasts, or bacteria in high-productivity fermentation systems.

7.6 OPPORTUNITIES VERSUS ECONOMICS

The success of any pigment produced by fermentation will depend upon its acceptability in the marketplace, regulatory approval, and the size of the capital investment required to bring the product to market. As of this writing (1992) colorants produced by industrial fermentation are mainly limited to β-carotenes produced by algae, but there are others. For example, riboflavin can be produced by fungal, bacterial, and yeast fermentation. Riboflavin is used to a certain extent in foods, but its use as a nutritional supplement is more important. *Monascus* pigments are produced and used in Japan to pigment meats, bean curd, fish, and rice wines, but they are not permitted for human food use in the United States. Considerable interest exists in these pigments; Francis [21] has noted that over 30 patents have been issued on *Monascus* pigment application. Shikonin, a naphthaquinone pigment, is also produced by plant cell culture in Japan.

Taylor [22] has expressed doubts about the successful commercialization of fermentation-derived food colors because of the high capital investment requirements for fermentation facilities and the expensive and lengthy toxicity studies required by regulatory agencies. We believe, however, that regulatory and public perception of biotechnology-derived products is changing from doubts about potential problems to optimism about economic opportunities. For example, in 1990 the FDA approved the first recombinant-DNA-derived food ingredient [121,122]. Pfizer successfully cloned the gene for calf chymosin, an enzyme used in cheese production, into *E. coli*. The enzyme produced by *E. coli* fermentation is identical in functionality to the enzyme obtained from extract of calf stomach. Interestingly, the development of this calf enzyme by biotechnology supplants other products of biotechnology. That is, when the supply of calves for slaughter decreased and the costs of natural chymosin escalated, microbial chymosins (or rennets) were developed. However, the microbial enzymes, for example, those from *Mucor pusillus*, were not as desirable because of lower yields, less desirable flavor profiles, and less thermal stability [123]. Australia has also approved the use of recombinant-DNA–derived calf chymosin in cheese production. The demonstration that biotechnology-derived food ingredients such as chymosin and natural pigments are identical in structure and functionality to those ingredients derived by more traditional means will be instrumental in the acceptance of biotechnology by both the public and regulators. The role of regulatory agencies in bringing biotechnology products to market is discussed in part in Chapter 2.

To better understand the market penetration that fermentation-derived food colors might achieve, it is important to understand the size of the natural color market, the prices of these pigments, and the costs to produce pigments fermentatively. Ilker [27] has estimated the world food color market to be growing at about 10 percent per year. The United States market in 1990 was approximately $135 million. Klaüi [124] has estimated that the total world market for food colorants was $320 million in 1987. Of this, one-third, or approximately $120 million, was for natural colors. This in turn was broken down into $35 million for natural extracted colors, $35 million for nature-identical (i.e. synthetic), and $50 million for process-derived pigments (e.g., caramel). On the other hand, Francis [125] has anticipated that interest in natural pigments will wane by the mid-1990s due to increased acceptance of synthetic colors. However, his view is controversial. When Paulus [126] polled 43 experts in the European Economic Community, they predicted that by the year 2000, 50 to 75 percent of the synthetic food colorants in use in 1989 would be replaced by natural pigments.

The traditional fermentation processes discussed here and plant cell culture offer great opportunities for the production of natural pigments. Research in plant cell culture includes the production of carotenoids [127], betalaines [16,17], anthocyanins [13], naphthaquinones [96,97], and anthraquinones [128]. In plant cell culture, perhaps the immediate opportunities lie in anthocyanin production. Ilker [27] quotes a price of $1250 to $2000/kg for pure anthocyanins and $4000/kg for the naphthaquinone shikonin. These prices are stimulating research in plant cell tissue cultures. However, at a price of $1000/kg for algal fermentation-derived β-carotene [52], which is substantially higher than the price of synthetic β-carotene, fermentation-derived β-carotene is only competitive in small niche markets. The relatively high price is due to the fact that production cost for a low volume fermentation product is determined mainly by capital investment and labor, which are expensive and fixed. For a high-volume product, cost is primarily related to raw materials and utilities [129]. Goldstein [130] has tried to explain this fact by organizing fermentation/biotechnology products into three categories: high value, intermediate value, and commodity. High-value products like therapeutic peptides with United States market sizes under 1000 kg/yr can cost in excess of $100,000/kg. Furthermore, the selling prices of some biotechnology-derived compounds such as tissue plasminogen activator and human growth hormone are 100- to 1000-fold more than predicted. Added downstream processing costs may account for the difference [131]. On the other hand commodity products such as food acidulants, glutamic acid, and ethanol with United States market sizes between 1 million and 1 billion/kg per year can be produced for $1/kg or less. Intermediate value products include amino acids, nucleotides and vitamins. The U.S. market size for these products is 10,000 to 10 million kg/yr and they are produced for $10 to $100/kg [130].

The future market penetration and growth of fermentation-derived food colors will depend, for the most part, on technologies involving strain development and fermentation design. Those pigments associated with plants, such as anthocyanins, await the development of plant cell culture techniques that can compete with extraction from natural vegetable sources. However, the successful marketing of algal fermentation-derived and vegetable-extracted β-carotene, both as a nutritional supplement and a food color, reflects the presence and importance of niche markets in which consumers are willing to pay a premium for "all natural ingredients." If fermentation is to access and economically compete in these niche markets for natural colors, production costs for these fermentation derived pigments will need to be brought in line with those incurred by extraction and synthetic processes. For the time being, fermentation-produced pigments will probably target those pigments with very low overhead, for example, algal β-carotene, or those with high synthetic costs.

ACKNOWLEDGMENTS

We would like to thank Ms. Carole Stewart for her typing skills and for preparing the tables. We would also like to thank Dr. Thomas Skatrud for his help in proofreading the manuscript.

REFERENCES

1. Clydesdale, F.M., in *Developments in Food Colors* (J. Walford, ed.), Vol. 2, Elsevier Applied Science, London and New York, 1984.
2. Farris, R. E., in *Kirk-Othmer Encyclopedia of Chemical Technology* (M. Grayson and D. Eckroth, eds.), 3rd ed., Vol. 8. Wiley, New York, 1979.
3. Walford, J., in *Developments in Food Colors* (J. Walford ed.), Vol. 1. Elsevier Applied Science, London 1980.
4. Anonymous, *Dairy and Food Sanit.* **4,** 131 (1984).
5. Anonymous, *Food Technology*, **34,** 77 (1980).
6. MacKinney, G., and Little, A.C., in *Color of Foods.* AVI Publ. Co., Westport, CT, 1962.
7. Counsell, J.N., in *Developments in Food Colors* (J. Walford, ed.), Vol. 1. Elsevier Applied Science, London, 1980.
8. Counsell, J.N., and Knewstubb, C.J., *Food* Aug., p. 18 (1983).
9. Lampila, E. E., Wallen, S.E., and Bullerman, L.B., *Dairy Food Sanit.* **4,** 90 (1984).
10. Emodi, A., *Food Technol.* **32,** 38 (1978).
11. Gordon, H.T., Johnson, L.E., and Borenstein, B., *Cereal Foods World* **30,** 274 (1985).
12. Bauernfeind, J.C., Rubin, S.H., Surmatis, J.D., and Ofner, A., *Int. J. Vitam. Res.* **40,** 391 (1970).
13. Francis, F.J., *Crit. Rev. Food Sci. Nutr.* **28,** 273 (1989).
14. Coulson, J., in *Developments in Food Colors* (J. Walford ed.), Vol. 1. Elsevier Applied Science, London, 1980.
15. Pasch, J.H., von Elbe, J.H., and Sell, R.J., *J. Milk Food Technol.* **38,** 25 (1975).
16. Sakuta, M., Takagi, T., and Komamine, A., *Physiol. Plant.* **71,** 455 (1987).
17. Sakuta, M., Takagi, T., and Komamine, A., *Physiol. Plant.* **71,** 459 (1987).
18. Spears, K., *Trends Biotechnol.* **6,** 283 (1988).
19. Timberlake, C.F., and Henry B.S., *Endeavour, new ser.* **10,** 31 (1986).
20. Riboh, M. *Food Eng.* **49,** 66 (1977).

21. Francis, F.J., *Food Technol.* **41**, (4) 62 (1987).

22. Taylor, A.J., in *Developments in Food Colors* (J. Walford ed.), Vol. 2, Elsevier Applied Science, London and New York, 1984.

23. Phillip, T., *Food Technol.* **38**(12), 107 (1984).

24. Chao, R.R., Mulvaney, S.J., Sanson, D.R., Hsieh, E.H., and Tempeston, M.S., *J. Food Sci.* **56**, 80 (1991).

25. Freund, P.R., Washam, C.J., and Maggion, M., *Cereal Foods World* **33**, 553 (1988).

26. Anonymous, *Certified Food Colors.* Warner-Jenkinson Co., St. Louis, MO, 1984.

27. Ilker, R., *Food Technol.* **41**, 70 (1987).

28. Donnelly, D.M., and Sheridan, M.H., *Phytochemistry* **25**, 2303 (1986).

29. Nelis, H. J., and de Leenheer, A.P., *Appl. Environ. Microbiol.* **55**, 2505 (1989).

30. Anonymous, *Riboflavin*, Tech. Bull. No. 8200. Warner-Jenkinson Co., 1984.

31. Florent, J., in *Biotechnology* (H. S. Rehm and G. Reed, eds.), Vol. 4. Verlag Chem., Weinheim, 1986.

32. Yoneda, F., in *Kirk-Othmer Encyclopedia of Chemical Technology* (M. Grayson and D. Eckroth, eds.), 3rd ed. Vol. 24. Wiley, New York, 1984.

33. Demain, A.L., *Annu. Rev. Microbiol.* **26**, 369 (1972).

34. Perlman, D., in *Microbial Technology* (H. Peppler and D. Perlman, eds.), 2nd ed., Vol. 1. Academic Press, New York, 1988.

35. Heefner, D.L., Weaver, C.A., Yarus, M.J., Burdzinski, L.A., Gywre, D.C., and Foster, E.W., International Patent Publication WO 88/09872 (1988).

36. Kawai, K., Matsuyama, A., and Takao, S., European Patent Application 89/109,875.8 (1989).

37. Bigelis, R., in *Biotechnology* (G. J. Jacobson and S. O. Jolly, eds.), Vol. 7B. Verlag Chem., Weinheim, 1989.

38. de los Angeles Santos, M., Iturriaga, E. A., and Eslava, A.P., *Curr. Genet.* **14**, 419 (1988).

39. Kiyohara, H., Watanabe, T., Imai, J.,Takizawa, N., Hatta, T., Nagao, K., and Yamamoto, A., *Appl. Microbiol. Biotechnol.* **33**, 671 (1990).

40. Endo, A., *J. Antibiotics*, **23**, 334 (1980).

41. Ban, Y.S., and Wong, H.-C., *Physiol. Plant.* **46**, 63 (1979).

42. Moll, H.R., and Farr, D.R., U.S. Patent 3,993,789 (1976).

43. Wong, H.-C., and Koehler, P.E., *J. Food Sci.* **48**, 1200 (1983).

44. Broder, C.V., and Koehler, P.E., *J. Food Sci.* **45**, 567 (1980).

45. Wong, H.-C., Lin, Y.C., and Koehler, P.E., *Mycologia* **73**, 649 (1981).

46. Lin, C.-F., and Hzuka, H., *Appl. Environ. Microbiol.* **43**, 671 (1982).

47. Wong, H.-C., and Koehler, P.E., *J. Food Sci.* **46**, 956 (1981).

48. Evans, P.J., and Wang, H.Y., *Appl. Environ. Microbiol.* **47**, 1323 (1984).

49. Mak, N.K., Fong, W.F., and Wong-Leung, Y.L., *Enzyme Microb. Technol.* **12**, 965 (1990).

50. Arden, G.B., Oluwole, J.O.A., Polkinghorne, P., Bird, A.C., Barker, F.M., Norris, P.G., and Hawk, J.L.M., *Hum. Toxicol.* **8**, 439 (1989).

51. Terao, J., *Lipids* **24**, 659 (1989).

52. Borowitzka, L.J., and Borowitzka, M.A., in *Algal and Cyanobacterial Biotechnology*, (R.C. Cresswell, T.A.V. Rees, and N. Shah, eds.). Wiley, New York, 1989.

53. Benemann, J.R., in *Algal and Cyanobacterial Biotechnology* (R.C. Cresswell, T.A.V. Rees, and N. Shah, eds.). Wiley, New York, 1989.

54. Ben-Amotz, A., and Avron, M., in *Algal and Cyanobacterial Biotechnology* (R.C. Cresswell, T.A.V. Rees, and N. Shah, eds.). Wiley, New York 1989.

55. Armstrong, G.A., Alberti, M., Leach, R., and Hearst, J.E., *Mol. Gen. Genet.* **216,** 254 (1989).

56. Perry, K.L., Simonitich, T.A., Harrison-Lavoie, K.J., and Liu, S.T., *J. Bacteriol.* **168,** 607, (1986).

57. Misawa, N., Nakagawa, M., Kobayashi, K., Yamano, S., Izawa, Y., Nakamura, K., and Harashima, K., *J. Bacteriol.* **172,** 6704 (1990).

58. Tada, M., Tsubouchi, M., Matsuo, K., Takimoto, H., Kimura, Y., and Takagi, S., *Plant Cell Physiol.* **31,** 319 (1990).

59. Murillo, F.J., and Cerda-Olmedo, E., *Mol. Gen. Genet.* **148,** 19 (1976).

60. Bejarano, E.R., and Cerda-Olmedo, E., *Phytochemistry* **28,** 1623 (1989).

61. De la Concha, A., and Murillo, F.J., *Planta* **161,** 233 (1984).

62. Nelson, M.A., Morelli, G., Canattoli, A., Romano, N., and Masino, G., *Mol. Cell. Biol.* **9,** 1271 (1989).

63. Ninet, L., and Renaut, J., in *Micobial Technology* (H. Peppler, and D. Perlman, eds.), 2nd ed., Vol. 1. Academic Press, New York, 1979.

64. Ratledge, C., in *Biotechnology* (H.J. Rehm, and G. Reed, eds.), Vol. 4. Verlag Chem., Weinheim, 1986.

65. Sandmann, G., Woods, W.S., and Tuveson, R.W., *FEMS Microbiol. Lett.* **71,** 77 (1990).

66. Armstrong, G.A., Alberti, M., and Hearst, J.E., *Proc. Natl. Acad. Sci. U.S.A.* **87,** 9975 (1990).

67. Marrs, B., *J. Bacteriol.* **146,** 1003 (1981).

68. Borowitzka, M.A., in *Micro-algal Biotechnology* (M.A. Borowitzka, and L.S. Borowitzka, eds.). Cambridge University Press, Cambridge, 1988.

69. Borowitzka, M.A., and Borowitzka, L.J., eds. *Micro-algal Biotechnology*. Cambridge University Press, Cambridge, 1988.

70. Borowitzka, M.A., and Borowitzka, L.J., in *Algal Biotechnology* (T. Stadler, J. Mollion, M.C. Verdus, Y. Karamanos, H.V. Morvan, and D. Cristiaen, eds.), Elsevier Applied Science, London and New York, 1988.

71. Lers, A., Biener, Y., and Zamir, A., *Plant Physiol.* **93,** 389 (1990).

72. Moulton, T.P., Borowitzka, L.J., and Vincent, D.J., *Hydrobiologia* **151/152,** 99 (1987).

73. Johansen, J.E., Svec, W.A., Liaaen-Jensen, S., and Haxo, F.T., *Phytochem.* **13** 2261 (1974).

74. Karrer, P., and Jucker, E., *Carotenoids*. Elsevier, New York, 1950.

75. Czeczuga, B., and Osorio, H.S., *Isr. J. Bot.* **38,** 115 (1989).

76. Johnson, E.A., and Lewis, M.J., *J. Gen. Microbiol.* **115** 173 (1979).

77. Okay, S., *Bull. Soc. Zool. Fr.* **74,** 11 (1949).

78. Brush, A.H., *FASEB J.* **4,** 2969 (1990).
79. Johnson, E.A., Lewis, J.J., and Grau, C.R., *Poult. Sci.* **59,** 1777 (1980).
80. Johnson, E.A., Conklin, D.E., and Lewis, M.J., *J. Fish. Res. Board Can.* **34,** 2417 (1977).
81. Torrissen, O.J., Hardy, R.W., and Shearer, D.D., *Aquat. Sci.* **1,** 209 (1989).
82. An, G.H., Schuman, D.B., and Johnson, E.A., *Appl. Environ. Microbiol.* **55,** 116 (1989).
83. Okagbue, R.N., and Lewis, M.J., *Appl. Microbiol. Biotechnol.* **20,** 33 (1984).
84. Haard, N.F., *Biotechnol. Lett.* **10,** 609 (1988).
85. Miranda, M., *personal communication* 1991.
86. Spencer, G., U.S. Patent 4,871,551 (1989).
87. Droop, M.R., *Arch. Mikrobiol.* **21,** 267 (1955).
88. Gudin, C., in *Algal Biotechnology* (T. Stadler, J. Mollion, M.C. Verdus, Y. Karamanos, H. Morvan, and D. Cristiaen, eds.). Elsevier Applied Science, London and New York, 1988.
89. Richmond, A., in *Micro-algal Biotechnology* (M.A. Borowitzka and L.J. Borowitzka, eds.). Cambridge University Press, Cambridge, 1988.
90. Vonshak, A. and Guy, R., in *Algal Biotechnology* (T. Stadler, J. Mollion, M.C. Verdus, Y. Karamanos, H. Morvan, and D. Cristiaen, eds.). Elsevier Applied Science, London, New York, 1988.
91. Mohn, F.H., in *Micro-algal Biotechnology* (M.A. Borowitzka and L.J. Borowitzka, eds.). Cambridge University Press, Cambridge, 1988.
92. Ensley, B.D., Ratzkin, F.J., Osslund, T.D., Simon, J.J., Wackett, L.P., and Gibson, D.T., *Science* **222,** 167 (1983).
93. Ensley, B.D., European Patent Application 83,110,688.5 (1984).
94. Mermod, N., Harayama, S., and Timmis, K.N., *Bio/Technology* **4,** 321 (1980).
95. Richardson, W.H., Schmidt, T.M., and Nealson, K.H., *Appl. Environ. Microbiol.* **54,** 1602 (1988).
96. Fugita, Y., Takahashi, S., and Yamada, Y., *Agric. Biol. Chem.* **49,** 1755 (1985).
97. Curtin, M.E., *Bio/Technol.* **1,** 649 (1983).
98. Parisot, D., Devys, M., and Barbier, M., *Microbios* **64,** 31 (1990).
99. Holme, T., and Zacharias, B., *Biotechnol. Bioeng.* **7,** 405 (1965).
100. Kahlon, S.S., and Malhotra, S., *Enzyme Microb. Technol.* **8,** 613 (1986).
101. Kumar, P.K.R., and Lonsane, B.K., *Biotechnol. Bioeng.* **30,** 267 (1987).
102. Kumar, P.K.R., and Lonsane, B.K., *Appl. Microbiol. Biotechnol.* **28,** 537 (1988).
103. Bailey, J.E., and Ollis, D.F., *Biochemical Engineering Fundamentals*, 2nd ed. McGraw-Hill, New York, 1986.
104. Atkinson, B., and Mavituna, F., *Biochemical Engineering and Biotechnology Handbook.* Nature Press, New York, 1983.
105. Pirt, S.J., *Principles of Microbe and Cell Cultivation.* Halsted Press, New York, 1975.
106. Oswald, W.J., in *Micro-algal Biotechnology* (M.A. Borowitzka, and L.J. Borowitzka, eds.). Cambridge University Press, Cambridge, 1988.
107. Aiba, S., Humphrey A., and Millis, N., *Biochemical Engineering.* Academic Press, New York and London, 1973.

108. Crueger, W., and Crueger, A., *Biotechnology: A Textbook of Industrial Microbiology*. Sinuaer Associates, Sunderland, MA, 1989.

109. Winkler, M.A., in *Principles of Biotechnology* (A. Wiseman, ed.). Surrey University Press, New York, 1988.

110. Whitaker, A., *Process Biochem.* **15**, 4 (1980).

111. Pirt, S.J., *J. Appl. Chem. Biotechnol.* **24**, 415 (1974).

112. Terry, K.L., and Raymond, L.P., *Enzyme Microb. Technol.* **7**, 474 (1986).

113. Richmond, A., and Becker, E.W., in *CRC Handbook of Microalgal Mass Culture* (A. Richmond is editor), CRC Press, Boca Raton, FL, 1986.

114. Richmond, A., *Hydrobiologia* **151/152,** 237 (1987).

115. Borowitzka, M.A., *Microbiol. Sci.* **3**, 372 (1986).

116. Chaumont, D., Thepenier, C., Gudin, C., and Junjas, C., in *Algal Biotechnology* (T. Stadler, J. Mollion, M.C. Verdus, Y. Karamanos, H. Morvan, and D. Cristiaen, eds.). Elsevier Applied Science, London and New York, 1988.

117. Stein, J., *Handbook of Phycological Methods*. Cambridge University Press, Cambridge, 1973.

118. Craig, R., Reichelt, B.Y., and Reichelt, J.L., in *Micro-algal Biotechnology* (M.A. Borowitzka and L.J. Borowitzka, eds.). Cambridge University Press, Cambridge, 1988.

119. Endo, H., and Shirota, M., *Ferment. Technol. Today* **4**, 533 (1972).

120. Farrow, W.M., and Tahenkin, B., U.S. Patent 3,280,502 (1966).

121. Salvage, B., *Food Business* May 7, p. 19 (1990).

122. Duxbury, D.D., *Food Process.* **51,**(6), 46 (1990).

123. Duxbury, D.D., *Food Process.* **45**(5), 44 (1988).

124. Klaui, H.M., *Natcol Q. Inf. Bull.* **3**, 2 (1989).

125. Francis, F.J., in *Developments in Food Colors*, (J. Walford, ed.). Vol. 2., Elsevier Applied Science, London and New York, 1984.

126. Paulus, K., *Natcol Q. Inf. Bull.* **3**, 2 (1989).

127. Yun, J.W., Kim, J.H., and Yoo, Y.J., *Biotechnol. Lett.* **12**, 905 (1990).

128. Robins, R.J., Payne, J., and Rhodes, M.J.C., *Phytochemistry* **25**, 2327 (1986).

129. Van Brunt, J., *Bio/Technology* **4**, 395 (1986).

130. Goldstein, W.E., *Biotechnol. Bioeng. Symp.* **17**, 16 (1986).

131. Knight, P., *Bio/Technology* **7**, 777 (1989).

Plant Tissue Culture

OM SAHAI

ESCAgenetics Corporation, San Carlos, California

Plants are a valuable source of natural colors, natural flavors, and aroma chemicals. The technology of plant tissue culture offers an alternative cost-effective means to produce such ingredients under controlled industrial conditions. Using this technology, a number of plant-derived products, including shikonin, taxol, ginseng, geranium, and vanilla, have either been commercialized or are quickly approaching commercialization. This chapter discusses the technology and economics of plant tissue culture, and reviews prior work on tissue culture production of food colors, flavors, and aroma chemicals.

8.1 PLANT-DERIVED FLAVORS, FRAGRANCES, AND COLORS

Plants are a valuable source of natural colors, natural flavors, nonnutritive sweeteners, spices, and complex aroma chemicals. According to a recent estimate, the U.S. market for plant-derived chemicals excluding carbohydrates is on the order of $10 billion per year at the consumer level and expected to increase at the rate of 3 percent annually through 1995 [1]. The market for natural fruit flavors is estimated at $200 million per year [2]. In the past decade, there has been an increasing and steady shift in consumer preference toward the use of food ingredients derived from natural sources, with the consumer willing to pay a premium for natural plant-derived products as opposed to their synthetic counterparts obtained through organic chemistry. Lately this trend appears to be catching on in the perfumery industry as well [3]. In the area of food colors, there has been a renewed interest in plant-based materials due to removal from commerce of many FD&C dyes, including Red No. 2 and Red Lake No. 3, with others such as Red Dye No. 3 and Yellow No. 5 facing increasing scrutiny from both regulatory agencies and consumer groups [4].

Bioprocess Production of Flavor, Fragrance, and Color Ingredients, Edited by Alan Gabelman,
ISBN 0-471-03821-0 © 1994 John Wiley & Sons, Inc.

8.1.1 Product Possibilities

From the pioneering work of White and Gautheret, who in 1939 showed that groups of undifferentiated plant cells can be multiplied indefinitely in a nutrient medium in culture, to the numerous recent reports confirming that compounds characteristic of the whole plant can be produced in significant quantities in tissue culture, the technology of plant cell culture has finally arrived at a point to offer a commercial alternative to extraction of products from whole-plant sources [5–7]. Table 8.1 lists some examples of plant-derived colors, flavors, and fragrances that can be targeted for production via plant tissue cultures. They include products useful for both the food and the perfumery industries: red colors such as anthocyanins and betacyanins; yellow colors such as betaxanthines and crocin; fruit flavors such as those characterizing strawberry, cherry, and raspberry; vegetable flavors like celery and asparagus; vanilla and licorice flavorings; essential oils derived from mint, garlic, and patchouly; nonnutritive sweeteners such as monellin and stevioside; and spices such as extracts of rosemary and sage.

Many of these products consist of a complex mixture of components, with one to six compounds contributing to the major flavor (or fragrance) impact. Table 8.2 gives examples of chemical constituents of select products, and includes current selling prices and approximate market sizes. The cost/capacity relationship of the products ranges from those that are high in value but low in volume (e.g., jasmine oil) to others that are low in value but high in annual usage (e.g., peppermint oil).

8.1.2 Rationale for Tissue Culture Production

The business rationale for producing these natural ingredients from plant tissue cultures is based on the premise of supply, quality, and cost. For example, many of the spices and essential oils are subject to fluctuations in yield, quality, and price due to diseases, pests, and adverse climatic conditions in addition to

TABLE 8.1 Examples of Products Amenable to Use of Plant Tissue Culture Processes

Food colors	Anthocyanins, betacyanins, crocin, bixin, β-carotene, betaxanthines, curcumin
Fruit flavors	Strawberry, grape, cherry, raspberry, banana, mango, peach, pineapple, melon
Vegetable flavors	Capsicum, celery, tomato, asparagus
Other flavors	Vanilla, chocolate, licorice
Essential oils	Spearmint, peppermint, vetiver, rose, garlic, onion, patchouly, jasmine, geranium, sandalwood, eucalyptus, chamomile, neroli
Sweeteners	Stevioside, thaumatin, monellin, miraculin
Spices	Cinnamon, cardamon, rosemary, sage, turmeric

TABLE 8.2 Key Chemical Components and Market Sizes of Select Flavors and Oils Derived from Plants

Product	Characteristic Components	Minor Components	Current Selling Price	Estimated World Market ($ MM/YR)
1. Peppermint Oil	Menthol Menthone	Isomenthol Isomenthone Methyl acetate Pulegone Piperitone Limonene	$30–40/kg	100
2. Rose Oil	1-Citronellol Geraniol Eugenol Nerol Phenylethyl alcohol		$7000–8000/kg	10–20
3. Jasmine Oil	Benzyl acetate Jasmone Jasmolactone Methyl jasmonate		$1000–5000/kg	2–5
4. Garlic Oil	Allinin Methyl cysteine sulfoxide		$80/kg	4–10
5. Sandalwood Oil	α- and β-Santalol		$170–180/kg	5–10

TABLE 8.2 (*Continued*)

Product	Characteristic Components	Minor Components	Current Selling Price	Estimated World Market ($ MM/YR)
6. Vanilla flavor	Vanillin	Vanillic acid 4-OH-Benzaldehyde Vanillyl alcohol 3,4-Dihydroxy benzaldehyde 3,4-Dihydroxy benzoic acid	~$70–90/kg beans or $2,500/kg flavor solids	100
7. Cherry flavor	Benzaldehyde		$200/kg	40–50
8. Raspberry flavor	4-OH-Phenyl-2-butanone	α- and β-Ionones 2-Hexenyl acetate cis-3-hexene-1-ol Ethyl acetate Acetaldehyde	$1–2/kg fruit or $10–$20,000/kg[a] flavor solids	4–5[b]
9. Saffron	Crocin Picro-crocin	α-, β-, and γ-Crocetin Crocins 2, 3, 4	$300/kg dry flowers or $7,500/kg active solids	15–20

[a]Estimated on basis of 100 ppm net flavor solids per kg fresh wt. of fruit.
[b]Author's estimate.

political and sociological considerations in the producing countries. A plant tissue culture-based industrial process guarantees control over these parameters at the point of usage, from one batch to the next and from one year to another. Other potential products such as fruit flavors are present in extremely low concentrations in the intact fruit, ranging from a few parts per million to about 100 ppm, and therefore are too costly to extract. As examples, on a fresh weight basis, strawberries contain 2 to 8 ppm of their characterizing flavor components, bananas 12 to 18 ppm, tomatoes 3 to 5 ppm and Concord grape 2 to 3 ppm. A simple cost calculation shows that if the above flavor chemicals were to be extracted from the whole plant source, the raw material cost alone would amount to $10,000 to $100,000/kg of pure flavor material [8].

The technical rationale for developing a plant tissue culture process is based upon the unique totipotency of plant cells. Theoretically, a cell derived from any plant has the potential to grow and differentiate into the complete plant. One could therefore argue that although strawberry flavor components are only found in the intact strawberry fruit, a cell derived from any part of the strawberry plant has the genetic potential to produce any or all of the strawberry flavor components in cell culture. As discussed in depth later in this chapter, the basic feasibility of producing characterizing components of many flavors and oils has been shown in plant cell cultures. Additionally, a number of tissue culture systems have been shown to yield products at concentrations comparable to or several fold higher than the intact plant, reaching up to 30 percent on a cell dry wt. basis (or 6 to 7 g product/L culture broth) in some cases. The latter is comparable to yields found in many antibiotic-producing microbial fermentation processes. Table 8.3 details product yields and associated process data for select high-product-yielding plant tissue culture systems [8a,8b].

8.1.3 Market Considerations

Various market considerations dictate the selection of potential targets for production via plant tissue cultures. The key factors include the overall market size in dollars and in production volume, the product selling price and the likelihood of penetration or expansion of the market with a biotechnology-based product. First is the issue of natural versus synthetic. A natural product, even with a chemical composition (or structure) identical to that of a synthetic product, is expected to command a higher selling price than the synthetic product. Examples include benzaldehyde, the characterizing compound in cherry flavor; methyl anthranilate, the characterizing compound in Concord grape flavor; and vanillin, the characterizing compound in natural vanilla flavor. Synthetic benzaldehyde sells for $1.50 to $1.75/kg; natural benzaldehyde commands a selling price of over $200/kg. Synthetic methyl anthranilate sells for $5.75/kg; natural methyl anthranilate is expected to be worth at least $400 to 500/kg. Synthetic vanillin sells for approximately $15/kg; natural vanillin is expected to sell for several times this value.

The second consideration is the positioning of the biotechnology-based prod-

TABLE 8.3 Examples of High Product Yields in Plant Cell Culture

Product	Plant Species	Product Yield in Culture (% dry wt.)	(mg/L)	Product Yield in Intact Plant (% dry wt.)	Cell Doubling Time[a] (hrs)	Cell Density[a] (gm/L)	Batch Cycle Time[b] (days)	Reference
Shikonin	*Lithospermum erythrorhizon*	23	4000	1.5	~62	17	23	[21,62]
Berberine	*Coptis japonica*	16	3500	7.5	~52	27	15	[8a,8b,73]
Anthraquinones	*Morinda citrofolia*	10	7000	7.5	N/A[c]	70	N/A[c]	[12]
		18	2500	2.2	~67	14	14	[7,28,51]
Rosmarinic Acid	*Coleus blumei*	25	5600	3.00	~50	22	14	[7]

[a]These values were approximated from the literature.
[b]Batch cycle time includes period for cell propagation and compound production.
[c]Not available.

uct versus the existing products, including the lower priced synthetic product and the higher priced whole plant-derived product. A good case in point is natural vanilla flavor. The annual worldwide consumption of vanilla beans is about 1350 tons, with a value of more than $100 million, whereas the worldwide usage of the synthetic counterpart, vanillin, is approximately 6500 tons, also with a value of $100 million. Use of the concept of "flavor units" puts the materials on a common basis. The vanilla bean has 1.2 flavor units per pound, whereas synthetic vanillin has 17 flavor units per pound. The cost per flavor unit for natural vanilla is approximately $31, about 85 times that for synthetic vanillin. The annual world consumption of natural vanilla is 3.6 million flavor units, compared to 243.0 million flavor units for synthetic vanillin. Natural vanilla obviously has special consumer appeal that permits it to retain a large dollar share of the total vanilla flavor market, although its share of the flavor unit market is relatively small. If 1.5 percent or 3.6 million flavor units of synthetic vanillin were converted to natural vanilla, the volume would equal that of natural vanilla and the sales dollar value would certainly be a multiple of the synthetic counterpart. The positioning of a plant tissue culture-based "Natural Vanilla Flavor" product between the cheap synthetic vanillin and the much more expensive natural vanilla derived from the bean therefore represents a solid business opportunity [9].

The third, and perhaps the most important consideration, is the relationship between the product value and the expected annual usage. As discussed in the next section, a number of process factors, including the type of process, product

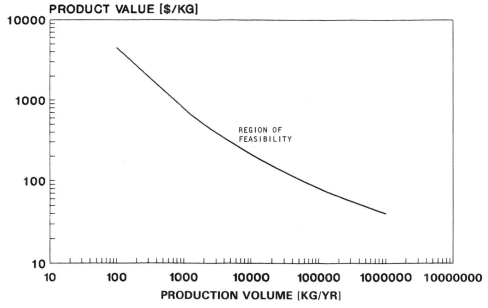

FIG. 8.1 Generalized plot to select candidates for production via plant tissue culture.

yields, batch cycle times, cell growth rates, and the nature of product purification affect the economics of a plant tissue culture-based production process. Making certain assumptions on the current state of plant tissue culture technology, Figure 8.1 presents a generalized plot that may be used to select (or reject) candidates for production via plant cell culture based on the relationship between product value and volume. The area above the curve indicates the region of economic feasibility. If a product lies close to the curve, then a more detailed economic analysis may be necessary, both for estimating manufacturing costs and for estimating the risks and the time-frames involved, including the costs of research and development leading to product commercialization [10].

8.2   TECHNOLOGY CONSIDERATIONS

8.2.1   Establishing Cell Cultures

Establishing plant cells in culture follows a relatively straightforward protocol [11]. Basically, it involves decontaminating the parent plant, taking sections of the appropriate tissue (that is, roots, leaves, stems, or fruits) and placing the cut tissue on a solid medium in a petri dish under totally aseptic conditions. A typical induction medium contains sugars, salts, micronutrients, vitamins, and phytohormones. The quality of the phytohormones (that is, the type and concentration of auxins such as 2,4-dichlorophenoxy acetic acid (2,4-D) or naphthalene acetic acid (NAA) and cytokinins such as benzyladenine or kinetin) are varied to induce a mass of disorganized cells or cell clumps called *callus*. The callus clumps are broken up, suspended in a liquid medium of an appropriate defined composition and agitated on a rotary shaker to form fine suspensions. The individual cells in suspension typically have linear dimensions ranging from 20 to 100 μm. The nature of the cell aggregates in these cultures can vary from predominantly single cells to aggregates of tens and thousands of cells, contingent upon the plant species used and the composition of the growth medium.

8.2.2   Characteristics of Plant Cells

From a manufacturing perspective, a plant tissue culture process for production of colors, food flavors, or fragrances would be similar to a conventional microorganism-based aerobic fermentation process, comprised of a bioreactor section and a downstream processing section. In the bioreactor section, cells grow and produce the product, whereas in the downstream processing section the culture broth is processed to recover and purify the product. However, there are sufficient differences between plant cells and microbial cells to impact both bioreactor design and process economics. The major points of difference include the following: (1) cell size, (2) cell doubling time, (3) batch cycle time, (4) cell settling characteristics, (5) sensitivity to mechanical shear, (6) metabolic activity, (7) oxygen requirements, (8) nature of product accumulation, and finally, and most importantly, (9) product yield.

A typical plant cell is 200,000 to a million times larger than a microbial cell, and the metabolic processes in plant cells are substantially slower. In contrast to *E. coli* that may double every 20 to 40 minutes or a penicillium mold that may double every 4 to 8 hours, plant cells typically exhibit doubling times on the order of 20 to 120 hours. Therefore, while a batch fermentation process based on a fast-growing bacterium may only last for 24 to 36 hours, and, based on a mold for 5 to 7 days, the batch cycle time for a plant cell culture process could easily range from 15 to 45 days, thus tying up expensive fermentation equipment for long periods of time and impacting both investments in capital and costs of manufacturing. In addition, this places more rigorous demands on long-term asepsis of the bioreactor. Because of lower metabolic rates, however, the oxygen requirement for plant cells is 10 to 100 times lower than for yeasts or bacteria (i.e., 2 to 10 mg O_2/g cells per hour, compared to 200 to 300 mg O_2/g cells per hour for fast-growing yeasts).

Plant cells settle rapidly in suspension and therefore need to be kept sufficiently mixed and agitated in a bioreactor. Appropriate agitator design is a basic requirement since plant cells are susceptible to damage by the mechanical shear typically imposed by Rushton impellers in a microbial fermenter. Also, in contrast to molds and bacteria, a majority of plant products are retained intracellularly within the cell vacuole. This limits the level of product accumulation within a reactor. The highest reported level of product accumulation in plant cells is on the order of 6 to 7 g/L (i.e., rosemarinic acid production in *Morinda citrifolia* and berberine production in *Coptis japonica*). This contrasts with 10 to 15 g/L for penicillin, 80 to 100 g/L for monosodium glutamate and 120 to 150 g/L for citric acid, all of which are extracellular products. The normal cell density obtainable in plant cell cultures is on the order of 20 to 30 g/L (dry basis), although recently a cell density on the order of 70 g/L has been reported. The latter result was attributed to an optimal bioreactor design in conjunction with a fed-batch nutrient feeding strategy [12].

8.2.3 Process Schemes

The bioreactor production scheme for plant tissue culture products, as it is for their microbial counterpart, is based on the nature of product accumulation (i.e., whether the product is intracellular or extracellular) and by the production kinetics (i.e., whether production is growth associated or non-growth associated). Most secondary plant products show non-growth-associated or mixed growth-associated production kinetics. Thus, for an intracellular nongrowth-associated product like shikonin or anthraquinones, a two-stage batch or continuous process may be used [12,13]. In such a process, the first stage is operated under conditions that favor biomass production, then in the second stage, conditions are shifted to favor product formation. The feasibility of multistage continuous culture with plant cells has been demonstrated at laboratory scale by at least two groups [13,14]. For a stable extracellular product like capsaicin, a two-stage batch (or continuous) suspension culture-based process separating biomass growth from product formation may also be appropriate; alternatively, a continuous immobilized cell process following batch production of cell biomass may be more cost effective [15]. Unstable extracellular products may require a product capture step linked to the production reactor, using adsorbents such as amberlite XAD resins or water-immiscible phases such as Miglyol [16]. Table 8.4 summarizes the process strategies applicable to plant tissue culture production of food colors, flavors, and aroma chemicals.

8.2.4 Scale-Up Issues

In the past 10 to 15 years, plant cell cultures have been shown to grow in bioreactors of up to 75,000 L in capacity [17–23]. Airlift fermenters, bubble columns, and impeller agitated fermenters (using low-shear marine, Ekato intermig or paddle impellers) have been used. Table 8.5 gives examples of large-scale suspension culture reactors used with plant cell cultures.

Plant cell cultures tend to foam readily in reactors when sparged with air deep in the main body of the culture. This foaming leads to accrual of large bodies of cell mass on the vessel wall at the liquid-air interface. Siliconizing of glass vessels, modification of bioreactor configuration, use of chemical or mechanical defoamers, lowering of overall gas flow by blending oxygen with air, and reduction of the calcium content of the medium have been suggested as appropriate solutions to this problem [12,24,25]. In addition to control of foaming, mechanical shear and oxygen transfer efficiency are important considerations for operation and scale-up of suspension culture reactors. Studies relevant to scale-up parameters for plant cell cultures are briefly discussed below.

Kato et al. investigated the effect of initial volumetric oxygen transfer efficiency (or k_La value based on air–water system) on growth of tobacco cell suspensions in a 15-L flat blade turbine fermenter [20]. The k_La was varied by changing the aeration rate from 0.25 to 1.0 vvm at fixed agitation speeds of 50 and 100 rpm. It was found that over a 6-day cultivation period, the final biomass concentration increased proportionally with the initial k_La over the range of 5 to 10/hour, but the biomass value leveled off as the k_La was increased further to 25/hour. Agitation speeds of 50 or 100 rpm were well tolerated by the cells; however, higher speeds (i.e., 150 to 200 rpm) led to uncontrollable bulking, foaming, and wall growth in the vessel. In a separate study, Smart and Fowler [23] found that biomass yield with *Catharanthus roseus* cultures in an airlift reactor increased as k_La was increased from 5 to 15/hour, but cell yield declined substantially as k_La was further increased to 37.5/hour. They attributed these results either to high levels of dissolved oxygen levels in the broth or to stripping of volatiles from the headspace of the fermenter and speculated that successful growth of plant cells in bioreactors may be possible only within a narrow window of k_La values.

TABLE 8.4 Process Strategies for Plant Tissue Culture-Derived Products

Product Type	Production Kinetics	Cell Type	Process Strategy
Extracellular	Nongrowth associated	Suspension culture	Batch/continuous two-stage process
		Immobilized cells	Continuous process
	Growth associated	Suspension culture	Batch/continuous one-stage process
Intracellular	Nongrowth associated	Suspension culture	Batch/continuous two-stage process
	Growth associated	Suspension culture	Batch/continuous one-stage process

TABLE 8.5 Examples of Large-Scale Suspension Culture Reactors used for Plant Cell Cultures

Plant Species	Product	Bioreactor Capacity and Type	Ref.
Catharanthus roseus	Serpentine	100-L airlift	23
Coleus blumei	Rosemarinic Acid	300-L airlift	22
Lithospermum erythrorhizon	Shikonin	750-L agitated	21
Nicotiana tabacum	Biomass	20,000-L agitated	20
	Biomass	1,500-L bubble column	19
Panax ginseng	Saponins	20,000-L agitated	18
Echinacea purpurea	Biomass	750–75,000-L agitated	17
Rauwolfia serpentina	Biomass	750–75,000-L agitated	17
Panax ginseng	Biomass	750–75,000-L agitated	17

Maurel and Parcilleux [26] reported decreased rates of cell growth in an agitated fermenter with suspensions of *C. roseus* at high rates of aeration and suggested use of carbon dioxide in the inlet gas to counteract adverse effects on cell physiology. Kessell and Carr [27] examined the effect of fixed dissolved oxygen (DO) concentration on growth and differentiation of carrot cells in paddle-agitated bioreactors and observed that cell growth was unaffected by increasing DO above a critical level of 16 percent saturation. Below this level, cell growth as determined by increase in dry weight was linear over time, whereas above this level it was exponential. Low levels of controlled dissolved oxygen led to morphological differentiation of cell suspensions into embryos. As discussed in the next section, such morphological differentiation may result in improved product yields in specific cases. From the above studies, it appears that based on their response to aeration–agitation conditions in bireactors, plant cell cultures may be divided into two categories. The first set which includes cultures of *Nicotiana tabacum*, is relatively insensitive to overgassing, to high k_La values and to high dissolved oxygen levels, and the second set, which includes cultures of *C. roseus*, is more sensitive to changes in gas phase conditions.

The effect of fermenter design on cultivation of plant cells has been investigated by a number of groups [12,28]. For example, Wagner and Vogelmann [28] examined various fermenter designs for growth and metabolite production by plant cell cultures and found airlift fermenters to be superior to impeller-agitated fermenters because of their ability to provide good macro-mixing and oxygen transfer while subjecting the cells to a minimum of mechanical damage. Fujita and Tabata [12] compared cell growth and shikonin production by *Lithospermum erythrorhizon* suspensions in airlift reactors, agitated reactors with modified paddle impellers, and rotating drum reactors. They found that cells grew well in the agitated reactors but that the yield of shikonin was much lower in such reactors than in shake flask cultures. In airlift reactors, major operational problems were caused by foaming and adherence of cells to the wall of the reactor at cell densities exceeding 15 g dry weight/L. In a rotary drum reactor, however, there was no problem with foaming (or wall growth), no cell injury, and good shikonin production. Furthermore, no decrease in shikonin yield was observed as the reactor was scaled up from less than 10 L to up to 1000 L.

Based on all published work and from our own experience, good quality impeller-agitated fermenters, typically used for microbial fermentation, are well-suited to cultivation of a wide range of plant cell cultures after retrofitting with appropriate impellers and spargers. The fermenters can be operated using an aeration strategy involving either a fixed gas flow rate (as determined by k_La values in air–water system) or a fixed dissolved oxygen level [26]. In order to obtain an equivalent process result (i.e., an equivalent rate of cell growth and/or compound production) using the first strategy, the following operating parameters need to be kept a constant at various scales of operation—(1) k_La value as based on air–water system, (2) impeller tip speed, impeller type, ratio of impeller diameter to vessel diameter, (3) gas phase composition, (4) strategy for control of foaming and wall growth, and (5) gas flow rate as expressed in volume gas flow per volume culture per minute.

8.2.5 Entrapped Cell Cultures

Immobilized or entrapped cell cultures offer another alternative to production of plant-derived products. The use of such cultures has been proposed both for technical and for economic reasons. From a technical viewpoint, close cell-to-cell contact in a high-density environment with gradients in levels of gases, substrates, and product intermediates may be critical in inducing or enhancing production of targeted metabolites. From a cost viewpoint, if the product is naturally leaked into the medium, or is induced to do so by use of permeabilization procedures, then entrapment of cells facilitates total reuse of expensively grown biomass and continuous production of products by compact, cost-effective bioreactors.

Brodelius [29], Shuler et al. [30], and Prenosil and Pedersen [31] provide good reviews on plant cell immobilization. Brodelius [29] also lists a number of examples where improved product yields were obtained with immobilized cells as compared to freely suspended cells [29]. As these reviews indicate, three principal methods have proven useful for immobilizing plant cells. They include (1) gel entrapment, (2) entrapment in nongel matrices, and (3) entrapment between membranes or fibers. Of the gel materials, alginate, agar, agarose, chitosan, and α-carrageenan have been used successfully to immobilize plant cells, with good retention of cell viability and the ability to produce the product of interest in batch or in continuous mode. Examples include production of ajmalicine and serpentine by alginate, agar, and agarose entrapped cells of *C. roseus* and of anthraquinones by alginate entrapped cells of *M. citrifolia* [29]. Sustained production of targeted compounds over 6 to 7 months has been reported [32]. Knorr and Teutonico [33] have reported success with chitosan immobilization of *Amaranthus tricolor* cells.

Alternate nongel matrices used with plant cells include polyurethane foam, stainless steel mesh, and nylon mesh. Lindsey and Yeoman [15] have shown successful entrapment of chili pepper cells in polyurethane foam, with a 100-fold improvement in the yield of capsaicin as a result of such entrapment. Kirsop and Rhodes [34] have shown immobilization of hop cells (*Humulus lupulus*) in nylon mesh and stainless steel mesh. Both Shuler et al. [30] and Prenosil and Pedersen [31] report entrapment of plant cells between membranes (or stainless steel screens) and between fibers in a hollow fiber reactor system. In one such system with tobacco cells, a cell density of approximately 65 to 80 g/L was obtained over prolonged reactor operation, with morphological differentiation of cells into compact nodulelike structures and improved phenolic synthesis [35].

To take full advantage of the immobilized cell process concept, the product of interest needs to be secreted into the medium. In some cases, such as in capsaicin production from chili pepper cells and in indole alkaloid production from cells of *C. roseus*, the normal intracellularly stored product was found to be spontaneously released into the extracellular medium under conditions of immobilization. In other cases, permeabilizing agents such as dimethyl sulfoxide, chitosan, and cyclic variations in medium pH were effective [29,33].

A number of standard reactor configurations, including packed bed, fluidized bed, and low-shear impeller agitated reactors are suitable for use with gel-entrapped and non-gel–entrapped matrices. Hollow fiber reactors and flat plate reactors offer still other alternatives. The selection of an appropriate bioreactor format will in large part be dictated by obtainable product yields, reactor cost, ease of operation under aseptic conditions and scaleability.

8.2.6 Economics of Production

Both Goldstein et al. [36] and Sahai and Knuth [37] have reviewed the economics of plant tissue culture processes in great detail. The factors that affect the economics of production of an intracellular product in a batch fermentation process include: (1) cell doubling time, (2) batch cycle time, (3) intracellular product concentration, (4) reactor cell density, (5) product recovery efficiency and (6) annual production volume. Figure 8.2 shows the relationship between product manufacturing cost and intracellular product concentration for a typical product, assuming realistic values for bioreactor cell density ($X = 20$ g/L), cell doubling time ($t_d = 60$ hours), batch cycle time ($T_B = 15$ days), and product recovery efficiency ($\eta = 90$ percent), and an average level of complexity in product recovery and purification steps. The manufacturing cost is projected to range from \$1000/kg to approximately \$50/kg as the product concentration increases from 3 to 25 percent on a dry weight basis and the annual production volume changes from 3000 to 1,000,000 kg. The projected costs are lower if less conservative assumptions are used (e.g., $t_d = 20$ to 24 hours, $X = 40$ to 70 g/L, and $T_B = 10$ days), and recent work does in fact support the use of less conservative assumptions. For example, a number of

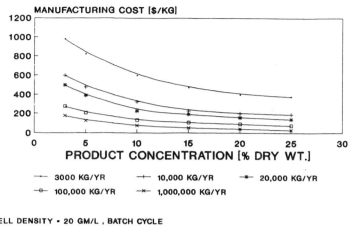

FIG. 8.2 Effect of product yield and annual production volume on manufacturing cost.

plant cell cultures have been shown to have doubling times of approximately 20 to 25 hours under optimal nutritional conditions, with the lowest known value of 15 hours reported for tobacco cells in shake flasks and in a 20,000-L fermenter [19]. Additionally, Fujita and Tabata [12] report obtaining cell densities of approximately 70 g/L in a bioreactor.

The economics of extracellular production via batch fermentation are comparable to the above with two exceptions. First, cost savings may be realized in recovery and/or purification steps because rupturing of cells is avoided. Second, cost savings may result from improved product yields, because the extracellular product concentration (expressed as grams of product per liter of broth) is no longer limited by reactor cell density. Furthermore, if the product is indeed a stable, extracellular entity (for example, capsaicin production in *Capsicum frutescens*), then alternate cost-effective process schemes, such as a continuous process incorporating immobilization (or retention) of cells in the reactor may be applicable. In such a process, cell suspensions still need to be propagated through fermenters of increasing scale prior to use in the final production stage. Therefore, cell doubling time and cell density may still be important to the economics of production, but the frequency of biomass production and the associated costs can be reduced substantially by improving the stability on the immobilized (or retained) biocatalyst. As indicated before, biocatalyst stability on the order of 6 to 7 months have thus far been shown for plant cell cultures [32,38]. Based on obtaining high volumetric productivity, good biocatalyst stability, and reasonable substrate-to-product conversion efficiency in continuous entrapped plant cell systems, manufacturing costs of approximately $25 to $50/kg in the production volume range of 10,000 to 100,000 kg/year may clearly become achievable in the future [36,37]. This assumes that entrapped cell reactors can be scaled up successfully from laboratory to production scale.

8.2.7 Yield Improvements

As discussed earlier, product yield is a key parameter affecting the economics of production in plant cell cultures. It is not surprising therefore that a large proportion of research effort in the past 10 to 15 years has been devoted to this area. The key strategies for yield improvement in plant cell culture systems may be categorized as follows: (1) use of high-product-yielding plants as starting material; (2) optimization of key nutritional factors, including the sources and concentrations of growth regulators, carbon, nitrogen, and phosphorus; (3) optimization of environmental factors, including sources and intensity of illumination, temperature, and gas-phase composition; (4) use of biochemical precursors and elicitors; (5) selection of high-product-yielding cell lines from the heterogeneous cell population by visual or chemical means or by resistance to certain growth-inhibitory substances; (6) induction of morphological differentiation, and (7) creation of artificial accumulation sites for the product. Each of these strategies is discussed briefly below. The reader is also referred to recent publications that review this area in great depth [39–43].

8.2.7.1 *Using High-Product-Yielding Parent Plants*   It is customary for an industrial plant tissue culture program to initiate tissue cultures from plants that have been shown to yield high levels of the product *in vivo*. If one assumes that the biosynthetic potential to form a secondary plant metabolite is genetically fixed, then the use of high-product-yielding differentiated plants as source material should maximize the possibility of obtaining high product yields in cell culture. Zenk et al. [44] were the first to conduct a comprehensive study to establish such a correlation. Using production of the alkaloids serpentine and ajmalicine from *C. roseus* as the model system, they showed that in fact high-product-yielding parent plants resulted in high-product-yielding cell cultures, whereas low-product-yielding parent plants led to low-product-yielding cell cultures. The findings of Zenk et al. [44] were further substantiated by the work of Kinnersley and Dougall [45], who showed that a similar correlation was valid for tissue cultures derived from high- and low-nicotine-yielding plants of *N. tabacum*. Note, however, that it is important to use all parts of the high-product-yielding plant to initiate cultures, not just the part with the highest concentration of product of interest, because the product may be synthesized in one part of the plant and then transported to another site for accumulation [46].

8.2.7.2 *Optimizing Nutrition*   The common plant tissue culture media, such as Murashige and Skoog (MS), Linsmaier and Skoog (LS) and Gamborg's B5 (B5), are nutrient rich and support prolific growth of a diverse range of cell cultures. It is widely acknowledged now that such nutrient-rich conditions favor

enzymes on the primary pathway toward protein and cell wall synthesis, and inhibit those controlling expression of secondary metabolites [47]. This hardly is surprising if one considers the parallel with microbial physiology, where nutrient-limiting and stress-inducing conditions have traditionally been used to promote synthesis of secondary metabolites such as antibiotics. This has prompted the concept of formulating the "production medium" as separate from that supporting routine cell propagation. In such a production medium, either the cells stop growing completely and switch immediately to product synthesis or additional cell growth occurs prior to product formation, with the bulk of production occurring as cells enter the stationary phase (that is, as cell growth slows down or stops completely). An example of the former is production of shikonin in cell cultures of *L. erythrorhizon* [48], whereas examples of the latter include production of indole alkaloids in suspensions of *C. roseus* and of anthraquinones in suspensions of *M. citrifolia* [12,28,44].

The key nutritional factors that play an important role in regulation of secondary metabolism in plant cell cultures include: (1) carbon source and amounts, (2) nitrogen source and amounts, (3) ratio of carbon to nitrogen, (4) levels of phosphorus, and (5) source and concentration of growth regulators. In addition, minor components may be important in specific cases, such as the role of copper in the production of shikonin in *L. erythrorhizon* cultures [21].

Carbon Source and Amounts. Sucrose is the most commonly used carbon source in plant cell cultures, and generally is used in the 2 to 3 percent range for routine cell propagation. An increase in sucrose concentration to the 4 to 14 percent range has been shown to improve yields of metabolites in a number of cases, including production of shikonin, indole alkaloids, diosgenin, and rosemarinic acid [44,49–51].

In contrast to microbial production of secondary metabolites, where the role of slowly metabolizable carbon sources in improving product yields has been clearly established, there is a limited body of data on the effect of carbon sources beyond sucrose, glucose, and fructose on production of secondary metabolites in plant cell cultures. In one such example, of the 14 carbohydrates tested at the 2 percent level, sucrose gave the highest product yield [52].

Carbon-to-Nitrogen Ratio. Alteration of the carbon-to-nitrogen ratio is a classical approach in microbial physiology to alter product yields. This parameter also has been shown to have a major effect on tissue culture production of phenolics, anthocyanins, and diosgenin [51,53,54].

Nitrogen Source and Amounts. The typical plant tissue culture medium, such as MS, LS, or B5, has both nitrate and ammonium as sources of nitrogen. However, the ratio of the ammonium to nitrate and overall levels of total nitrogen have been shown to markedly affect production of secondary plant products. For example, reduced levels of NH_4^+ and increased levels of NO_3^- promoted production of shikonin and betacyanins, whereas higher ratios of

NH_4^+ to NO_3^- increased production of berberine and ubiquinone [48,55–57]. Reduced levels of total nitrogen improved product yields in quite a few cases, including production of capsaicin in *C. frutescens*, anthraquinones in *M. citrifolia*, polyphenols in Paul's Scarlet Rose and anthocyanins in *Vitis* species [47,50,52,54]. The use of organic nitrogen sources such as casein hydrolysate, peptone, and yeast extract were shown to inhibit production of shikonin in *L. erythrorhizon*, but to promote synthesis of rosemarinic acid in suspensions of *Coleus blumei* [44,49].

Phosphate Levels.   The phosphate concentration in the medium can have a major effect on production of secondary plant products. Sasse et al. [58] have given a number of examples to show that a medium limited in phosphate either induces or stimulates both the product and the levels of key enzymes leading to the product. Thus reduced phosphate levels induced production of ajmalicine and phenolics in *C. roseus*, of caffeoyl putrescenes in *N. tabacum*, and of harman alkaloids in *Peganum harmala*. Similar results were obtained in a separate study for the production of betacyanins in callus cultures of *Beta vulgaris* [55]. In contrast, increased phosphate was shown to stimulate synthesis of digitoxin in *Digitalis purpurea* and of betacyanins in *Chenopodium rubrum* and *Phytolacca americana* [55,59].

Growth Regulators.   Both the source and concentration of growth regulators (i.e., auxins, cytokinins, gibberellin, and abscisic acid) have been shown to have a profound effect on product yields in plant cell cultures. The growth regulator, 2,4-D, very commonly used for good and stable growth of callus and cell suspensions, has been shown to inhibit production of secondary metabolites in a large number of cases. In such cases, total elimination of 2,4-D or replacement of 2,4-D by NAA or indole acetic acid (IAA) has been shown to enhance production. Examples include the production of the following: anthocyanins in suspensions of *Populus* and *Daucus carota*; betacyanin in suspensions of *Portulaca*; nicotine in suspensions of *N. tabacum*; shikonin in suspensions of *L. erythrorhizon*; anthraquinones in suspensions of *M. citrofolia* and phenolics in suspensions of *N. tabacum* [52,55,60–62]. In a limited number of cases, stimulation by 2,4-D has been observed, as for example, in carotenoid synthesis in suspensions of *Daucus carota* [63].

Cytokinins have different effects depending on the type of metabolite and species concerned. Thus, kinetin stimulated production of anthocyanins in *Haplopappus gracilis* but inhibited formation of anthocyanins in *Populus* and of carotenoids in carrot cell cultures [60,63]. Gibberellic acid and abscisic acid are reported to suppress production of anthocyanins and betacyanins in a number of cultures [55,60].

8.2.7.3 Optimizing the Environment
The key factors in the culture environment that affect product synthesis include: temperature, illumination, and gas-phase composition.

Temperature.   Plant cell cultures are normally maintained in the temperature range of 24°C to 28°C. Very few studies have examined the optimum temperature for product formation as distinct from cell growth. Watts et al. reported that synthesis of limonene in celery cultures was induced by maintaining the cells at refrigeration temperature (that is, at 4°C) for the first 5 days and then shifting to the normal 25°C [64]. Ikeda et al. observed a higher yield of ubiquinone in tobacco cells at 32°C when compared to either 24°C or 28°C [56]. Courtois and Guern [65] observed a 12-fold higher production of crude alkaloids in cell cultures of *C. roseus* at 16°C as compared to the normal 27°C.

Illumination.   With a few exceptions, such as in cultures of *Daucus carota* and *Vitis* hybrids, anthocyanin accumulation in cell cultures has been shown to be strongly stimulated by light [60]. Illumination was found to affect the composition of sesquiterpenes in callus cultures of *Matricaria chamomilla* [42]. Farnescene and bisabolol were synthesized when the cultures were illuminated, but the biosynthesis was shifted to spathulenol at the expense of these two compounds in conditions of total darkness. In another example (i.e., callus cultures of *Citrus limon*), exclusion of all light promoted accumulation of monoterpenes [42].

Gaseous Environment.   Ambid and Fallot [66] investigated the effect of the composition of the gaseous environment on production of volatiles by fruit suspension cultures. They reported that addition of carbon dioxide stimulated synthesis of monoterpenes by Muscat grape suspensions and induced formation of linalool. With apple cultures, use of high levels of carbon dioxide induced production of flavor components characteristic of apple. Kobayashi et al. [67] reported that use of carbon dioxide at the 2 percent level was critical to prevent cell browning and sustain berberine production in suspensions of *Thalictrum minus* in bubble column reactors.

8.2.7.4 *Inducing Morphological Differentiation*   A number of plant species that fail to accumulate compounds in callus or suspension cultures have been shown to do so under conditions of morphological differentiation. The level of differentiation can vary from simple aggregation of cells with formation of specialized glands, as in cultures of *Mentha piperita*, to formation of actual roots, leaves, or shoots, as in tissue cultures of eucalyptus, onion, pyrethrum and *Digitalis* [42].

8.2.7.5 *Creating Artificial Accumulation Sites for the Product*   A low accumulation of characterizing compounds in cell culture in a number of cases may not be due to a lack of key biosynthetic enzymes, but rather due to feedback inhibition, enzymatic or nonenzymatic degradation of the product in the medium, or volatility of substances produced. In such cases, it should be possible to increase net production by addition of an artificial site for product accumulation, for example, by use of a second phase. Beiderbeck and Knoop

[16] have examined the concept of the two-phase culture with plant cells and have shown two types of second phases to be effective: (1) lipophilic second phase, such as Miglyol 812, liquid paraffin, or Lichroprep RP8—suitable for lipophilic products, and (2) solid nonpolar phases, such as XAD-resins and activated charcoal (AC). As examples, addition of AC led to 20 to 60-fold improvement in yields of coniferyl aldehyde in *Matricaria chamomilla* suspensions, and addition of Miglyol or silica gel RP8 stimulated anethol production by cell cultures of *Pimpinella anisum*. Also, highly volatile monoterpenes from *Thuja occidentalis* such as α-pinene, β-pinene, myrcene and limonene were produced in cell suspensions only in the presence of Miglyol [16,42].

8.2.7.6 *Using Precursors and Elicitors*   Attempts to induce or increase the production of plant metabolites by supplying precursors or intermediate compounds have been effective in many cases. However, as Ibrahim [68] notes, unsuccessful attempts to induce or increase product yields in specific cases may be due to our lack of knowledge concerning the time of addition of such compounds, their uptake, and their compartmentation in relation to the enzymes involved in their utilization. Thus, addition of L-phenylalanine led to improvement in rosemarinic acid yield in cultures of *Coleus blumei*, hydroxy cinnamoyl esters in apple cell cultures, polyphenol in tobacco cells, and naphthoquinones in *Lithospermum* cell cultures [68]. Similarly, anthocyanin synthesis in carrot cell cultures was restored by addition of dihydroquarcetin, naringen, or inodoctyl. Furthermore, addition of leucine led to enhancement of the volatile monoterpenes, α- and β-pinene in callus cultures of *Perilla frutescens*, whereas addition of geraniol to Lady Seton Rose cultures led to accumulation of nerol and citronellol [42].

Plants produce secondary metabolites in nature as a defense mechanism against attacks by pathogens. Plants have been found to elicit the same response as the pathogen itself when challenged by compounds of pathogenic origin (i.e., elicitors). The role of elicitors in inducing or enhancing accumulation of secondary metabolites in plant cell cultures has received increasing attention in the last 10 years or so. Eilert presented a good review of this area [69]. As discussed in this review, the range of biotic elicitors useful in plant cell cultures extends from chitosan and purified yeast extract to autoclaved spores of *Botrytis*, *Rhodotorula*, *Phytophthora*, *Fusarium*, and *Pythicum*. Depending on the culture conditions, the cell line used and the product of interest, the use of elicitors has resulted in a 10- 1,000-fold improvement in product yields. However, one should note that elicitors derived from pathogenic sources may have limited usefulness for commercial manufacturing of food ingredients.

8.2.7.7 *Selecting High-Product-Yielding Cell Lines*   Plant cells in culture form a very heterogenous population. The yields observed in suspensions often represent the mean of a population of cells that shows wide variation in product-forming ability on an individual basis. This therefore offers a source material to establish high-product-yielding cell lines. Alternatively, cell suspensions can be treated with mutagens, such as γ-irradiation or ethyl methane

sulfonate (EMS) to increase the frequency of variant cells in culture. Methods that have been used for selection include: (1) visual selection, based on pigmented cell lines, (2) selection for resistance to certain growth-inhibitory substances, and (3) selection by direct chemical analysis. Widholm [70], Zenk [7] and Tabata et al. [71] provide good reviews of prior work in this area. As discussed in these reviews, for colored compounds such as anthocyanins, shikonin, berberine, and betacyanins, visual selection of cells has been very effective in improving yields. For noncolored compounds, chemical selection procedures based on rapid and sensitive immunoassays were shown to be effective in selection of high-yielding strains of *C. roseus* (serpentine and ajmalicine) and *N. tabacum* (ubiquinone-10) [44,72]. Selection for resistance to 4-methyl-tryptophan (4-MT) with *Peganum* cells was shown to produce strains with 100-fold higher levels of serotonin than the wild type [70].

8.2.8 Stability and Preservation of High-Yielding Cell Lines

For commercial exploitation of plant cell cultures for production of food and perfumery ingredients, it is important that the productivity potential of the high-yielding cell lines be maintained for long periods of time over repeated subcultures. Zenk [7] noted that, with regard to their production stability, there are at least two kinds of cell cultures. One kind produces specific compounds in high yields and almost all cells are producing. Such cultures are not clonally selected, but are high yielding from the beginning and are stable for many years. Examples include diosgenin production in *Dioscorea deltoidea*, ginsengosides in *Panax ginseng*, anthraquinones in *M. citrifolia*, visnagin in *Amni visnaga* and rosemarinic acid in *Coleus blumei*. Regarding *M. citrifolia*, Zenk [7] recalled that this culture had been maintained in his laboratory for over 7 years (or around 180 transfers) without any change in productivity.

A second type of cell culture is represented by *C. roseus* (indole alkaloids) and *Nicotiana rustica* (nicotine), where high-yielding cell lines were obtained by clonal selection only and the strains showed considerable biochemical instability, with rapid decline in product yield on subculture. The problem was resolved by repeated clonal selection of high-yielding cells; similar strategies were used to obtain stable cell lines in *C. japonica* (berberine) and *L. erythrorhizon* (shikonin) [12,73].

An alternative to repeated clonal selection is freeze preservation of high-yielding cell lines. Kartha [74] gave an excellent review of this critical developing field and noted that cryopreservation of cultures has been effective in retaining the ability for compound synthesis in *C. roseus* (indole alkaloids), *D. carota* (anthocyanins), *D. lanata* (cardenolides), *Lavandula vera* (biotin), and *Dioscorea deltoidea* (diosgenin) [74].

8.3 REVIEW OF PRIOR TISSUE CULTURE WORK

Over the past 30 years or so, with the exception of plant pigments, the bulk of plant tissue culture research had been focused on producing pharmaceutical compounds in culture. Only recently has the focus shifted to the area of food additives and cosmetics. Shikonin, the first commercial product of plant tissue culture work, was originally conceived as a pharmaceutical (that is, as an antiinflammatory agent) but found use in cosmetics as a biolipstick [75,76]. The product, developed by Mitsui Chemicals, is currently being marketed by Kanebo Ltd. Kanebo is also working on commercializing a geranium perfume using plant cell culture [76]. ESCAgenetics Corporation (San Carlos, CA) is currently involved in scale-up of its process to produce vanillin and related vanilla flavor components from *Vanilla* cells in culture [77].

The following is a brief summary of prior work conducted on production of food colors, food flavors, and volatile oils in plant cell culture.

8.3.1 Food Colors

The four classes of compounds comprising the food colors include anthocyanins, betalaines (that is, betacyanins and betaxanthines), carotenoids (that is, β-carotene, bixin, norbixin), and crocin. Of these, by far the largest number of studies have been done on anthocyanins and betalaines. Table 8.6 gives examples of prior tissue culture work on food colors and the highest product yield obtained in these cultures.

Seitz and Hinderer [60] have given an excellent summary of prior tissue work on anthocyanins. In this review, they note that some 27 different species have been shown to produce anthocyanins in plant cell culture, ranging from grape and carrot to petunia and eucalyptus. Seven different anthocyanidins were detected in these cultures: cyanidin (17 species), delphinidin (five species), malvidin (four species), petunidin (three species), pelargonidin (two species), peonidin (one species), and hirsutidin (one species). The authors concluded that plant cells cultivated *in vitro* retain the capacity to produce the same spectrum of anthocyanidins as that produced *in vivo*. By far the highest concentration of anthocyanins has been shown to accumulate intracellularly in cell cultures of grape (that is, at levels approaching 16 percent dry weight). The

TABLE 8.6   Examples of Tissue Culture Production of Food Colors

Product	Product Yield in Tissue Culture (% Dry Wt.)	(mg/l)	Plant Species	Ref.
Anthocyanins	13–16	830	*Vitis* hybrids	53
Betacyanins	1	100	*Chenopodium rubrum*	79
Carotenoids	0.04–0.07[a]	—	*Lycopersicon esculentum*	80
Crocin	0.3[b]	—	*Crocus sativus*	82

[a]Estimated based on cell fresh weight to dry weight ratio of 10–20.
[b]Estimated assuming (i) crocin concentration in tissue culture is 17% of that in parent plant, and (ii) parent plant is 2 percent crocin on a dry wt. basis.

key nutritional factors that affect anthocyanin accumulation in *Vitis* cultures include the concentration of the carbon source, the nitrogen-containing salts, and phosphate. High levels of sucrose (approximately 8 percent) and low levels of phosphate and nitrate enhance anthocyanin production in culture [53]. Callebaut et al. [78] showed that relatively high concentrations of anthocyanins (i.e., 3 percent dry weight) can be produced and sustained through repeated subcultures of *Ajuga* species using relatively cheap cheese whey as the sole carbon source.

Bohm and Rink [55] have given a detailed review on plant tissue culture production of betalaines. They conclude that five of the ten betalain-producing plant families (i.e., the *Centrospermae*), including species from *Amaranthus*, *Beet*, *Chenopodium*, *Phytolacca*, and *Portulaca*, produce similar compounds in culture. Both betacyanins, the red pigments and betaxanthines, the yellow pigments, were produced in culture. The highest concentration of betacyanins (1 percent dry weight) was obtained in cell cultures of *Chenopodium rubrum* [79].

Fosket and Radin [80] reported the production of carotenoids in cell cultures of tomato (that is, *Lycopersicon esculentum*). They found that dark-grown tomato suspension cultures contain low levels of carotenoids, of which lycopene and β-carotene were the most abundant. However, the addition of the bioregulator 2,4-chlorophenyl-thio-triethylamine (CPTA) to the culture medium resulted in a 60-fold increase in total carotenoids over a 14-day period. The level of carotenoids accumulated in this culture was on the order of 0.04 to 0.08 percent dry weight.

Two sets of investigators have recently reported the production of crocin, the active coloring material in saffron, in plant tissue cultures [81,82]. Nawa and Ohtani [81] detailed production of a number of yellow pigments in callus cultures derived from *Gardenia jasminoides* and identified crocin as the major component. They noted that the level of pigments in the callus was approximately 12 to 16 percent of that in the parent plant tissue. Sarma [82] reported production of crocin (the color) and picro-crocin (the characteristic bitter taste of saffron) in tissue cultured stigmas of *Crocus sativus*. The levels of crocin and picro-crocin in tissue culture were lower (by 6 and 9 times, respectively) compared to the parent plant. The material produced via tissue culture was subjected to sensory analysis, and was clearly identifiable as saffron by the sensory panel. The reader is referred to Chapter 7 of this book for an in-depth discussion of pigment generation in other biological systems.

8.3.2 Food Flavors

In contrast to food colors, a limited number of studies have been carried out on production of food flavors via plant cell culture with varying degrees of success. The flavors investigated include citrus, apple, strawberry, black currant, grape, licorice, vanilla, celery, cocoa, onion, garlic, and chili pepper. The following is a brief summary of these investigations.

8.3.2.1 Citrus Flavor Bricout and Paupardin [83] and Drawert and Bergern [84] examined callus and suspensions derived from a number of *Citrus* species and reported that none of the volatile components characteristic of citrus were produced in culture. However, Cresswell et al. [85] recently demonstrated production of sabinene and octanol as key constituents in callus and suspension cultures of *Citrus limon*. Tisserat et al. [86] approached tissue culture production of citrus somewhat differently. Instead of dealing with dedifferentiated callus and suspensions, these investigators cultivated juice vesicles of the lemon fruit on solid medium *in vitro* and showed that the key lemon flavonoids characteristic of tree-grown lemons, hesperidin and eriocitrin, can be produced in the differentiated cultures.

8.3.2.2 Apple Flavor Ambid and Fallot [66] investigated production of apple flavor in cell culture. They found that apple cell suspensions grown under standard culture conditions do not produce volatile compounds characteristic of the fruit, with the exception of ethanol and ethyl acetate. However, in the presence of fatty acid precursors and a high carbon dioxide concentration (i.e., 20 percent on a volumetric basis in the gas phase), certain compounds characteristic of the natural aroma of the fruit were formed, including isobutanol, 1-hexanol, ethyl butyrate, and ethyl caproate.

8.3.2.3 Strawberry Flavor Hong et al. [87] examined flavor production in callus and suspension cultures of *Fragaria ananasa* and were unable to detect volatile components characteristic of the intact fruit. However, addition of flavor precursors such as short-chain fatty acids and α-keto acids resulted in production of ethyl butyrate, butyl butyrate, butanal, and butanol. At ESCAgenetics, we have examined a wide germplasm base of *Fragaria* species in cell culture and have shown feasibility of *de novo* production of a number of strawberry flavor components, including ethyl acetate, ethyl butyrate, methyl propionate, isopropyl acetate and ethyl hexanoate in callus and cell suspensions [88].

8.3.2.4 Black Currant Flavor In a recent study, Enevoldsen et al. [89] reported production of α-terpineol and limonene in cell suspension cultures of black currant. Although limonene was present in trace quantities, α-terpineol accumulated up to a level of 100 ppm on fresh tissue weight basis.

8.3.2.5 Grape Flavor Ambid and Fallot [66] reported that when suspension cultures derived from Muscat grapes were cultivated in the presence of air, α-terpeniol and nerol were obtained in small amounts as the major monoterpenes. However, increased concentration of carbon dioxide in the gas phase stimulated production of the above compounds and also induced the production of a new compound, linalool. Ambid et al. [90] further showed that the above suspension cultures could convert exogenous citral into the corresponding monoterpenic alcohols, nerol and geraniol, and the ester, geranyl acetate. We have examined callus and suspension cultures derived from Concord grapes and have found *de novo* production at the parts per million level of methyl octanoate, 1-hexanol, trans-2-hexenal, and methyl anthranilate. Also, some of the Concord grape suspensions were able to convert exogeneously fed anthranilic acid to methyl anthranilate at conversion efficiencies approaching 10 percent.

8.3.2.6 Licorice Flavor There is a controversy in tissue culture literature as to whether glycyrrhizin, the key flavor chemical in licorice, can be produced in cell culture. In 1973, Tamaki et al. reported production of glycyrrhizin by callus and suspension cultures of *Glycyrrhiza* sp. in substantial quantities (i.e., at levels approaching 3 to 4 percent dry weight) [91]. However, Hayashi et al. [92] recently reported that despite all attempts, they were unable to produce detectable amounts of glycyrrhizin in callus and cell suspensions of *Glycyrrhiza glabra*.

8.3.2.7 Vanilla Flavor We have examined production of key components of natural vanilla in cell suspensions derived from *Vanilla fragrans* and *Vanilla phaeantha*. Vanillin as well as some of the other phenolic constituents of cured vanilla beans, including 4-hydroxybenzaldehyde, vanillic acid, and 4-hydroxybenzoic acid, were produced in culture [93]. Initial studies showed accumulation of 100 mg vanillin per liter culture over a 14-day period in a production medium containing cells at a low density (4.5 g dry weight cells/L) and powdered activated charcoal as an absorbent [93]. Thus, with little or no increase in cell mass during the course of vanillin production under these conditions, the vanillin yield obtained was 2.2 percent on a dry weight basis. This compares with 1 to 3 percent vanillin content in cured vanilla beans. Subsequent optimization work improved the cell doubling time from 106 hrs to 50 hrs and the yield of vanillin from the initial 100 mg/L to over 1000 mg/L. The latter is equivalent to 8 percent vanillin on a dry weight basis. Some of the slower growing cell lines (cell doubling time—106 hrs) have produced vanillin levels of the order of 1900 mg/L under high cell density conditions (27 g dry weight cells/L). However, these cultures proved to be unstable and unsuitable for propagation in bioreactors. The vanillin production process was successfully scaled up from shake flasks to an impeller agitated 72L fermenter. Scale-up beyond this point is currently in progress [77].

8.3.2.8 Celery Flavor The main flavor components characterizing celery (or *Apium graveolens*) are the terpenes limonene and selinene, as well as certain phthalides. Collin [94] and Watts et al. [95] carried out extensive work on production of celery flavor components in cell culture. They reported in 1984 that suspensions derived from *Apium* sp. did not produce celery flavor components under normal conditions of cultivation [95]. However, when the auxin 2,4-D (or NAA) was replaced by 3,5-D, highly aggregated green cultures were formed, with significant synthesis of phthalides. In a recent review, Collin [94] indicated that the levels of phthalides in celery suspensions obtained subsequently declined upon subculturing, whereas the levels of limonene increased to levels approaching those found in the intact plant.

Follow-on work showed that aggregated cells in a 2,4-D medium could also produce limonene, 3-butyl phthalide, and sedanenolide. In an attempt to increase the genetic variability of celery cultures and to select for high-yielding clones, a number of suspensions with celery aroma showing a wide variation in levels and composition of terpenes and phthalides were selected [95]. Again, on subculturing, the high-yielding cell clones showed a progressive decline in production of flavor compounds. Collin [94] concluded that slow growing, aggregated, greening or partially differentiated cell cultures of celery may be ideal for production of aroma components. Immobilization as a means to induce differentiation and stabilize production was suggested.

8.3.2.9 Cocoa Flavor

8.3.2.9 Cocoa Flavor   Limited success has been obtained in producing cocoa flavor components in de-differentiated cell cultures. Townsley [96] did not observe production of flavor precursors in culture derived from *Theobromo cacao*, but reported that a cocoa aroma was produced when the cultures were maintained at a roasting temperature. Jalal and Collin [97] showed that the polyphenol compounds in callus cultures were different and somewhat fewer in number than in the explant tissues [97].

8.3.2.10 Onion Flavor   Initial studies conducted in the 1970s showed that undifferentiated callus cultures derived from onion contain fewer than 10 percent of the flavor compounds present in the intact bulbs [98–100]. Methyl cysteine sulfoxide was detected in small amounts, whereas ethyl-, propyl-, and *trans*-propenyl-L-cysteine sulfoxide were absent in callus culture. However, differentiation of onion callus into roots and shoots was reported to yield types and levels of onion flavor precursors comparable to those found in the intact plant [99,101]. A recent paper by Collin and Musker [102] indicates that under the right nutritional or hormonal conditions, undifferentiated callus cultures do produce key onion flavor precursors, including *trans*-propenyl-L-cysteine sulfoxide. Apparently the use of specific growth regulators, picloram and benzylaminopurine (BAP), leads to induction of flavor precursors in culture.

8.3.2.11 Garlic Flavor   The key flavor components in garlic (i.e., *Allium sativum*) include allinin and methyl cysteine sulfoxide. Malpathak and David [103] reported that callus cultures from garlic produce large amounts of allinin and that using γ-irradiation or the mutagen EMS (ethyl methane sulfonate) led to substantially improved allinin yields. A Japanese patent application by Ajinomoto describes production of allinin and methyl cysteine sulfoxide by garlic cell cultures [104]. Both irradiation (continuous or discontinuous) and presence of sulfate ions in the medium are considered essential to efficient production of the two flavor compounds *in vitro*.

8.3.2.12 Chili Pepper Flavor Yeomen's group at the University of Edinburgh has done extensive work on the production of capsaicin, the key flavor component of chili pepper, by cell cultures of *Capsicum frutescens* [15,47]. These researchers showed that initially only trace quantities of capsaicin were produced in undifferentiated cell suspensions, but when a product precursor such as vanillylamine or isocapric acid was added to the medium, or when cell growth was restricted (i.e., by immobilization, by lowering of sucrose and nitrate levels in the medium, or by use of inhibitors of protein synthesis), then an improved accumulation of capsaicin followed. Capsaicin levels comparable to those found in the intact fruit were obtained by following such strategies. Additionally, the predominantly intracellular capsaicin was secreted into the medium.

8.3.3 Volatile Oils

Quite a few reports are available on the production of volatile oils in plant cell culture [104a]. The following discussion summarizes tissue culture production of the following oils: chamomile oil, mint oil, rose oil, Eucalyptus oil, Coriander oil, geranium oil, jasmine oil, and aniseed oil. Mulder-Krieger et al. [42] and Collin [94] provide a good survey of this area.

8.3.3.1 Chamomile Oil Production of chamomile oil in plant tissue culture has been investigated by a number of researchers. Szoke et al. [105] reported that callus and cell suspensions derived from *Matricaria chamomilla* accumulated volatile components of chamomile oil at levels of 0.14 and 0.06 percent on a dry weight basis, respectively, as compared to 0.3 to 0.6 percent found in the intact plant. Reichling et al. obtained 0.2 percent of the oil in callus that had differentiated [106,106a]. Bisson et al. [107] examined the use of a two-phase system for production of chamomile oil by cell suspension cultures of *M. chamomilla*. They found that use of a lipophilic phase, such as Miglyol, or an adsorbent, such as RP-8 (commonly used in reversed phase liquid chromatography) stimulated accumulation of key volatile components of the oil.

8.3.3.2 Mint Oil The characterizing flavor components of peppermint oil include menthol and menthone. The minor components include isomenthol, isomenthone, methyl acetate, pulegone, piperitone, and limonene. Extensive research has been done by various groups since the 1960s on production of key components of mint oil via plant cell culture. Earlier work by Lin and Staba [108] and Becker [109] showed that undifferentiated callus cultures of *Mentha piperita* lacked the ability to produce the essential oils, and that structural differentiation was needed to induce product synthesis. However, later investigations by Bricout and Paupardin [110] and Kireeva et al. [111] showed that certain components of mint oil were produced in undifferentiated callus cultures, but that the composition of the oil was different from that of the intact plant. Kireeva et al. [111] found levels of oil in the callus comparable to those found in the plant (i.e., 3.6 percent on a dry weight basis), with equal amounts of the product in the cells and in the medium. The product composition in the callus was 8 percent menthol, 5 percent menthone, and 15 percent pulegone, compared with the plant containing 49 percent menthol, 27 percent menthone, and 6 percent pulegone. Bricout and Paupardin [110] found pulegone and menthofuran as the primary components of mint oil in cell cultures of *M. piperita*, with only trace amounts of menthol and menthone present. These investigators also showed that the addition of colchicine was effective in increasing yields of mint oil constituents in culture. They found that this was closely linked to an increase in the number of neoformed secretory glands in the callus material.

8.3.3.3 Rose Oil There are conflicting reports on production of rose oil in tissue cultures of *Rosa damascena*. Kireeva et al. [113] showed that callus cultures of *R. damascena* produce essential oils, aromatic alcohols, and glycoside-bound terpenoids similar to those found in the intact plant. On the other hand, Banthorpe and Barrow [114] could not detect any monoterpenes in callus and suspensions of *R. damascena*, although the critical biosynthetic enzymes were present in the tissue culture systems.

8.3.3.4 Eucalyptus Oil Gupta and Mascarenhas [115] found that undifferentiated callus cultures of *Eucalyptus citriodora* do not contain the monoterpenes found in the intact plant, but that organogenesis or differentiation of callus into roots stimulates oil production. Similar results were obtained by Yamaguchi and Fukuzumi [116].

8.3.3.5 Coriander Oil Sardesai and Tipnis [117] found that callus cultures of *Coriandrum sativum* produced geraniol, but none of the other monoterpenes associated with the flavor principles of coriander oil was detectable [117].

8.3.3.6 Geranium Oil Although a number of papers are available on the growth of geranium cells in tissue culture, none of the reports deals specifically with the production of geranium oil flavor components [118]. However, Kanebo Ltd. of Osaka claimed success in producing geranium fragrance components in cell culture [76].

8.3.3.7 Jasmine Oil Two sets of reports are available on production of jasmine oil by plant cell cultures. Banthorpe et al. [119] reported production of nerol, linalool, citronellal, citronellol and citral in callus cultures of *Jasminum officinale*. Tomoda et al. [120] reported that benzyl acetate, a key jasmine oil component was present in cell culture, but that the aroma of the callus culture of *Jasminum sambac* was different from that of the intact plant.

8.3.3.8 Aniseed Oil Reichling et al. [106] demonstrated accumulation of key components of aniseed oil in undifferentiated callus and suspension cultures of *Pimpinella anisum*. The accumulation of the oil in callus and suspensions ranged from 1 to 4 mg/100 g fresh weight, compared to 100 to 400 mg oil/100 g fresh weight in the intact plant. The major chemical components produced in cell culture included anethol, myristicum, β-bisabolene, and pseudo-isoeugenol-2-methyl butyrate. RP-8, an adsorbent commonly used with reversed phase liquid chromatography, stimulated accumulation of anethol in cell cultures of *P. anisum* [42].

8.4 CONCLUSIONS AND FUTURE PROSPECTS

The technology of plant tissue culture has come a long way in the last decade. A number of processes have been commercialized or are approaching commercialization, including shikonin, taxol, ginseng, geranium and vanilla. Broadening of this product base will require a greater participation by the flavor and fragrance industry and the willingness to invest in long-term research and development. The ability to select the right product with an appropriate value–volume relationship, to improve product yields, and to scale up the system to multithousand liter levels are critical.

For high-value products, current tissue culture technology is adequate; however, for medium- or low-value products, innovations in technology and in strategy are needed. These could entail ''forced'' secretion of predominantly intracellular products and use of high-density biocatalytic reactors for continuous production of the secreted products, or improvements in critical process parameters that affect costs of production of intracellular products. Alternate approaches to reduce costs may include using a multipurpose facility, retrofitting existing fermenters, producing a number of products from the same cell line, or using simple, cheap bioreactors.

The future prospects for application of plant tissue culture technology to production of colors, flavors, and fragrances are indeed bright as stated so eloquently in the following editorial comment from the New York Times [121]:

Vanilla plants grow mainly in Madagascar and are not cheap. Thus scientists in California are trying to cultivate the essence of vanilla in a laboratory, using living cells taken from the plant. Chocolate without the cocoa plant, orange juice without trees and morphine without the poppy are all contemplated in the race to develop cell culture technology.

The search for rare substances that added flavor and color to life also propelled Christopher Columbus, Vasco da Gama and Ferdinand Magellan across vast, mysterious oceans. The prizes then were silk and precious stones from China, pepper and cinnamon from the fabled Spice Islands. Today's explorers are scientists whose search is confined mostly to quiet laboratories, when they fill test tubes instead of cargo holds and scan computer screens, not watery blue horizons. Yet their realm remains vast and exciting, limited only by the bounds of human imagination.

REFERENCES

1. Biotech Forum Europe, *Plant Biotechnology to Play Major Role in Plant-Derived Chemicals*, Vol. 8, No. 1, p. 8, Biotech Forum Europe, 1991.
2. McCormick & Wild, Natural flavors continue dramatic gains in flavor market. *Food Eng.* **63**(6), 34 (1991).
3. Chemical Week, Cosmetics: Raw materials are more than skin deep. *Chem. Week* **1345**,(20) 24, 41–42 (1989).
4. Moore, L., The natural vs. certified debate rages on. *Food Eng.* **8**, 69–72 (1991).
5. White, P.R., Potentially unlimited growth of excised plant callus in an artificial medium. *Am. J. Bot.*, **26**, 59–64 (1939).
6. Gautheret, R.J., Sur la possibilité de Réaliser la culture indefinié des tissus de tubercules de carotte. *C.R. Hebd. Seances Acad. Sci.* **208**, 118–121 (1939).
7. Zenk, M.H., The impact of plant cell culture on industry. In *Frontiers of Plant Tissue Culture* (T.A. Thorpe, ed.), pp. 1–13. University of Calgary Press, Calgary, Alberta, Canada 1978.
8. Moshy, R.J., Nieder, M.H., and Sahai, O.P., Biotechnology in the flavor and food industry of the USA. In *Biotechnology Challenges for the Flavor and Food Industry* (R.C. Lindsay and B.J. Willis, ed.), pp. 145–163. Elsevier, New York, 1989.
8a. Matsubara, K., Kitani, S., Yoshioka, T., Morimoto, T., Fujita, Y., and Yamada, Y., High density culture of *Coptis japonica* cells increases berberine production. *J. Chem. Technol. Biotechnol.* **46**(1), 61–70 (1989).
8b. Ikuta, A., Isoquinolines. In *Cell Culture and Somatic Cell Genetics of Plants* (F. Constabel and I. Vasil, eds.), Vol. 5, 289–314. Academic Press, San Diego, 1988.
9. Moshy, R.J., Impact of biotechnology on food product development. *Food Technol.* **39**(10), 113–118 (1985).
10. Goldstein, W.E., Economic factors in relationship to specialty chemical products by biocatalysis. *Biotechnol. Bioeng. Symp.* **17**, 763–776 (1986).
11. Seabrook, J.E.A., Laboratory culture. In *Plant Tissue Culture as a Source of Biochemicals* (E. John Staba, ed.), pp. 1–20. CRC Press, Boca Raton, FL, 1980.
12. Fujita, Y., and Tabata, M., Secondary metabolites from plant cells. Pharmaceutical applications and progress in commercial production. In *Plant Tissue and Cell Culture* (C.E. Green, D.A. Somers, W.P. Hackett, and D.D. Biesboer, ed.), pp. 169–185. Alan R. Liss, New York, 1987.
13. Sahai, O.P., and Shuler, M.L., Multistage continuous culture to examine secondary metabolite formation in plant cells: Phenolics from *Nicotiana tabacum*. *Biotechnol. Bioeng.* **26**, 27–36 (1984).
14. Tal, B., Roken, J.S., and Goldberg, I., Factors affecting growth and product formation in plant cells grown in continuous culture. *Plant Cell Rep.* **2**, 219–222 (1983).
15. Lindsey, K., and Yeoman, M.M., The synthetic potential of immobilized cells of *Capsicum frutescens* Miller *annuum*. *Planta* **152**, 495–501 (1984).

16. Beiderbeck, R., and Knoop, B., Two-phase culture. In *Cell Culture and Somatic Cell Genetics of Plants* (F. Constabel and I. Vasil, ed.), Vol. 4, pp. 255–266. Academic Press, San Diego, 1987.

17. Ritterhaus, E., Ulrich, J., and Westphal, K., Large-scale production of plant cell cultures. *Int. Assoc. Plant Tissue Cult. Newsl.* **61**, 2–10 (1990).

18. Ushiyama, K., Oda, H., and Miyamoto, Y., Large-scale tissue culture of *Panax Ginseng* root. In *Sixth International Congress of Plant Tissue and Cell Culture* (D. Somers, B. Gengenbach, D. Biesboer, W. Hackett, and C. Green, ed.), Abstr., 252. University of Minnesota, Minneapolis, 1986.

19. Noguchi, M., Matsumato, T., Hirata, Y., Yamamoto, K., Katsuyama, A., Kato, A., Azechi, S., and Kato, K., Improvement of growth rates of plant cell cultures. In *Plant Tissue Culture and its Biotechnological Application* (W. Barz, E. Reinhard, and M.H. Zenk, ed.), pp. 85–94. Springer-Verlag, Berlin, 1977.

20. Kato, A., Kawazoe, S., Iizima, M., and Shimizu, Y., Continuous culture of tobacco cells. *J. Ferment. Technol.* **54**, 82 (1976).

21. Tabata, M., and Fujita, Y., Production of shikonin by plant cell cultures. In *Biotechnology in Plant Science* (M. Zaitlin, P. Day, and A. Hollaender, ed.), pp. 207–218. Academic Press, Orlando, FL, 1985.

22. Rosevear, A., Putting a bit of color into the subject. *Trends Biotechnol.* **2**(5), 145–146 (1984).

23. Smart, N.J., and Fowler, M.W., Effect of aeration on large-scale culture of plant cells. *Biotechnol. Lett.* **3**, 171–176 (1981).

24. Takayama, S., Misawa, M., Ko, K., and Misato, T., Effect of cultural conditions on the growth of *Agrostemma githago* cells in suspension culture and the concomitant production of an anti-plant virus substance. *Physiol. Plant.* **41**, 313–320 (1977).

25. Tulecke, W., and Nickell, L.G., Methods, problems and results of growing plant cells under submerged conditions. *Trans. N.Y. Acad. Sci.* [2] **22**, 196–198 (1960).

26. Maurel, B. and Pareilleux, A. Effect of carbon dioxide on the growth of cell suspensions of *Catharanthus roseus. Biotechnol. Lett.* **7**(5), 313–318 (1985).

27. Kessell, R.H.J., and Carr, A.H., The effect of dissolved oxygen concentrations on growth and differentiation of *Daucus carota* tissue. *J. Exp. Bot.* **23**, 996–1007 (1972).

28. Wagner, F., and Vogelmann, H., Cultivation of plant tissue cultures in bioreactors and formation of secondary metabolites. In *Plant Tissue Culture and its Biotechnological Application* (W. Barz, E. Reinhard, and M.H. Zenk, ed.), pp. 245–252. Springer-Verlag, Berlin and New York, 1977.

29. Brodelius, P., Immobilized plant cells and protoplasts. In *Handbook of Plant Cell Culture* (D.A. Evans, W.R. Sharp, and P.V. Ammirato, ed.), Vol. 4, pp. 287–315. Macmillan, New York (1987).

30. Shuler, M.L., Sahai, O.P., and Hallsby, G.A., Entrapped plant cell tissue cultures. *Ann. N.Y. Acad. Sci.* **413**, 373–382 (1983).

31. Prenosil, J., and Pedersen, H., Immobilized plant cell reactors. *Enzyme Microb. Technol.* **5**, 323–331 (1983).

32. Rosevear, A., and Lambe, C.A., Immobilized plant cells. *Adv. Biochem. Eng.* **31,** 37–58 (1985).

33. Knorr, D., and Teutonico, R.A., Chitosan immobilization and permeabilization of *Amaranthus tricolor* cells. *J. Agric. Food Chem.* **34,** 96–97 (1986).

34. Kirsop, B.H., and Rhodes, M.J.C., Plant cell cultures as sources of valuable secondary products. *Biologist* **29**(3), 134–140 (1982).

35. Shuler, M.L., and Hallsby, G.A., Bioreactor considerations for chemical production from plant cell cultures. In *Biotechnology in Plant Science* (M. Zaitlan, P. Day, and A. Hollaender, eds.), pp. 191–205. Academic Press, New York 1985.

36. Goldstein, W.E., Lasure, L.L., and Ingle, M.B., Product cost analysis. In *Plant Tissue Culture as a Source of Biochemicals* (E.J. Staba, ed.), pp. 191–234. CRC Press, Boca Raton, FL, 1980.

37. Sahai, O., and Knuth, M., Commercializing plant tissue culture processes: Economics, problems and prospects. *Biotechnol. Prog.* **1,** 1–10 (1985).

38. Furuya, T., Yoshikawa, T., and Taira, M., Biotransformation of codeinone to codeine by immobilized cells of *Papaver somniverum. Phytochemistry* **23**(5), 999–1001 (1984).

39. Constabel, F., and Vasil, I.K., *Cell Culture and Somatic Cell Genetics of Plants,* Vol. 4. Academic Press, San Diego, 1987.

40. Constabel, F., and Vasil, I.K., *Cell Culture and Somatic Cell Genetics of Plants,* Vol. 5. Academic Press, San Diego, 1988.

41. Staba, E.J., ed., *Plant Tissue Culture as a Source of Biochemicals.* CRC Press, Boca Raton, FL, 1980.

42. Mulder-Krieger, T., Verpoorte, R., Svendsen, A., and Scheffer, J., Production of essential oils and flavors in plant cell and tissue cultures—A review. *Plant Cell, Tissue Organ Cult.* **13,** 85–154 (1988).

43. Barz, W., Reinhard, E., and Zenk, M.H., *Plant Tissue Culture and its Biotechnological Application.* Springer-Verlag, Berlin and New York, 1977.

44. Zenk, M.H., El Shagi, H., Arens, H., Stockigt, J., Weiler, E.W., and Deus, B., Formation of the indole alkaloids serpentine and ajmalicine in cell suspension cultures of *Catharanthus roseus.* In *Plant Tissue Culture and its Biotechnological Application* (W. Barz, E. Reinhard, and M.H. Zenk, eds.), pp. 27–43 Springer-Verlag, Berlin, 977.

45. Kinnersley, A.M., and Dougall, D.K., Correlation between nicotine content of tobacco plants and callus cultures. In *W. Alton Jones Cell Science Center*, Annu. Rep., pp. 7–8. Lake Placid, NY, 1981.

46. Guern, J., Renaudin, J.P., and Brown, S.C., The compartmentation of secondary metabolites in plant cell cultures. In *Cell Culture and Somatic Cell Genetics of Plants* (F. Constabel and I. Vasil, eds.), Vol. 4, pp. 43–76, Academic Press, San Diego, 1987.

47. Yeoman, M.M., Miedzybrodzka, M.B., Lindsey, K., and McLaughlin, W.R., The synthetic potential of cultured plant cells. In *Plant Cell Cultures: Results and Prospectives* (F. Sala, B. Parisi, R. Cella, and O. Cifferi, eds.), pp. 327–343. Elsevier North-Holland, Amsterdam, 1980.

48. Fujita, Y., Hara, Y., Ogino, T., and Suga, C., Production of shikonin derivatives by cell suspension cultures of *Lithospermum erythrorhizon*. I. Effects of nitrogen sources on the production of shikonin derivatives. *Plant Cell Rep.* **1,** 59–60 (1981).

49. Mizukami, H., Konoshima, M., and Tabata, M., Effect of nutritional factors on shikonin derivative formation in *Lithospermum* callus cultures. *Phytochemistry* **16,** 1183–1186 (1977).

50. Davies, M.E., Polyphenol synthesis in cell suspension cultures of Paul's scarlet rose. *Planta* **104,** 50–65 (1972).

51. Tal, B., Gressel, J., and Goldberg, I., The effect of medium constituents on growth and diosgenin production by *Dioscorea deltoidea* cells grown in batch cultures. *Planta Med.* **44,** 111–115 (1982).

52. Zenk, M.H., El Shagi, H., and Schulte, V., Anthraquinone production by cell suspension cultures of *Morinda citrifolia*. *Planta Med. Suppl.*, pp. 79–101 (1975).

53. Kubek, D.J. and Shuler, M.L., The effect of variations in carbon and nitrogen concentrations on phenolics formation in plant cell suspension cultures. *J. Nat. Prod.* **43,** 87–96 (1980).

54. Yamakawa, T., Kato, S., Ishida, K., Kodama, T., and Minoda, Y., Production of anthocyanins by *Vitis* cells in suspension culture. *Agric. Biol. Chem.* **47,** 2185–2191 (1983).

55. Bohm, H. and Rink, E. Betalains. In *Cell Culture and Somatic Cell Genetics of Plants* (F. Constabel and I. Vasil, eds.), Vol. 5, pp. 449–463. Academic Press, San Diego, 1988.

56. Ikeda, T., Matsumoto, T., and Noguchi, M., Effects of inorganic nitrogen sources and physical factors on the formation of ubiquinone by tobacco plant cells in suspension culture. *Agric. Biol. Chem.* **41,** 1197–1201 (1977).

57. Nakagawa, K., Konagai, A., Fukui, H., and Tabata, H., Release and crystallization of berberine in the liquid medium of *Thalictrum minus* cell suspension cultures. *Plant Cell Rep.* **3,** 254–257 (1984).

58. Sasse, F., Knobloch, K., and Berlin, J., Induction of secondary metabolism in cell suspension cultures of *Catharanthus roseus*, *Nicotiana tabacum* and *Peganum harmala*. In *Proceedings of the 5th International Congress of Plant Tissue and Cell Culture* (A. Fujiwara, ed.), pp. 343–344. Abe Photo Printing Co., Ltd., Tokyo, 1982.

59. Hagimori, M., Matsumoto, T., and Obi, Y., Studies on the production of *Digitalis* cardenolides by plant tissue cultures. *Plant Cell Physiol.* **23,** 1205–1211 (1982).

60. Seitz, H.U., and Hinderer, W., Anthocyanins. In *Cell Culture and Somatic Cell Genetics of Plants* (F. Constabel and I. Vasil, eds.). pp. 49–76. Academic Press, San Diego, 1988.

61. Sahai, O.P., and Shuler, M.L., Environmental parameters influencing phenolics production by batch cultures of *Nicotiana tabacum*. *Biotechnol. Bioeng.* **26,** 111–120 (1984).

62. Tabata, M., Naphthoquinones. In *Cell Culture and Somatic Cell Genetics of Plants* (F. Constabel and I. Vasil, eds.), Vol. 5, pp. 99–111. Academic Press, San Diego, 1988.

63. Mok, M.C., Gabelman, W.H., and Skoog, F., Carotenoid synthesis in tissue cultures of *Daucus carota*. *J. Am. Soc. Hortic. Sci.* **101**, 442–449 (1976).

64. Watts, M.J., Galpin, I.J., and Collin, H.A., The effect of growth regulators, light and temperature on flavor production in celery cultures. *New Phytol.* **98**, 583–591 (1984).

65. Courtois, D., and Guern, J., Temperature response of *Catharanthus roseus* cells cultivated in liquid medium. *Plant Sci. Lett.* **17**, 473–482 (1980).

66. Ambid, C., and Fallot, J., Role of the gaseous environment on volatile compound production by fruit cell suspensions cultured *in vitro*. In *Flavour '81*. (P. Schreier, ed.), pp. 529–538, de Gruyter, Berlin and New York, 1981.

67. Kobayashi, Y., Fukui, H., and Tabata, M., Effect of carbon dioxide and ethylene on berberine production and cell browning in *Thalictrum minus* cell cultures. *Plant Cell Rep.* **9**, 496–499 (1991).

68. Ibrahim, R.K., Regulation of synthesis of phenolics. In *Cell Culture Somatic Cell Genetics of Plants* (F. Constabel and I. Vasil, eds.), Vol. 4, pp. 77–95. Academic Press, San Diego, 1987.

69. Eilert, U., Elicitation: Methodology and aspects of application. In *Cell Culture and Somatic Cell Genetics of Plants*, (F. Constabel, and I. Vasil, eds.), Vol. 4, pp. 153–196 Academic Press, San Diego, 1987.

70. Widholm, J., Selection of mutants which accumulate desirable secondary compounds. In *Cell Culture and Somatic Cell Genetics of Plants* (F. Constabel and Vasil, eds.), Vol. 4, pp. 125–137. Academic Press, San Diego, 1987.

71. Tabata, M., Ogino, T., Yoshioka, K., Yoshikawa, N., and Hiaraoka, N., Selection of cell lines with higher yield of secondary products. In *Frontiers of Plant Tissue Culture 1978* (T.A. Thorpe, ed.), pp. 213–222. University of Calgary Press, Calgary, Alberta, Canada, 1978.

72. Matsumoto, T., Ikeda, T., Kenno, N., Kisaki, T., and Noguchi, M., Selection of high ubiquinone—10 producing strains of tobacco by cell cloning techniques. *Agric. Biol. Chem.* **44**, 967–969 (1980).

73. Sato, F., and Yamada, Y., High berberine producing cultures of *Coptis japonica* cells. *Phytochemistry* **23**, 281–285 (1984).

74. Kartha, K., Cryopreservation of secondary metabolite—producing plant cell cultures. In *Cell Culture and Somatic Cell Genetics of Plants* (F. Constabel and I. Vasil, eds.), Vol. 4., pp. 217–227. Academic Press, San Diego, 1987.

75. Curtin, M.E., Harvesting profitable products from plant tissue culture. *Biotechnology* **10**, 649–657 (1983).

76. Biotechnology Newswatch, *After Shikonin Lipstick, in-vitro Geranium Perfume?* McGraw Hill, New York, 1985.

77. ESCAgenetics Annual Report. ESCAgenetics Cooporation, San Carlos, California, 1993.

78. Callebaut, A., Voets, A., and Motte, J., Anthocyanin production by plant cell cultures on media based on milk whey. *Biotechnology Lett.* **12**(3), 215–218 (1990).

79. Berlin, J., Sieg, S., Strack, D., Bokern, M., and Harms, H., Production of betalains by suspension cultures of *Chenopodium rubrum* L. *Plant Cell, Tissue Organ Cult.* **5**, 163–174 (1986).

80. Fosket, D., and Radin, D., Induction of carotenogenesis in cultured cells of *Lycopersicon esculentum*. *Plant Sci. Lett.* **30**, 165–175 (1983).

81. Nawa, Y., and Ohtani, T., Water soluble carotenoid pigments in callus cells from fruit of *Gardenia jasminoides* Ellis. In *Abstracts from the 7th International Congress of Plant Cell and Tissue Culture*. International Association for Plant Cell Culture, Amsterdam, 1990.

82. Sarma, K., Chemical and sensory analysis of saffron produced through tissue cultures. In *Abstracts from the 7th International Congress of Plant Cell and Tissue culture*. International Association for Plant Cell Culture, Amsterdam, 1990.

83. Bricout, J., and Paupardin, C., Sur la composition de l'huile essentielle de tissus de pericarpe de citron (*Citrus limon* Obseck) cultivées *in vitro*. *C.R. Hebd. Seances Acad. Sci., Ser. D* **278**, 719–722 (1974).

84. Drawert, F., and Bergern R.G., Uber die Biogenese von Aromastoffen bei Pflanzen und Fruchten. XIX. Mitt: Vergleich der Biosynthese von Aromastoffen in Segment-, Kallus- und Suspensionskulturen von Citrusarten. *Chem. Mikrobiol., Technol. Lebensm.* **7**, 143–147 (1982).

85. Cresswell, R., The production of flavor components from citrus tissue cultures. In *Abstracts from the 7th International Congress of Plant Cell and Tissue culture*. International Association for Plant Cell Culture, Amsterdam, 1990.

86. Tisserat, B., Vandercook, C., and Berhow, M., Citrus juice vesicle culture: A potential research tool for improving juice yield and qualilty. *Food Technol.* **2**, 95–100 (1989).

87. Hong, Y., Huang, L., Reineccius, G., Harlander, S., and Labuza, T., Production of aroma compounds from strawberry cell suspension cultures by addition of precursors. *Plant Cell, Tissue Organ Cult.* **21**, 245–251 (1990).

88. Dziezak, J., (1986). Biotechnology and flavor development: Plant tissue cultures. *Food Technol.* **40**(4), 122–129 (1986).

89. Enevoldsen, K., Joersbo, M., and Andersen, J., Establishment of black current cell suspension cultures for production of flavor compounds. In *Abstracts from the 7th International Congress of Plant Cell and Tissue culture*. International Association for Plant Cell Culture, Amsterdam, 1990.

90. Ambid, C., Moisseeff M., and Fallot, J., Biogenesis of monoterpenes, bioconversion of citral by a cell suspension culture of muscat grapes. *Plant Cell Rep.* **1**, 91–93 (1982).

91. Tamaki, E., Morishita, I., Nishida, K., Kato, K., and Matsumoto, T., Process for preparing licorice extract-like material for tobacco flavoring. U.S. Patent 3,710,512 (1973).

92. Hayashi, H., Fukui, H., and Tabata, M., Examination of triterpenoids produced by callus and cell suspension cultures of *Glycyrrhiza glabra*. *Plant Cell Rep.* **7**, 508–511 (1988).

93. Knuth, M., and Sahai, O., Flavor composition and method. U.S. Patent 5,057,424 (1991).

94. Collin, H.A., Flavors. In *Cell Culture and Somatic Cell Genetics of Plants* (F. Constabel and I. Vasil, eds.), Vol. 5, pp. 569–585. Academic Press, San Diego, 1988.

95. Watts, M.J., Galpin, I.J., and Collin, H.A., The effect of greening on flavor production in celery tissue cultures. *New Phytol.* **100,** 45–56 (1985).

96. Townsley, P.M., Chocolate aroma from plant cells. *J. Inst. Can. Sci. Technol. Aliment.* **7,** 76–78 (1974).

97. Jalal, M.A.F., and Collin, H.A., Polyphenols of mature plant, seedling and tissue cultures of *Theobroma cacao. Phytochemistry,* **16,** 1377–1380 (1977).

98. Davey, M.R., Mackenzie, I.A., Freeman, G.G., and Short, K.C., (1974). Studies of some aspects of the growth, fine structure and flavor production of onion tissue grown *in vitro. Plant Sci. Lett.* **3,** 113–120 (1974).

99. Freeman, G.G., Whenham, R.J., Mackenzie, I.A., and Davey, M.R., Flavor components in tissue cultures of onion (*Allium cepa* L.). *Plant Sci. Lett.* **3,** 121–125 (1974).

100. Selby, C., and Collin, H.A., Clonal variation in growth and flavour production in tissue cultures of *Allium cepa* L. *Ann. Bot.* (*London*) **40,** 911–918 (1976).

101. Turnbull, A., Galpin, I.J., Smith, J.L., and Collin, H.A., Comparison of the onion plant (*Allium cepa*) and onion tissue culture. IV. Effect of shoot and root morphogenesis on flavor synthesis in onion tissue culture. *New Phytol.* **87,** 257–268 (1981).

102. Collin, H.A., and Musker, D., *Allium* compounds. In *Cell Culture and Somatic Cell Genetics of Plants* (F. Constabel and I. Vasil, eds.), Vol. 5, pp. 475–493. Academic Press, San Diego, 1988.

103. Malpathak, N., and David S., Effect of gamma irradiation and ethyl methane sulfonate on flavor formation in garlic *Allium sativum* L. Cultures. *Indian J. Exp. Bot.* **28**(6), 519–521 (1990).

104. Ajinomoto Corporation, Production of allinin and methylcysteine sulfoxide by garlic cell culture. Japanese Patent 1,257,487 (1989).

104a. Tudge, C., Drugs and dyes from plant cell cultures. *New Sci.* **12,** 25 (1984).

105. Szoke, E., Verzar-Petri, G., Shavarda, A.G., Kuzovkina, I.N., and Smirnov, A.M., The difference in etheric oil composition between excised roots, callus and cell suspension cultures of chamomile (*Matricaria chamomilla* L.). *Izv. Akad. Nauk SSSR, Ser. Biol.* **6,** 943–949 (1979).

106. Reichling, J., Becker, H., Martin, R., and Burkhardt, E., Vergleichende Untersuchungen zur Bildung und Akkumulation von Atherischen Ol in der Intaken Pflanze und in Zellkulturen von *Pimpinella anisum. Z. Naturforsch.* **40,** 465–468 (1985).

106a. Reichling, J., Bisson, W., and Becker, H., Vergleichende Untersuchungen zur Bildung und Akkumulation von Etherischem Ol in der Intaken Pflanze und in der Callus-kultur von *Matricaria chamomilla. Planta Med.* **48,** 334–337 (1983).

107. Bisson, W., Biederdeck, R., and Reichling, J., Die Produktion Atherischer Ole Durch Zell-suspensionen der Kamille in Einem Zweiphasensystem. *Planta Med.* **47,** 164–168 (1983).

108. Lin, M.L., and Staba, E.J., Peppermint and spearmint tissue cultures. I. Callus formation and submerged culture. *Lloydia,* **24,** 139–145 (1961).

109. Becker, U.H., Untersuchungen zur Frage Der Bildung Fluchtiger Stoffwechselprodukte in Callus Kulturen. *Biochem. Physiol. Pflanz.* **161,** 425–441 (1970).

110. Bricout, M.J., and Paupardin, C., (1975). Sur la composition de l'huile essentielle de *Mentha piperita* L. cultivée *in vitro*: Influence de quelques facteurs Sur sa Synthèse. *C.R. Hebd. Seances Acad. Sci., Ser. D* **281**, 383–386 (1975).

111. Kireeva, S.A., Mel'nikov, V.N., Reznikov, S.A., and Meshcheryakova, N.I., Essential oil accumulation in a peppermint callus culture. *Fiziol. Rast. (Moscow).* **25**, 564–570 (1978).

112. Bricout J., Garcia-Rodriguez, M.J., and Paupardin, C., Action del la colchicine sur la synthèse d'huile essentielle par des tissus de *Mentha piperita* cultivées *in vitro. C.R. Hebd. Seances Acad. Sci., Ser. D* **286**, 1585–1588 (1978).

113. Kireeva, S.A., Burgorskii, P.S., and Reznikova, S.A., Tissue cultures of damask rose and accumulation of terpenoids in them. *Fiziol Rast. (Moscow)* **24,** 824–831 (1977); *Chem. Abs.* **87**, 130499b.

114. Banthorpe, D.V., and Barrow, S.E., Monoterpene biosynthesis in extracts from cultures of *Rosa damascena. Phytochemistry,* **22**, 2727–2728 (1983).

115. Gupta, P.K., and Mascarenhas, A.F., Essential oil production in relation to organogenesis in tissue cultures of *Eucalyptus citriodora* Hook. *Basic Life Sci.* **22**, 299–308 (1983).

116. Yamaguchi, T., and Fukuzumi, T., Volatile compounds from callus cells of woody plants. In *Proceedings of the 5th International Congress of Plant Tissue and Cell Culture* (A. Fujiwara, ed.), pp. 287–288. Japanese Association for Plant Tissue Culture, Tokyo, 1982.

117. Sardesai, D.L., and Tipnis, H.P., Production of flavoring principles by tissue culture of *Coriandrum sativum. Curr. Sci.* **38**, 545 (1969).

118. Pedro, L., Souse, M.J., Novais, J.M., and Pais, M.S.S. Callus and suspension cultures from geranium. *Biotechnol. Lett.* **12**(6), 439–442 (1990).

119. Banthorpe, D., Branch, S., Njar V., Osborne, M., and Watson, D., Ability of plant callus cultures to synthesize and accumulate lower terpenoids. *Phytochemistry* **25**(3), 629–636 (1986).

120. Tomoda, G., Matsuyama, J., and Iikubo, H., Tissue culture of fragrant plants. *Bull. Fac. Agric., Tamagawa Univ.* **16**, 16–22 (1976).

121. New York Times, Spice islands of the mind. *N.Y. Times*, June 29 (1987).

Genetic Engineering and Other Advanced Technologies

ALAN GABELMAN

Tastemaker
Cincinnati, Ohio

Biotechnology is one of the major advancing technologies of the 20th century. Impressive progress has been made since cloning experiments were begun in the 1970s; indeed, we have reached the point where creation of recombinant organisms is routine, and the technology is inexpensive enough to interest low-margin industries like the food industry. For example, food ingredient suppliers have already commercialized recombinant deoxyribonucleic acid (rDNA)-derived chymosin, a product that saves an average-sized cheese plant about $1000/day. A second example is Calgene's Flavr-Savr™ tomato, which can be vine-rippened for better taste because it is genetically engineered to stay firm longer than traditional tomatoes. Numerous other examples of genetically engineered food, flavor, and agricultural products are presented in this chapter.

In addition to genetic engineering, considerable progress has been made in related areas of biotechnology. For example, biotechnologists are learning to modify the properties of a protein by changing its amino acid sequence, a technique known as protein engineering. *Furthermore, using genetic engineering, protein engineering, and other techniques, scientists are studying the modification of entire biochemical pathways to obtain results like enhanced production of valuable metabolites. In years to come, these technologies will progress to the point where biological production of novel food ingredients with made-to-order properties will be possible. Current progress in several advanced technologies, including protein engineering, flux mapping, pathway engineering, and nonaqueous enzymatic catalysis, is discussed in this chapter.*

Bioprocess Production of Flavor, Fragrance, and Color Ingredients, Edited by Alan Gabelman,
ISBN 0-471-03821-0   © 1994 John Wiley & Sons, Inc.

9.1 INTRODUCTION

Genetic engineering, otherwise known as the new biotechnology, can be considered one of three major advancing technologies of the century (the other two are electronics and advanced materials). Tremendous advances have been made in the relatively short amount of time that has passed since the first DNA cloning experiments were performed in the early 1970s. That is, expression of foreign DNA in bacteria is now routine, and expression in higher organisms, including plants and even animals, is now possible.

The so-called ''first wave'' of products manufactured using this new biotechnology has come from the pharmaceutical industry. This is not surprising because these products command a market and a profit margin sufficient to support the research and development costs. Commercial pharmaceuticals produced using the new biotechnology include tissue plasminogen activator, insulin, pituitary growth hormone, EPO, hepatitis B vaccine, and others.

However, our knowledge and proficiency with genetic engineering has progressed to the point where the cost of developing new products has decreased substantially, so that now the second wave of new biotechnology products has begun to appear. This second wave includes those products that command smaller profit margins than pharmaceuticals, for example, food products. The first such product manufactured using a genetically engineered organism is chymosin, an enzyme used in cheese production. A number of others have followed, and many more are in the pipeline; specific examples are discussed later in this chapter.

A sales forecast for biotechnology products is presented in Table 9.1. This forecast was prepared by Consulting Resources Corporation, a Boston-area consulting firm specializing in biotechnology [1]. The table indicates that the expected impact of biotechnology on food and flavors is substantial. That is, sales of specialty products, which include enzymes, flavors, and fragrances, are expected to grow at a rather astounding rate of 30 percent per year for the

TABLE 9.1 U.S. Biotechnology Sales Forecasts: 1992–2002[a]

Key Sectors	Sales in 1992	Sales Forecasts		1992–2002 Sales Growth (ppa)
		1997	2002	
Human therapeutics	2250	4800	9200	15
Human diagnostics	1050	1700	2500	9
Agriculture	70	375	1400	35
Specialties	95	400	1300	30
Nonmedical diagnostics	10	100	250	38
Total	3475	7375	14,650	15

[a]Figures are given in million 1992 dollars.

Source From Consulting Resources Corporation [1].

next several years, reaching $400 million in 1997 and $1.3 billion in 2002. Furthermore, the rate of growth of these biotechnology-based specialty products is expected to be double that of biotechnology products in general; specialties represent 2.7 percent of the 1992 sales, and that number is expected to increase to 8.9 by 2002. According to the Technology Management Group (New Haven, CT), in 1988 over 75 U.S. firms were already using advanced technologies such as genetic engineering in pursuit of better flavors and fragrances [2].

These figures imply that we are only beginning to feel the impact of a far-reaching technology. That is, scientists have achieved an advanced level of skill in the genetic manipulation of organisms to obtain desired results. Impressive progress has also been made in related areas of biotechnology. For example, in the field known as *protein engineering*, investigators are modifying the amino acids sequences of enzymes and other proteins to obtain increased and/or altered activity, improved resistance to heat, extremes of pH and other harsh conditions, and other advantages. Advances in this area will eventually lead to the ability to design enzymes from scratch that have the desired catalytic activity and are stable under the desired conditions of use.

Genetic engineering and protein engineering can be used to modify entire biochemical pathways of living organisms. This powerful approach, known as *pathway engineering*, will not only allow enhanced yields of existing materials, but will also lead to cost-effective production of novel food and flavor ingredients with unique properties. Advances in the relationship between structure and function of food and flavor ingredients will allow food technologists to specify a chemical structure depicting an ingredient with a specific desired effect in the food, then use pathway engineering to develop an organism capable of producing the ingredient. The availability of such novel ingredients will literally revolutionize the food industry, requiring food technologists and other food industry professionals to adjust to a whole new way of thinking. The food technologists of today are trained to work with existing ingredients, but such limitations will no longer be necessary. For example, it would never occur to the traditional food technologist that nonstaling flour would be available, yet such a product is currently in development [3].

Admittedly, routine production of novel food ingredients in this manner will not occur in the next few years. However, advanced technologies such as those discussed in this chapter will in fact have far-reaching effects on the food, flavor, and fragrance industries in the near term. For example, many important raw materials are currently extracted from plants that grow in parts of the world subject to political instability and climatic variation, causing uncertain supply, inconsistent quality, and high prices. The technologies discussed in this chapter will allow these materials to be mass produced inexpensively and with consistent quality, completely changing sources of raw materials for the flavorist and the food technologist.

This chapter discusses genetic engineering and some related technologies, including protein engineering, flux mapping, pathway engineering, and nonaqueous enzymatic catalysis. The discussion is not intended to be an exhaustive review of these topics, but rather an overview of the various technologies, emphasizing practical examples rather than in-depth scientific treatment. Examples from the flavor area are cited wherever possible; however, because at this time flavor applications are somewhat limited, food and agricultural applications are covered as well, with the intention of demonstrating the types of things one can accomplish with these techniques. There are a number of exciting technologies with food and flavor applications that are not covered; indeed, an entire book could be devoted to the topic of advanced technologies rather than just one chapter. A partial list of advanced technologies of interest to the food industry, including those covered in this chapter and many that are not, is given in Table 9.2.

Although fragrances are not emphasized in this chapter, the potential impact of these technologies on the fragrance industry should not be overlooked. For example, fragrance companies are interested in genetically engineered plants that produce more enzymes involved in fragrance production or are able to efficiently convert a fragrance precursor into the desired fragrance with a minimum of by-products. Fragrance opportunities exist not only in perfumes, but also in soaps, toiletries, skin care products, sunscreens, shampoos, household cleaners, and pet foods; the current world market exceeds $2 billion [2].

TABLE 9.2   **Advanced Technologies with Food and Flavor Applications**

Genetic engineering
Protein engineering
Flux mapping
Pathway engineering
Nonaqueous enzymatic catalysis
Nonaqueous enzymatic catalysis using lyotropic liquid crystals
Plant tissue culture (see Chapter 8)
Synthetic enzymes
Enzyme mimetic systems
Cyclodextrins
Immobilization of cells or free enzymes
Supercritical fluid extraction
Protoplast fusion
Catalytic antibodies
Mathematical modeling of fermentations and other biological processes
Biosensors
Biomaterials
Encapsulation
Controlled release
Two-phase aqueous extraction

9.2 GENETIC ENGINEERING—BACKGROUND

Organisms have undergone genetic mutations since the appearance of life on the planet. These mutations result from errors that are occasionally introduced into the genetic code during reproduction. Such mutations are rare, but considering the amount of time that has passed since life first appeared on Earth, even infrequent occurrences have happened many times. According to Darwin's well-known work, since the dawn of time mutations providing an organism with some survival advantage have been propagated, and in this way life on Earth has become more complex and diversified.

Mutations responsible for natural evolution result in biological systems that are optimized for the survival of the organism; this is not always consistent with our objectives as food industry professionals. For example, in nature there is usually no need for high concentrations of metabolites, or for enzymes that are stable at high temperatures or extreme pHs. The food industry is interested in the genetic modification of organisms that lead to biological systems with desirable properties not provided by nature.

Scientists have learned to accelerate the genetic modification process as it occurs in nature, so that genetically modified organisms can be produced in a reasonable time frame rather than eons. For example, in microbiology one can use a classic mutation–selection strategy to obtain a modified organism that overproduces a desired metabolite, perhaps because the gene coding for the enzyme which breaks down that metabolite is altered. With this technique, one starts with a culture of a wild-type microorganism having characteristics similar to the desired ones. Mutations are then induced with a suitable mutagenic agent (chemicals, radiation, etc.), with the hope that one of the many mutated organisms will have the desired characteristics. Another example of accelerated genetic modification is the well-known hybridization strategies used to develop new varieties of plants.

Although such traditional techniques substantially reduce the time required to obtain a useful genetic modification (from eons to months or years), the process is still long, inefficient, imprecise, random, and uncontrollable. That is, one must create many genotypes then screen for the desired one, assuming it even exists. With traditional plant breeding and selection, for example, thousands of genes are shuffled randomly with every cross, with the hope that an offspring with the desired characteristics will be found. For this reason up to 20 years may be required to develop a new commercial plant variety, and even then its characteristics are limited to what can be derived from the genetic material of the parents [4,5].

These disadvantages can now be eliminated by the use of genetic engineering, which can bring about the desired genetic changes quickly, precisely and in a highly controlled manner. Furthermore, the source of genetic material is virtually unlimited, because in principle, genetic material can be taken from any organism. To continue with the plant breeding example, often the results include hardier, more flavorful, and more nutritious varieties of fruits and vegetables.

The discussion below is a review of the basic principles of biotechnology and genetic engineering, followed by a number of examples taken from the food and flavor industry. For a more detailed discussion of the fundamentals of biotechnology the reader is referred to several references [6–10]. Factors influencing productivity of fermentations using recombinant microorganisms are discussed by Zabriskie and Arcuri [11], and an introduction to developments in biotechnology intended for the non-specialist is presented by Marx [12].

9.3   BASICS OF BIOTECHNOLOGY

9.3.1   DNA

The code for the genetic make-up of an organism is contained in its DNA. The DNA molecule is a chain of nucleotides, each of which is composed of a nitrogenous base, a pentose (2'-deoxy-D-ribose), and phosphoric acid, linked as shown in Figure 9.1. The nitrogenous bases include the pyrimidines cytosine and thymine and the purines adenine and guanine. The related nucleic acid, RNA (ribonucleic acid), is composed of nucleotides having the same structure except the pentose is D-ribose and uracil is used instead of thymine. The specific sequence of nucleotides found in the DNA of a given organism is found in every cell of that organism and is characteristic of that organism.

Indeed this specificity, along with advances in nucleic acid biochemistry, has formed the basis of a powerful new technique known as DNA fingerprint-

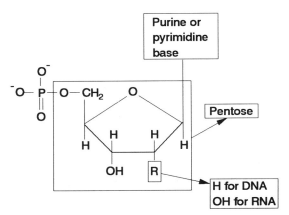

FIG. 9.1   Chemical structure of nucleotides, the building blocks of DNA and RNA.

ing. One of the uses of this technique which has captured the attention of the media is in police work. That is, DNA isolated from a sample of blood, semen, hair, or other biological tissue found at the scene of a crime is compared to the DNA of the suspect. If a match is found then there is an exceedingly high probability that the tissue did in fact come from the suspect, although the technique can give false results if the correct procedures are not carefully followed [13,14].

The structure of DNA was first proposed by James D. Watson and Francis Crick at Cambridge University (England) in 1953. Based on X-ray diffraction data and observed characteristic base pairing, they proposed that DNA consists of two helical chains coiled around a common axis to form a right-handed double helix. The helices are joined by hydrogen bonding between complementary bases pairs (*i.e.*, adenine to thymine and guanine to cytosine). The structure of DNA and its role in protein synthesis and transmittal of genetic information has been studied extensively in the ensuing years.

Replication of DNA, that is, transmission of genetic information to successive generations, is a complex process requiring over 20 enzymes, but one that is highly precise and only rarely results in errors in the transmitted information. (However, as explained above, it was these occasional errors that provided prehistoric organisms with survival advantages, eventually leading to the complexity and diversity of life on Earth today.) The replication process begins with the unwinding of the double helix to form two separate strands. Each of these strands forms a template for the replicated DNA, and the new complementary strands are formed with the assistance of enzymes known as *polymerases*. Once the new complementary strands are in place, rewinding occurs and the process is complete.

The DNA molecule contains the genetic information required to synthesize (or *express*) enzymes and other proteins that form the basis of the organism's metabolism. Each of the 20 amino acids found in proteins is coded by a sequence of three nucleotides (called a *codon*), so that entire polypeptides or proteins are coded by segments of DNA or genes. The information is transferred to the site of protein synthesis (the ribosomes) by messenger RNA (mRNA), which is formed by a process known as *transcription*. In this process, the segment of DNA to be transcribed is unwound and a complementary strand of mRNA is formed. The process is illustrated schematically in Figure 9.2. The reader is referred to Lehninger [15] or another biochemistry text for a more detailed description of protein synthesis.

It is important to emphasize that the genes contained in the DNA molecule do not code for specific characteristics directly, but for proteins, that is enzymes. These enzymes catalyze the biochemical reactions that allow the organism to function and are responsible for various inherited characteristics. Of course not all of the proteins contained in the genetic code are synthesized all of the time. The regulatory mechanisms of genetic transcription and protein synthesis are very complex and well beyond the scope of this review, but in

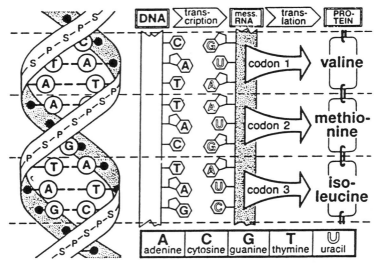

FIG. 9.2 DNA transcription and protein synthesis. *Source* Reprinted from Over-beeke [35], with permission.

simple terms enzymes can be divided into two general categories: constitutive and induced. Constitutive enzymes are always present, whereas induced enzymes are those that are synthesized only when needed.

For example, the presence of some chemical above a threshold concentration may trigger the synthesis of an induced enzyme that catalyzes the degradation of that chemical before its concentration reaches a toxic level. Such a triggering mechanism may involve an interaction between the chemical and a particular site on the DNA coding for the enzyme; for example, enzyme synthesis may be blocked unless that site is occupied by the chemical. As the concentration of the chemical increases, the probability of attachment to the site increases, so that enzyme synthesis occurs when needed, i.e., when the chemical concentration becomes unacceptably high.

This is just one example of the myriad of complex interactions that control protein synthesis and enzyme function, in turn leading to regulation of metabolite concentrations and biochemical reactions in living systems. The power of biotechnology comes from the ability to achieve some desirable result by understanding and to some degree controlling these interactions.

The ability to perform the types of genetic manipulations described in this chapter arose from a combination of scientific developments that occurred during the 1970s. These include the ability to introduce DNA into bacteria (and later other higher organisms), and the ability to cut DNA into fragments using highly specific endonucleases and to join the resulting fragments using DNA ligase [16]. These advances allow DNA from virtually any source to be propagated by transferring the DNA into one of the bacterial workhorses listed in Table 9.3, which offer advantages such as ease of growth and industrial har-

TABLE 9.3 Bacterial Workhorses of Genetic Engineering

Escherichia coli
Bacillus subtilis
Bacillus licheniformis
Aspergillus niger

diness. The recipient of the foreign DNA is called the *host organism*, and the modified DNA of the host organism is known as *recombinant DNA*.

9.3.2 Vectors

The carrier of the foreign DNA into the host organism is called the *vector*. One type of vector enjoying widespread use in current research is known as a *plasmid*, that is, a circular piece of DNA. A schematic representation of the use of a plasmid to genetically modify the bacterium *Escherichia coli* is shown in Figure 9.3 [162]. The main advantage of using a plasmid is that it is not part of the chromosome(s), so that dealing with the complexity of the chromosome(s) can be avoided, yet the plasmid contains the complete DNA of the organism and is self-replicating. A further advantage is that genetic markers for antibiotic resistance are often found on plasmids, and these markers are easily selectable. This is important because it provides a way to determine if offspring of the transformed organism retain the plasmid; that is, organisms that grow on a nutrient medium containing the relevant antibiotic are known to contain the plasmid.

However, antibiotic-resistant cultures probably will not be permitted in food fermentation systems because of the possible transfer of such resistance to bacteria found in the human gut [17]. Other markers more suitable for food use are being developed, including resistance to nisin (an antimicrobial agent recently approved for use in processed cheese spreads) [18], lactose-fermenting ability [19], and resistance to bacteriocins [20].

Other types of vectors also exist (see Table 9.4); a complete review of the subject is beyond the scope of this chapter, but some techniques are covered briefly. One approach in particular has attracted considerable attention lately and that is the use of mechanical acceleration of DNA-coated particles into cells. Once the particles are inside, the DNA migrates from the particle and becomes part of the chromosome of the host organism, where it functions like native DNA. W. R. Grace (Boca Raton, FL) and Du Pont (Wilmington, DE) recently signed a cross-licensing pact covering Du Pont's Biolistic gene delivery technology [21]. In a related development, a novel microprojectile accelerating system in which DNA remains in solution, avoiding the need to bind it to particles, was recently reported by Sautter et al. [22].

Another approach is known as *protoplast transformation*; here the cell wall is removed enzymatically in the presence of an osmotic stabilizing agent, then

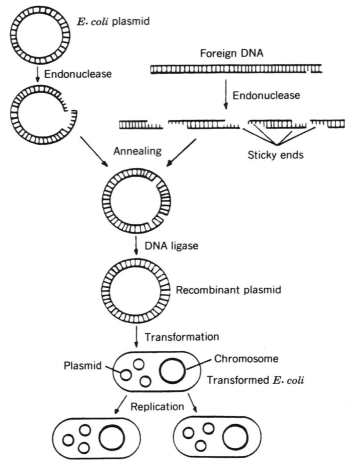

FIG. 9.3 Transformation of foreign DNA into *E. coli* using a plasmid. *Source* Reprinted from Levin et al. [162] with permission.

TABLE 9.4 Vectors Used by Biotechnologists

Plasmids
Mechanical acceleration of DNA-coated particles
Protoplast transformation
Electroporation
Viruses

DNA uptake is facilitated by polyethylene gl.   (ill another alternative is *electroporation*, which is the use of electrical pulse. to reversibly permeabilize bacterial cells and allow uptake of DNA [17]. Viruses are also used as vectors.

9.3.3   Necessary Prerequisites and Suitable Organisms

The first step in genetic engineering is to understand the biochemical reactions that occur upon transformation of some substrate (say sugar) to some metabolite (say ethanol), and to understand the enzymes involved and how they are regulated. That is, to use genetic engineering for production of flavors or other food ingredients one needs the following information [23]:

1. The structure of the compound of interest
2. Its position within a metabolic pathway
3. The identity of each enzyme in the pathway
4. A knowledge of each enzyme's regulation

The second step is to identify the genes that code for the enzymes catalyzing the biochemical reactions of interest. One can then manipulate the organism's DNA to control enzyme synthesis in a way that leads to production of high levels of the desired metabolite. Alternatively, and in many cases preferably, one can introduce the gene of interest into one of several bacteria amenable to large-scale production in a industrial setting, the so-called workhorses of the fermentation industry (see Table 9.3). This is important because many organisms containing genes of interest are not suitable for economic large-scale production.

However, even though the genetic material of these bacteria is relatively simple to understand and manipulate, these lower organisms do not always possess the ability to carry out more complex procedures such as posttranslational glycosylation. On the other hand, higher organisms (yeast, mammalian cells, and higher plants) may have the required abilities, but these organisms are more difficult to manipulate because they have more DNA and regulation of gene transcription is more complex. Yet in spite of the complexity, excellent progress has been made. For example, genetically engineered versions of Baker's yeast (*Saccharomyces cerevisiae*) have been developed for various applications, and some of these are discussed below. Other yeasts, which are also suitable candidates for genetic recombination and have the potential to join the list of industrial workhorses, are listed in Table 9.5; reviews are given by Buckholz and Gleeson [24], Stewart and Russel [25], and Rank et al. [26].

In fact, even though yeasts are more complex than bacteria, the former offer several advantages [27]:

1. *S. cerevisiae* and many other yeasts are nonpathogenic.
2. The fermentation technology is well established with strains that are highly stable.

TABLE 9.5   Yeasts with Potential to Become Genetic Engineering Workhorses

Saccharomyces cerevisiae
Hansenula polymorpha
Kluyveromyces lactis
Pichia pastoris
Schizosaccharomyces pombe
Schwanniomyces occidentalis
Yarrowia lipolytica

3. Product recovery is simplified because yeast produce fewer extracellular proteins, and the ones they do make are glycosylated and possess biological activity.

Other investigators have suggested that the filamentous fungi potentially offer several distinct advantages as recombinant hosts [28,29].

9.3.4   Problems Yet to Be Solved

Although there are a number of examples of successful genetically engineered products in both the food and pharmaceutical industries, some problems still need to be worked out. These are discussed below and summarized in Table 9.6.

9.3.4.1   *Secretion of Products*   Often products of genetically engineered organisms are not secreted but instead remain inside the cell. Additional processing is required to remove the product from the cell, for example, mechanical shear (homogenizers), lytic enzymes, osmotic shock, or a combination of these techniques. Enzymes produced by *E. coli* are sometimes stored in intracellular inclusion bodies; these enzymes can be removed using detergents, but this leads to denaturation and requires the enzymes to be renatured, which reduces yield and increases cost [30]. Some investigators are allowing *E. coli* proteins to be secreted by removing a signal peptide from them, using the signal peptidases present in the organism. Another method of protein release

TABLE 9.6   Unsolved Problems in Genetic Engineering

Many recombinant proteins are not secreted.
Complexity of the genetic material of higher organisms is problematic.
Transformed plasmids can be lost upon cell division.
Native plasmids can interfere with transformed ones.
Structural deletions can occur in plasmids.
Foreign proteins are sometimes degraded by the host organism.
Consumer perception of biotechnology products is not always favorable.

involves rupturing the cells; unfortunately this method also releases unwanted proteins and other intracellular components into the medium, which may complicate purification.

9.3.4.2 Complexity of Higher Organisms As mentioned above, higher organisms are required for synthesis and secretion of complex molecules, and further advances in the ability to understand and manipulate such organisms are necessary.

9.3.4.3 Problems with Plasmids Upon cell division, plasmid DNA may not divide equally between daughter cells, and plasmidless cells have a survival advantage because they do not need to expend energy during product formation [30]. (One way to determine if daughter cells contain the plasmid of interest is to use antibiotic markers present on the plasmid, as explained in Section 9.3.2.) Furthermore, plasmids native to a given organism sometimes interfere with transformed plasmids because of incompatibility, or, if the plasmids are similar in size, interpretation of results may be difficult [17]. Structural deletions in plasmid DNA can also occur. Unfortunately, even if these problems do not occur and the recombinant gene is successfully expressed, degradation of foreign proteins can be a problem [30].

9.3.4.4 Consumer Perception In addition to the technical problems, consumer acceptance is critical to the success of any biotechnology product. Regarding genetically engineered foods, the public must be convinced that they pose no threat to the environment, they do not adversely affect food safety, they will not harm food animals, and they will have no deleterious effect on farm revenue. Even if there is no scientific basis for concerns of this nature, a genetically engineered food product will fail if the public perceives such concerns, so food companies are well advised to adopt a proactive strategy to address public concerns [31]. The issue of consumer perception, including the recent controversy involving milk and genetically engineered bovine somatotropin (BST), is discussed in more detail in Chapter 2.

9.3.5 Summary

The problems discussed here currently limit the use of genetic engineering, especially in areas like foods and flavors, where products do not command markets in the same league as pharmaceuticals. However, as progress is made toward solving the above problems, the technology will become less expensive to apply and a broader range of products will benefit. Indeed, a number of successful applications have already been achieved in the food and flavor industry, several more are close to fruition, and others cover the range from laboratory curiosity to real commercial potential; a number of examples are discussed in the pages that follow. In spite of the title of this book, more of these examples deal with foods in general rather than flavors, fragrances, or colors per se, mainly because the number of applications relating specifically to these latter topics is limited at this time (but see Section 9.4.7). However, the examples given achieve the objective of presenting the types of things one can do with this technology. Reviews of perfumery applications are given by Gocho [32] and Klausner [33].

9.4 APPLICATIONS OF GENETIC ENGINEERING IN THE FOOD AND FLAVOR INDUSTRY

Most of the food examples discussed below are summarized in Table 9.7; flavor and enzyme examples are listed in Tables 9.8 and 9.9, respectively.

TABLE 9.7   Genetic Engineering Examples from the Food Industry

Product	Problem	Solution
Chymosin	Product derived from young calves is expensive and nonkosher.	Introduce gene into microorganism. Microbial product is inexpensive and kosher.
Cheese starter cultures	Traditional cultures eventually lose lactose-fermenting ability because genes are on plasmid.	Insert genes onto chromosome to stabilize lactose-fermenting ability.
Locust bean gum (LBG)	Guar gum can be converted to the more desirable LBG using α-galactosidase; this enzyme is found in plants but at low levels.	Introduce gene for α-galactosidase into microorganism.
Xanthan gum	Improved properties are desired.	Genetically engineered version with improved properties is under development.
Beer	Diacetyl formation leads to off-flavors, requiring 5 weeks additional processing time to remove.	Introduce gene allowing formation of acetoin instead of diacetyl into Brewer's yeast.
Beer	Improved clarity and filterability are desired.	Add genes for β-glucanase to Brewer's yeast.
Light beer	Lower calorie content is desired.	Add genes for amylases, allowing starch to be used as a substrate.
Wine	Reduced acidity is desired.	Introduce gene for malolactic enzyme into wine yeast strains, allowing conversion of malic to lactic acid.
Baker's yeast	More consistent product quality is desired in the face of doughs with varying sugar content.	Introduce maltose permease and maltase genes into yeast to obtain more consistent product.

TABLE 9.7   (*Continued*)

Product	Problem	Solution
High-nucleotide yeast extracts and autolysates	Yeasts do not have the proper nucleases to generate flavor-enhancing 5′-nucleotides.	Introduce genes for proper nucleases into yeast strains.
Phenylalanine	Higher titers and minimal by-product formation are desired.	Use genetic manipulation to relieve feedback inhibition and minimize formation of by-products.
Whey disposal	Cheese whey is a major disposal problem in the face of increasingly stringent environmental regulations.	Genetically manipulate yeasts to allow use of lactose as a substrate.
Ascorbic acid	Reduced cost and more consistent product quality are desired.	Introduce gene for 2,5-DKG reductase into producing organism to achieve these advantages.
Thaumatin	This sweet protein is found in the fruit of an African plant; a microbial source is desired.	Thaumatin has been expressed in Baker's yeast with promising results.
Debittered citrus juices	Bitter compounds are formed during juice extraction.	Remove bitterness biologically using an organism containing added gene for debittering enzyme.

TABLE 9.8   Flavor Chemicals that Are Candidates for Production by Recombinant Organisms

Flavor	Use
Diacetyl	Butter flavors
Benzaldehyde	Cherry, almond flavors
Vanillin	Vanilla flavors
Meaty beefy peptide	Meat flavors
Ethyl butyrate and other esters	Fruit flavors; flowery aroma
Terpenes	Characteristic grape aroma and odor of oils
L-Menthol	Peppermint flavors and aromas
5′-Nucleotides	Savory flavors

TABLE 9.9 Enzymes That Have Been the Subject of Genetic Engineering Studies

Enzyme Type	Use
Protease	To generate peptides for cheese flavors
Lipase	Cheese flavors
Amylase, glucoamylase	Flour supplement; starch saccharification
Pectinase	Fruit juice clarification
Phospholipase	Modification of egg yolk; emulsifiers
Human manganese superoxide dismutase	Food preservation
Glucose isomerase	High fructose corn syrup production
Glucose oxidase	Desugaring of eggs; removal of oxygen from moist food products and beverages
2,5-Diketogluconic acid reductase	Vitamin C production
β-Galactosidase	Breakdown of lactose (for lactose-intolerant individuals)
Pullulanase	Hydrolysis of pullulan, amylopectin, and glycogen; used in starch and brewing industries

Additional information is provided by Wheat and Wheat [31], who give a complete list of genetically engineered food products. A review of the application of genetic engineering methods to the production of foods and food additives was recently published by Leuchtenberger [34]. A timeline showing when some genetically engineered food products are expected to reach commercialization is given in Figure 9.4.

9.4.1 Chymosin

The first genetically engineered food product to be commercialized was chymosin (also known as rennin), an enzyme used in cheese production to facilitate coagulation of milk protein and subsequent curd formation. Traditionally, the enzyme has been obtained from the stomach of young calves; problems with that source include scarce supply (hence high price), lack of kosher status, and objections from animal rights groups.

Several years ago a number of companies realized that genetic engineering could be used to solve this problem, hence the appearance of four patent applications within a 6-month period in 1982 [35]. Pfizer successfully introduced the chymosin gene into *E. coli*, which can be cultured inexpensively at industrial scale, and now sells chymosin derived from this genetically engineered bacterium. The Pfizer product (known as Chymax™) is approximately 50 percent less expensive than traditional chymosin, which saves an average-sized cheese plant approximately $1000/day [36]. Gist-brocades also offers

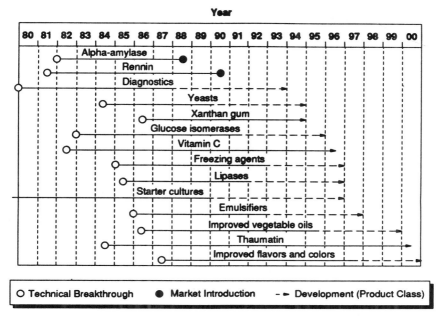

FIG. 9.4 Timeline for development of biotechnology products. *Source* Reprinted from Shamel [48], with permission. Data prepared by Consulting Resources Corporation. Copyright 1991 by Decision Resources, Inc., 17 New England Executive Park, Burlington, MA 01803, telephone 617-270-1200, fax 617-273-3048.

biotechnology-derived chymosin; they use the yeasts *S. cerevisiae* and *Klyveromyces lactis* as host organisms. Genencor offers a recombinant chymosin as well. Not only are these achievements noteworthy from a scientific standpoint, but the approval of these products for commercial sale is an important milestone in overcoming the regulatory obstacles to allowing the application of biotechnology to food products.

At least two studies [37,38] have demonstrated that there is no difference between cheddar cheese made with recombinant and standard chymosin, which of course is not at all surprising. Interestingly, one European producer of calf rennin has called for special labeling of cheese made with rDNA chymosin, citing "consumer reaction to the consumption of cheese if a problem arose with the use of rDNA chymosin. In the worst scenario, for example, if the use of a mutant organism led to fatalities from the consumption of cheese, the industry could be devastated." At a recent symposium sponsored by the Biotechnology Division of the Institute of Food Technologists, U.S. biotechnologists expressed deep concern over the proposed labeling, saying it could arouse unwarranted consumer fears and indefinitely delay food applications of recombinant DNA technology [39]. Additional discussion on public perception of the products of the new biotechnology can be found in Chapter 2.

9.4.2 Other Dairy and Cheese Applications

In the production of cheese, starter cultures containing various bacteria are used to ferment the lactose in milk to lactic acid, giving cheese its characteristic sharp flavor. Most starter cultures currently on the market consist of naturally occurring microorganisms. Loss of lactose-fermenting ability and viability in these cultures is an old problem, but recent advances in understanding of bacterial genetics has led to genetic engineering solutions. For example, genes coding for enzymes required for lactose utilization in *Streptococcus lactis* are located on a plasmid. Sometimes this plasmid is lost, hence the organism is no longer able to ferment lactose. Researchers have succeeded in stabilizing the lactose-fermenting ability by inserting the appropriate genes onto the chromosome itself [40–43]. Additional information on genetically engineered starter cultures can be found in reviews given by Klaenhammer [44], Yu et al. [45], Gasson et al. [46], and Barach [47]. Commercial availability is expected by the mid-1990s [48].

Other improvements in cheesemaking brought about by genetic engineering include accelerated ripening and increased bacteriophage (virus) resistance [49]. The Institute of Food Research (UK) is developing a genetically engineered bacterium that may prevent the growth of *Listeria* in cheese [50]. Genetically engineered lipases, used as flavor enhancers and in the production of enzyme-modified cheese, are expected to be commercialized in the late 1990s (see Section 9.4.11) [48,51]. See Harlander [52] and Jimenenz-Flores and Richardson [53] for additional dairy industry applications. See Chapter 6 for a more in-depth discussion of cheese and related topics.

9.4.3 Food Gums

Guar gum and locust bean gum (LBG) are polysaccharides consisting of a mannose backbone with galactose side groups. Guar gum has more galactose side groups than LBG, which renders guar gum less desirable regarding gelling and other properties. For this reason LBG is considerably more expensive than guar gum. The enzyme α-galactosidase can be used to convert guar gum to locust bean gum by removing galactose side groups (see Figure 9.5); unfortunately, this enzyme is found only at low concentrations in plants, so obtaining usable quantities is prohibitively expensive.

Clearly, the ability to produce large quantities of this enzyme in a microbial system would be desirable, and this has been achieved using genetic engineering. That is, the plant gene coding for α-galactosidase was introduced into various microbial hosts; several of these were able to synthesize the enzyme, including Baker's yeast, *Hansenula polymorpha*, and *Bacillus subtilis*. Baker's yeast, leading to an enzyme that was both excreted and correctly glycosylated, appeared to be the most promising for scale-up. *Hansenula polymorpha* also was promising; this organism has the advantage of being able to grow on methanol, an inexpensive carbon source, but also has the disadvantage of not being approved for food use. *Bacillus subtilis* was unable to glycosylate the

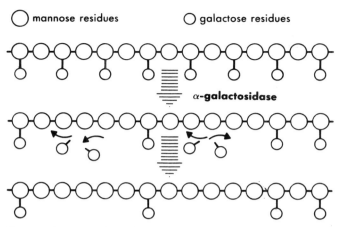

FIG. 9.5 Enzymatic conversion of guar gum to locust bean gum. *Source* Reprinted from Overbeeke [35], with permission.

enzyme, but full activity still was observed, which indicates that glycosylation is not required. However, *B. subtilis* also is not approved for food use, and a high level of residual protein was observed with this organism, which could present purification problems [35,40,54].

Xanthan gum, widely used as a thickener and viscosifier in salad dressings, ice cream, and a variety of other foods, is also the subject of genetic engineering research. This gum is produced by bacterial fermentation; a genetically engineered version of xanthan gum, expected to offer improved properties and economic benefits, is currently under development, with commercialization expected in the mid-1990s [48]. (Also see Section 9.4.8.)

9.4.4 Alcoholic Beverages

Diacetyl formation is a problem in brewing because the buttery flavor that results is undesirable in beer. Part of the relevant biochemical pathway is shown in Figure 9.6; α-acetolactate, a metabolic intermediate, sits at a branch point and can be converted to either diacetyl or acetoin. Unfortunately for the brewing industry, Brewer's yeast strains do not synthesize α-acetolactate decarboxylase (α-ALDC), the enzyme needed to form acetoin. The diacetyl that is formed can only be removed by a secondary fermentation, known as *lagering*, which takes 3 to 5 weeks. Cloning of α-ALDC from a bacterium (*Enterobacter aerogenes*) and subsequent transformation to Brewer's yeast resulted in decreased diacetyl production, eliminating the need for the secondary fermentation and reducing the total process time for beer production from 5 weeks to 1 week; no other beer characteristics were affected [35,40,55–60].

Another brewing application involves removal of β-glucans from the wort to improve filterability and clarity of the beer. This can be accomplished with β-glucanases, but these enzymes are not synthesized by Brewer's yeast strains.

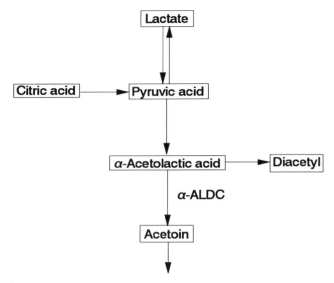

FIG. 9.6   Biochemical pathway with branches to diacetyl and acetoin.

The gene for a fungal β-glucanase was recently transformed into a Brewer's yeast strain, with the result of nearly complete removal of β-glucans [31,56]. Scientists are also developing genetically modified Brewer's yeast strains that (1) produce amylases allowing starch to become more accessible as a substrate (thereby reducing the calorie content of the beer; also see Section 9.4.11) [31, 56,61–63], and (2) are more tolerant of the toxic effects of ethanol [61].

Other alcoholic beverage applications have been reported by Berry [64], who discusses genetic approaches to eliminating off-flavors in beer, for example, phenolic off-flavors caused by the decarboxylation of ferulic acid present in the wort to 4-vinyl guaiacol, which is then reduced to 4-ethyl guaiacol. The enzyme responsible for the decarboxylation of ferulic acid could be eliminated using genetic engineering, although classical genetic approaches have also been successful [65,66].

Other examples involve wine and sake. Scientists are studying the cloning of the malolactic enzyme into yeast strains used in wine production, to allow conversion of malic to lactic acid, thereby decreasing the acidity of the wine. Currently the achievable concentration of enzyme is too low, but improvement to useful levels is expected in the future [67]. Fukuda and coworkers [68] have reported on a genetically engineered yeast that produces a sake with improved taste attributable to higher levels of β-phenethyl alcohol and tyrosine.

9.4.5   Other Yeast Applications

Baker's yeast has been used for production of bread and alcoholic beverages since ancient times. A genetically engineered strain (*S. cerevisiae* N52Ng) was approved for food use in the United Kingdom on March 1, 1990; the improved strain produces elevated levels of carbon dioxide by virtue of increased levels of two enzymes, maltose permease and maltase, which are responsible for dough leavening. The genes coding for these enzymes were obtained from a related strain of *S. cerevisiae*. The improved organism allows the baker to maintain more consistent product quality in the face of doughs with widely varying sugar concentrations. This development is an important milestone in the acceptance of biotechnology products for food use because it represents the first such approval in Europe. Improved genetically engineered baking and brewing strains are expected to be available in the United States by the mid-1990s [4,48,49].

Yeasts contain nucleases that are capable of hydrolyzing their RNA into component nucleotides. These nucleases are activated when the yeast encounters stressful conditions, for example, starvation, to prevent protein synthesis and thereby conserve energy. Certain nucleotides, notably guanosine 5′-monophosphate (5′-GMP) and inosine 5′-monophosphate (5′-IMP), possess flavor-enhancing properties and are important components of savory flavors (see Chapter 5). However, usually the 2′- rather than the 5′-nucleases dominate, and 2′-nucleotides possess no flavor-enhancing properties.

One could use genetic engineering to amplify the production of 5′-nucleases, which in turn could make possible yeast extracts and autolysates with improved flavor attributable to naturally high levels of 5′-nucleotides [69,70]. A related development is a genetically engineered *E. coli* which is able to produce 5′-GMP from xanthosine 5′-monophosphate (5′-XMP) and ammonia and/or L-glutamine [71].

Other yeast applications are as follows: (1) Liljestrom-Suominen et al. [72] reported on a genetically modified Baker's yeast with stable α-galactosidase activity, allowing it to more fully utilize available sugars when grown on molasses. (2) Davies [73] has proposed a genetically modified yeast with the ability to synthesize biotin, which would allow this valuable vitamin to be produced from food wastes. (3) Davies [73] has also proposed a more complex problem: the cloning of entire plant pathways into yeast as an alternative to plant tissue culture (e.g. the pathway for alkaloid biosynthesis). Such an approach would avoid some of the disadvantages of plant tissue culture (see Chapter 8), and allow valuable compounds to be produced by an organism that can be grown inexpensively and in large quantities. (4) Genetically engineered yeast strains that can grow on lactose are discussed in Section 9.4.8. Comprehensive reviews of the application of recombinant DNA technology to yeasts are given by Vakeria [74], Tubb [75], Hammond [76], von Wettstein [77], and Holmberg [78].

9.4.6   Amino Acids

Backman and coworkers [79] applied the principles of genetic engineering to a complex problem in their work with phenylalanine, used in the manufacture of the popular sugar substitute, aspartame. These investigators were interested in engineering an organism to produce high titers of phenylalanine at a high rate and with a minimum of by-products (including cell mass), so that purification problems would in turn be minimized.

Using *E. coli*, these workers first identified the rate-limiting steps of phenylalanine biosynthesis, then they increased the rates of those reactions by adding DNA coding for the enzymes catalyzing them (also see Section 9.8). The next step was to identify the mechanism of feedback inhibition by accumulated phenylalanine and eliminate it; this was done at the genetic level, and subsequent genetic manipulations largely relieved the inhibition, both at the enzyme expression and enzyme activity levels. Even though some of the genetic changes were carried on plasmids, the organism was shown to be completely stable with respect to both plasmid retention and ability to produce phenylalanine. Genetic manipulation was also used to minimize by-product formation, simplifying purification as mentioned above; purification was further simplified because the phenylalanine was excreted into the medium rather than retained inside the cell.

Other reports on enhanced production of amino acids have also appeared: Ajinomoto developed a process for efficient production of the aromatic amino acids (phenylalanine, tyrosine, and tryptophan) using a genetically engineered *Coryneform* bacterium [80]. Follettie and Sinskey [81] reported on genetic modification of *Corynebacterium glutamicum* for enhanced amino acid production. Venkat et al. [82] also reported on overproduction of aromatic amino acids using a genetically modified organism. Momose [83] developed a bacterium (*E. coli*) able to overproduce tryptophan by virtue of its genetically modified tryptophan synthetase. Finally, threonine levels increased almost fivefold when genes coding for key enzymes in the biosynthetic pathway were put on a plasmid and multiple copies were introduced into a bacterial host [40].

9.4.7 Flavors

A number of flavor compounds have been the subject of genetic engineering studies, including 5'-nucleotides, diacetyl, pyrazines, terpenes, lactones, aromatic aldehydes, meaty peptides, and others [35,50,84]. The 5'-nucleotides are discussed in Section 9.4.5; possibilities involving other flavor chemicals are covered in this section (see Table 9.8).

Diacetyl is used in butter flavors. Work on increasing diacetyl production by *Streptococcus lactis* using classical mutagenesis dates back to 1979. As mentioned earlier, the biochemical pathway is shown in Figure 9.6; α-acetolactic acid, a metabolic intermediate, can be converted to either diacetyl or acetoin. Genetic engineering efforts involve removing the gene coding for α-acetolactate decarboxylase (α-ALDC), thereby shutting off production of acetoin in favor of diacetyl [35]. This of course is the opposite of the problem encountered in brewing (see Section 9.4.4), where efforts are directed at suppressing diacetyl production in favor of acetoin because the former is responsible for an undesirable off-taste.

Several aromatic aldehydes are important flavor compounds, including benzaldehyde (cherry, almond flavors) and vanillin (vanilla flavors). One can envision a microbial route to natural benzaldehyde and vanillin by considering a degradative sequence of aromatic acids that proceeds via these aldehydes. One such sequence, found in certain fungi and yeasts, degrades phenylalanine via cinnamic acid and benzaldehyde to benzoic acid. A similar type of degradation occurs with ferulic acid, which is transformed via vanillin to vanillic acid.

Casey and Dobb [85] demonstrated accumulation of up to 800 mg/L of benzoic acid by *Trichosporon beigelii* and presented convincing evidence that the pathway described above was in fact followed, but were not able to actually isolate benzaldehyde. Apparently the oxidation of the aldehyde to the acid, catalyzed by a nicotinamide adenine dinucleotide (NAD)-requiring dehydrogenase, was too rapid for the aldehyde to accumulate. This problem might have a genetic engineering solution; in principle one could remove the gene coding for the dehydrogenase, thereby allowing accumulation of benzaldehyde. A similar approach might also be effective in allowing accumulation of vanillin.

An octapeptide found in beef has been identified as an important component of beef flavor. The peptide, designated BMP for beefy meaty peptide, is produced during the aging of meat after slaughter. An interesting possibility would be to determine the structure of the protein from which BMP is derived, then introduce the DNA coding for that protein into a suitable host organism, for example, a bacterium or a yeast. The modified organism could be used to produce commercial quantities of BMP, which could be used to improve the flavor of less expensive cuts of meat [86].

Certain esters impart fruity flavors and flowery aromas to foods. Some of these esters can be produced microbially; for example, organisms that produce the important flavor ester ethyl butyrate include *Pseudomonas fragi*, *Streptococcus diacetylactis* and *Lactobacillus casei*. Some of these organisms, such as *S. diacetylactis*, can be genetically manipulated to enhance ester production [69].

A number of terpenes are useful in flavors, primarily because they impart a characteristic grape aroma and odor of oils (see Chapter 4). Genetic engineering can be used to boost yields of the terpenes citronellon and linalool by *Kluveromyces lactis* [69]. Also, Javelot et al. [87] were able to enhance the terpene-producing ability of *S. cerevisiae* in wine fermentations using genetic engineering.

L-Menthol is widely used in not only in foods, but also in cosmetics, toiletries, and household detergents where a peppermint aroma is desired. Worldwide demand is over 3000 tons, which represents a $2 billion market. The plant *Mentha piperita* produces L-menthone, which is converted to L-menthol during full bloom, but at an efficiency of only 40 percent. Alternatively, the bacterium *Pseudomonas putida* YK-2 has been shown to convert L-menthone to L-menthol; genetic engineering could be used to improve the conversion efficiency, and also to obtain higher yields of a more desirable and purer form of L-menthol from a racemic mixture [69].

A patent was recently issued for a process to produce large amounts of the proteins responsible for flavor production in cocoa beans, using genetically engineered *S. cerevisiae* or *E. coli* [88].

9.4.8 Waste Disposal

Concern for the environment has increased markedly in recent years, and all indications are that this trend will continue. Regulations are becoming increasingly stringent; not only must chemical and food processors carefully control releases of chemicals into the environment, but they must also clean up chemicals released in the past (e.g., Superfund sites). These increasingly difficult environmental problems are encouraging chemical and food processors to look to new technologies for solutions, and biotechnology can clearly play a role. Some examples of treatment of food wastes using biotechnology are considered in this section.

One of the first steps of the cheesemaking process is coagulation of proteins from milk to form the curd. The remaining liquid, known as *whey*, is a waste product, and disposal of whey has become a major problem. One solution is to use whey as a carbon source for growth of commercial yeast. Tons of yeast per day are used in the baking, brewing, and other industries, so growth of this yeast on lactose, the sugar found in whey, would alleviate the whey disposal problem substantially. Although commercial Baker's and Brewer's yeast strains are unable to use lactose as a carbon source, researchers have succeeded in imparting this ability to these strains through genetic engineering [89]. Other organisms have also been genetically engineered to allow use of lactose as a carbon source for production of food and flavor materials, including xanthan gum [90], ethanol, and single cell protein [40,91].

Related waste disposal problems that could be handled effectively using genetically modified organisms include cellulosic waste from fruit and vegetable processing; blood, bone, and collagen from meat processing; shells from egg and nut processing; and starch from potato processing. Currently, there is not a strong economic incentive to develop genetically modified organisms directed at these waste streams, but that will change as the cost of the technology comes down and environmental regulations become increasingly stringent [49].

9.4.9 Lactic Acid Bacteria

These include a number of bacterial species that produce lactic acid from lactose or other sugars. Many of the lactic acid bacteria are important to the food industry. For example, members of the bacterial genus *Lactobacillus* are used in several commercially important food fermentation processes, where they contribute to the development of desirable flavor and nutritional profiles in foods and also discourage the growth of spoilage microorganisms. For these reasons, the improvement of lactobacilli using genetic engineering has been the objective of numerous studies; reviews are given by Chassy [92,93] and Gasson [94].

Other lactic acid bacteria have also been the subject of genetic engineering studies, for example, members of *Lactococcus* and *Streptococcus* have been genetically modified to produce proteolytic enzymes, bacteriocin (an antimicrobial agent) [95], β-galactosidase [96,97], α-amylase, chymosin, and pepsin, and to impart resistance to degradation by phages or citric acid. A bacteriocin-producing organism might be useful in controlling *Salmonella* in poultry when used as a feed supplement or for inhibiting proliferation of pathogens and spoilage organisms in fermented meats such as sausage [4,49,95]. Teuber [98] has also reported on strategies for genetic modification of Lactococci and Sandine [99] has discussed some practical aspects of genetic engineering research on lactic acid bacteria.

9.4.10 Ascorbic Acid

Erwinia herbicola produces 2,5-diketo-D-gluconic acid (2,5-DKG) from glucose, but lacks the enzyme (2,5-DKG reductase) necessary to convert 2,5-DKG to 2-keto-L-gluconic acid, the precursor to ascorbic acid (vitamin C). A gene coding for 2,5-DKG reductase was identified in a *Corynebacterium* sp. and expressed on a plasmid in *Erwinia*, thereby allowing a single recombinant organism to produce ascorbic acid [100–102]. Expected benefits over traditional production methods include reduced cost and a more consistent product. Commercialization is expected by the mid-1990s [48].

9.4.11 Enzymes

The food industry is the largest single user of enzymes, accounting for over 50 percent of enzyme sales of $445 million in 1987 [49]. A number of studies have been done on the use of genetic engineering to enable host organisms to produce greater amounts of enzymes, as well as improved versions of enzymes important to the food and flavor industry. A partial listing of these enzymes is given in Table 9.9; specific examples are discussed below in more detail.

High fructose corn syrup (HFCS) has partially or totally replaced sugar in many soft drinks. A genetically engineered version of one of the enzymes used to prepare HFCS from starch, α-amylase, has been on the market since 1988. This enzyme facilitates the conversion of starch to glucose, which in turn is converted to fructose using glucose isomerase. Improved genetically engineered versions of this latter enzyme are also forthcoming, probably by the mid-1990s [48,103,104].

Food industry uses of glucose oxidase (GO) include desugaring of eggs and removal of oxygen from moist food products and beverages. Unfortunately, GO obtained from traditional microorganisms is contaminated with other similar enzymes, which makes purification difficult and expensive. Researchers at Chiron have succeeded in introducing the polynucleotide sequence coding for GO in *Aspergillus niger* or *Penicillium amagasakiense* into *Saccharomyces* sp., allowing production of contaminant-free GO in a workhorse organism (Table 9.5) [104,105].

Robert-Baudouy [102] has published a review of the molecular biology of the bacterial genus *Erwinia*. These organisms synthesize several enzymes of interest to the food industry, including pectinases (fruit juice clarification) and 2,5-diketogluconic acid reductase (vitamin C production; see Section 9.4.10). Some *Erwinia* enzymes have been expressed in *E. coli*.

Other food-industry enzymes that are the subject of genetic engineering studies include protease (peptides for cheese flavors) [4,106]; lipase (cheese flavors; also see Section 9.4.2) [51]; pectinase (fruit juice clarification [4,107]; phospholipase (modification of egg yolk, emulsifiers); human manganese superoxide dismutase (food preservation) [108]; and β-galactosidase (lactose breakdown, for lactose-intolerant individuals) [49]. Reviews are given by Haas [109] and Pitcher [110].

9.4.12 Sugar Substitutes

Increased public attention to health, fitness and diet has led to the proliferation of sugar substitutes in recent years. Aspartame stands out as the most successful commercially, but others are either on the market already or are in developmental stages. One example is thaumatin, a sweet-tasting protein (100,000 times sweeter than sucrose) found in the fruit of the African plant *Thaumatococcus daniellii*. This protein became the first plant enzyme to be expressed in a microorganism; although yields were quite low in the initial work with *E. coli*, partly because of protein instability, more promising results have been obtained with Baker's yeast. Commercialization of a genetically engineered version is expected by the year 2000 [35,40,48,111–114].

Hallborn et al. [115] recently reported on a genetically engineered version of *S. cerevisiae* able to efficiently convert xylose to another promising sugar substitute, xylitol.

9.4.13 Debittering Citrus Juices

Intact citrus fruits contain precursors to bitter lemonoids and flavonoids, primarily limonin, nomilin, and naringen. In some types of fruit, these bitter compounds are formed as the juice is extracted, a problem that costs California food processors $8 million annually. There is a microbial solution to this problem; *Aspergillus niger* possesses the enzyme naringenase, which facilitates the conversion of naringen to its nonbitter form, naringenin. In recent work using immobilized cells, juice bitterness was reduced 75 percent and the cells were reused up to 20 times with no loss of activity. The economics of such a debittering process could be improved by using genetic engineering to increase the activity of debittering enzymes in *A. niger* and other organisms [69].

Hasegawa [116] has also reported on biological removal of limonoid bitterness in citrus fruits, and Bar-Peled et al. [117] have elucidated the biochemical mechanism of bitterness formation, making the goal of a genetically modified grapefruit with reduced bitterness more attainable.

9.4.14 Miscellaneous Applications

Genetic engineering has been proposed to improve single cell protein production from petroleum derivatives. For example, when catechol is used as a substrate, certain aromatic compounds are formed that are toxic to many organisms. Genetic engineering approaches to reducing the formation of these toxic aromatics have been suggested [118].

Genmark (Salt Lake City) is trying to produce healthier milk and leaner meat by manipulating the DNA of cows [119].

Researchers at Cornell University have found that when heated, honey clarifies fruit juices in the same manner as gelatin. Further work showed that the effect is attributable to one particular protein of the honey. If there were sufficient commercial interest, DNA coding for this protein could be obtained from the bee (or synthesized) and introduced into a suitable microorganism, which would allow the protein to be produced inexpensively in large quantities [120,121].

9.5 APPLICATIONS OF GENETIC ENGINEERING TO AGRICULTURE

The agricultural community faces a prodigious task in the coming years. To feed the United States alone, American farmers must deal with 10,000 possible infectious plant diseases, 2000 species of weeds, 1000 types of worms, and 10,000 varieties of ravenous insects. Approximately 800 million lbs of chemicals costing $1.5 billion are sprayed on crops each year; the long-term damage to the environment could be disastrous. Biotechnology will play an increasingly important role in solving these problems. As the world's population swells to 5.2 billion over the next two decades, biotechnology will indeed be asked to help feed the world [122]. This is evident from the sales forecast for biotechnology products given in Table 9.1, which projects that sales of agricultural products will grow at an impressive rate of 35 percent/year over the next 10 years.

Genetic manipulation of crops offers a number of potential benefits, some of which are listed in Table 9.10 [122a]. Robert Fraley, director of plant science technology at the Monsanto Company, has said, "Within two years, every major crop in the world will be manipulable using gene-transfer technology." Transgenic food-producing plants have already been engineered for species of tomato, potato, carrot, walnut, peas, lettuce, rapeseed, sunflower, cabbage, pear, sugarbeet, cucumber, asparagus, rice, soybean, and corn [4]. Reviews are given by Hardy [123], DeJong [124] and Fraley [125]; articles have also

TABLE 9.10 Potential Benefits of Genetic Engineering of Crops[a]

Improved yield.

Increased disease resistance, reducing dependence on insecticides and/or allowing the use of "natural" insecticides such as insecticidal proteins from *Bacillus thuringiensis*.

Development of faster maturing, drought-resistant, and temperature-tolerant plants, allowing more efficient use of land.

Increased herbicide resistance, including increased tolerance to more "environmentally friendly" herbicides, allowing a decrease in the use of less friendly ones.

Higher salt tolerance, allowing use of ocean water for irrigation.

Development of plants that can fix atmospheric nitrogen, eliminating the need for nitrogen-containing fertilizers.

Increased shelf life.

Improved taste and texture.

Higher protein content.

Enhanced production of natural products, e.g., amino acids, latex, enzymes, oils, starch, etc.

Improved metal tolerance.

Heat and cold tolerance.

---

[a]As described in Harlander [4] and Chakrabarty [122a].

been written by Knight [126] and Gasser and Fraley [127]. In this space I have given a brief overview, with most examples summarized in Table 9.11.

9.5.1 Tastier Tomatoes

An agricultural application close to market is being touted by Calgene (Davis, CA), who used genetic engineering to develop a tastier tomato. Everyone knows that a store-bought tomato cannot compare to a home-grown one when it comes to taste. The reason is home-grown tomatoes are allowed to ripen on the vine, whereas store-bought tomatoes must be picked green and allowed to ripen in transit, because if they were vine-ripened they would be rotten by the time they reached the consumer. The softening process leading to a rotten tomato involves the production of ethylene gas and polygalacturonase; the latter facilitates degradation of pectin, which provides part of the structural integrity of tomatoes as well as other fruits and vegetables. Softening is important in nature because it provides a means for the seeds to escape from the fruit and propagate the species, but of course it is undesirable to those of us who enjoy tasty tomatoes.

Calgene has been able to suppress production of ethylene gas and polyga-lacturonase using antisense technology; that is, the reverse of the gene coding for polygalacturonase was introduced into the organism to prevent synthesis of the enzyme [128–130]. The bioengineered tomato remains firm for a week

TABLE 9.11 Genetic Engineering Examples from Agriculture

Product	Problem	Solution
Flavr-Savr™ tomato	Tomatoes must be picked while still green so that they arrive at the grocery store before rotting.	A genetically engineered tomato stays firm up to a week longer than the traditional fruit; it can be vine-ripened, resulting in a desirable home-grown taste.
Disease-resistant crops	More disease resistance is desired.	Genetically engineer disease resistance into tomatoes, soybeans, potatoes, and other crops.
Nitrogen-fixing crops	Crops require nitrogen supplements (fertilizers) because they are unable to fix atmospheric nitrogen.	Introduce nitrogen-fixing genes from rhizobia into alfalfa, wheat, corn, and other plants.
Glyphosate-resistant crops	Improved resistance to powerful glyphosate-containing weedkillers like Roundup™ is desired.	Introduce glyphosate resistance from petunias into tobacco, tomato, canola, soybean, and other plants.
Crops that synthesize bioinsecticides	Crops that can synthesize bioinsecticides (perceived as environmentally friendly) are desired.	Genetically engineered versions of tomato, corn, and cotton plants can synthesize insecticidal protein found in *B. thuringiensis*.
Crops that are resistant to frost damage	Resistance to frost damage is desired.	Delete frost-promoting protein from bacterium *P. syringae*, then spray modified organism onto plants.
Vegetable oils	Specialty oils with higher nutritional content are desired.	Several improved oils have been produced using genetically engineered plants.
Crops with improved nutritional content	Crops with higher levels of certain amino acids are desired.	A variety of corn with higher levels of tryptophan has been developed.
Improved frozen vegetables	Reduced rancidity in frozen vegetables is desired.	Identify and delete lipoxygenase gene from the plant.
Potatoes with higher starch content	Higher starch content in potatoes is desired; this leads to improved texture, storage and cooking properties.	Increase starch content by introducing the gene for a bacterial enzyme into the potato plant.

longer than traditional tomatoes, so that it can be vine-ripened and still be structurally sound upon arrival at the grocery store. At this writing the genetically modified tomato, named Flavr-Savr™, is awaiting FDA approval [131]. However, the product has been field tested for 4 years and is widely acknowledged to be safe for human consumption. The U.S. retail market for fresh tomatoes is reportedly in excess of $5 billion per year [132].

Calgene may be challenged by Enzo Biochem following a recent announcement that Enzo Biochem will receive a patent for technology that essentially turns off genes that lead to decay or disease in plants, animals, and humans. The patent, filled in 1983, claims technology similar to that used by Calgene to develop the Flavr-Savr tomato, and may set up a legal battle between the companies [133]. Calgene is fighting another legal battle with ICI Americas, Inc., who also claims technology for firmer, more flavorful tomatoes that have less tendency to soften and are less likely to break. ICI recently announced agreements with Hunt-Wesson, a ConAgra subsidiary, to develop improved tomatoes through biotechnology for use in processed tomato products, and with Dole Fresh Vegetables, Inc. (Salinas, CA) to evaluate ICI's technology for improving the quality of fresh vegetables [134]. Finally, DNA Plant Technology Corporation also expects to offer a genetically engineered tomato in 1994; however, they are already producing speed bred tomatoes under the name "Vine Sweet" [134a].

9.5.2 Disease-Resistant Crops

In 1987, a genetically engineered tomato plant developed by Roger Beachy and colleagues at Washington University became the first genetically altered food plant ever to be field tested. Genetically altered plants survived a viral injection nicely, whereas the control group became infected and shriveled up. A disease-resistant cassava plant is being developed by Beachy in collaboration with French scientists. Beachy's research has served as a model for engineering disease resistance into plants such as soybeans, potatoes, canola, flax, and cotton, which are being field tested by the Monsanto Company [122]. In other work, several genetically engineered vegetables that may confer virus resistance are being tested by Upjohn's Asgrow seed division [135]. Developments like these could lead to substantial progress in reduction of world hunger.

9.5.3 Nitrogen Fixation

Members of the plant family known as legumes, including peas, beans, and clovers, have a symbiotic relationship with bacteria known as rhizobia, which reside on the roots and enable these plants to trap nitrogen from air for use as a nutrient. Researchers are removing the genetic material that allows rhizobia to fix nitrogen in this manner and transferring it to alfalfa, wheat, corn, and other plants. Such genetically engineered crops will potentially save up to $20 billion/year on synthetic nitrogen fertilizers, as well as make available the environmental benefits of reduced fertilizer usage [40]. These developments were reviewed by Cocking and Davey [119].

9.5.4 Glyphosate, Bioinsecticides, and Frost Tolerance

Glyphosate, manufactured by Monsanto under the tradename Roundup™, is a highly effective weed killer. However, because it kills most plants it contacts, Roundup™ usually cannot be used with commercial crops. Using genetic material obtained from petunias, researchers have developed genetically engineered versions of tobacco, tomato, canola, and soybean plants that are resistant to glyphosate [40].

Biologically derived pesticides have considerable market potential because the public perceives these as more environmentally friendly than chemical pesticides. In July of 1991, Mycogen (San Diego) became the first company to obtain approval to sell a bioengineered insecticide, MVP, a bacterium that secretes a protein harmful to caterpillars and other crop pests [131]. Genetically engineered versions of tomato, corn, and cotton plants have been developed that are able to synthesize the insecticidal protein found in the bacterium *Bacillus thuringiensis*, rendering the plants resistant to a variety of insect pests.

Crops have been made resistant to frost damage by genetically modifying *Pseudomonas syringae* to delete a protein that promotes ice nucleation, then spraying the engineered organism onto the plants to compete with natural flora [35,40].

9.5.5 Vegetable Oils

Genetic engineering has allowed modification of oil production in plants to obtain specialty oils, such as oils with higher nutritional content. Examples include:

Canola oil with increased stearic acid levels [136]

Sunflower seeds that produce an oil with a high level of the desirable oleic acid [121]

rapeseed and linseed oils with improved properties [135,137]

genetic research on palm oil, which has led to major improvements in quality and extraction yield [103]

Other examples are given by Hardin [138], Scowcroft [139], and Rattray [140]. Oils from genetically engineered plants are expected to be commercialized by the late 1990s [48].

9.5.6 Crops with Improved Flavor and Nutritional Content

Some important grains and legumes are deficient in certain essential amino acids. For example, soybean is low in sulfur-containing amino acids, and corn is low in lysine and methionine. To improve the nutritional value of these crops, one could identify the genes coding for proteins containing higher levels of these amino acids, then insert additional copies of these genes [35,40]. Along these lines, a variety of corn with higher levels of tryptophan has been developed using genetic engineering [141]. Other possibilities include elimination of antinutritional factors in plants (e.g., the trypsin inhibitor in soybeans), elimination of natural toxicants (e.g., cyanogenic glycosides in broccoli), and increasing the levels of desired vitamins [35,40].

Frozen vegetables sometimes develop rancidity due to the action of the enzyme lipoxygenase. A genetic engineering solution to this problem would be to identify and delete the gene coding for this enzyme [35,40]. Some progress has been made along these lines by Davies and coworkers [142], who improved the flavor of a soybean preparation by genetic removal of lipoxygenase-2.

9.5.7 Modification of Sugar and Starch Levels in Fruits and Vegetables

Texture, storage, and cooking properties of potatoes can be improved by increasing the amount of sugars and starch through genetic engineering [125]. For example, by expressing a bacterial ADP glucose pyrophosphorylase gene in potato tubers, investigators were able to increase starch content by 20 to 40 percent [143]. A second example involves expression of the sucrose phosphate synthase gene in transgenic tomato plants, which increased sucrose levels and reduced starch content [144].

9.5.8 Miscellaneous Applications

The potential of genetic engineering for development of improved wheat strains is discussed by Bushuk [145], who stated, ''Potential improvement in terms of yield and processing quality that may be realized by [modern techniques such as genetic engineering] is limited only by man's imagination.'' Bushuk talks of new wheat varieties with improved disease resistance, better stress tolerance, nitrogen-fixing ability, and prescribed quality.

DNA Plant Technology is using their ''Transwitch'' technology to change the colors of flowers, and might also apply it to enhancing colors of tomatoes, peppers, and chilies [135].

9.6 PROTEIN ENGINEERING

A protein is a chain of α-amino acids joined by peptide bonds, that are formed by the elimination of water between the amine group of one amino acid and the carboxyl group of another. There are 20 amino acids commonly found in proteins, and the structure and function of a particular protein is determined by the properties of its component amino acids. The number of amino acid sequences that are possible is enormous; for example, theoretically there are 20^{100} possible proteins containing 100 amino acids. This accounts for the extreme versatility of these molecules; indeed, proteins constitute over half the dry weight of most organisms and serve many functions. That is, there are enzymes, transport proteins, nutrient and storage proteins, contractile proteins, structural proteins, defense proteins, regulatory proteins, and so on [146].

Protein engineering is the alteration of the amino acid sequence to change the function or properties of the protein in some manner. For example, protein engineering might be used to increase and/or decrease enzyme activity toward certain substrates, change the pH and/or temperature optima, or render the enzyme more (or less) stable at extremes of temperature, pH, or other environmental conditions. Protein engineering might also be used to improve nutritional value by increasing the content of certain amino acids (e.g., lysine; see Section 9.5.6).

To apply the technique of protein engineering, one usually follows a procedure similar to the one shown in Figure 9.7. One must first determine the structure of the native protein, using approaches such as X-ray diffraction, nuclear magnetic resonance [147], and molecular modeling. Although solving the structure of a protein is no trivial task, the techniques for doing so have advanced considerably in recent years, particularly in the area of molecular modeling. For example, Lee [148] recently reported on the use of homology software to determine protein structure once the primary amino acid sequence is known. In this homology modeling, the protein whose structure is to be solved is compared to other similar proteins with known structure. The resulting preliminary structure assigned to the unknown protein is then optimized to minimize energy and relieve strain from misplaced amino acid side groups.

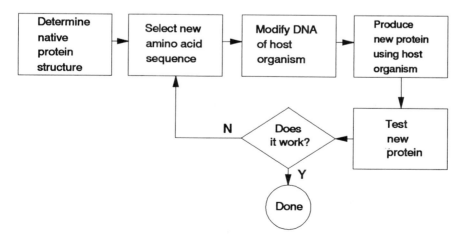

FIG. 9.7 Procedure for protein engineering.

Once the protein structure is known, the next step is to decide how it should be altered to achieve the desired change in properties. This is difficult because it requires an understanding of the relationship between protein structure and function, a problem that has received considerable attention in recent years but to date is largely unsolved, although significant progress has been made in elucidating the roles of particular milk proteins in cheesemaking and wheat flour in breadmaking [149]. Clearly the selection of an altered structure can be a formidable task because there are 19 amino acids to choose for each residue (in addition to the one present in the native protein), so that even if only a few residues are to be changed the number of possibilities is quite large.

Once the desired changes are determined, a suitable host organism is selected and genetic material coding for the altered protein structure is inserted into the organism. The altered protein is then produced by the organism and tested for the desired activity. If the desired activity is not achieved, another altered structure is selected, the appropriate genetic material is inserted into the host, and that protein is expressed. The process is repeated until the desired activity is obtained [149]. It is clear that genetic engineering is an integral part of protein engineering; that is, the latter builds upon advances in genetic engineering made in recent years.

In addition to the need for a working model of the structure–function relationship, the drawback of this approach is that many proteins of interest to the food industry are produced by higher organisms (animals, plants), and genetic manipulation of these higher organisms is difficult. One is almost forced to use a yeast or bacterium as the producing organism to express the altered protein; unfortunately, the properties of a plant or animal protein produced by a bacterial host cell may not be the same as the properties of the protein produced by the higher organism [149,150].

9.6.1 Examples of Protein Engineering

Most of the examples discussed below are summarized in Table 9.12.

9.6.1.1 Chymosin and Milk Proteins
The cheesemaking enzyme chymosin (see Section 9.4.1) cleaves a specific peptide bond in the milk protein κ-casein to initiate coagulation and curd formation. However, the enzyme also facilitates further hydrolysis, which is undesirable because it can lead to bitterness and other off-flavors. Perhaps alteration of the structure of chymosin could eliminate this secondary activity [149]. Another desirable change, which might be achieved through protein engineering, is reduced stability of chymosin, so that off-flavors are not generated during storage [151].

Jimenenz-Flores and Richardson [53,152] have reported on modification of milk proteins.

9.6.1.2 Subtilisin
Even though this example is not a food application, I have included it because it demonstrates the potential of the technology. Sub-

TABLE 9.12 Examples of Protein Engineering

Protein	Problem	Solution
Chymosin	Secondary activities lead to bitterness and other off-flavors.	Alter structure to eliminate undesirable activities, or to develop low-temperature sensitivity so that off-flavors are not generated during storage.
Subtilisin	Native form of this laundry-detergent enzyme is sensitive to oxidation by bleach.	Eliminate bleach sensitivity by changing the Met residue at position 222.
Protease	Improved thermal stability is desired.	Cognis evaluated over 150 modified proteases in less than a year, with a success rate of over 60%. Some showed an order-of-magnitude improvement in thermal stability.
T4 lysozyme	Improved thermal and pH stability are desired.	Insert a disulfide bond between two unpaired cysteine residues to achieve improved stability.
β-galactosidase	Yogurt sours with time, even at refrigeration temperature.	Modify the enzyme responsible for souring (β-galactosidase) so that it loses activity at reduced temperature.

tilisin is an enzyme secreted by many species of *Bacillus* and used in laundry detergent. The problem was the sensitivity of the native enzyme to oxidation by bleach [149]. According to peptide-mapping studies, a single amino acid residue, Met222, is responsible for this sensitivity. Estell et al. [153] prepared a series of engineered enzymes in which the other 19 amino acids were tried at the 222 position. Several of these had reduced bleach sensitivity, and some had enhanced activity as well.

Chen and Arnold [154] also investigated engineered versions of subtilisin but with a different objective—enhanced activity in polar organic media. As discussed in Section 9.9, polar organic solvents are often unsuitable for enzymatic catalysis because the stabilizing water layer is stripped from the surface of the enzyme. Chen and Arnold have shown that protein engineering can be effective in improving enzyme activity in such media. Using random mutagenesis, they developed an engineered version of subtilisin E that contained three amino acid substitutions and was 38 times more active than the native enzyme in 85 percent DMF (dimethyl formamide).

9.6.1.3 *Enzymes with Improved Stability* Cognis, an independent subsidiary of Henkel KGaA, Germany, has succeeded in improving the storage stability of a protease by modifying its structure. Scientists at Cognis selected

a reasonable mutational strategy based on a three-dimensional model of the protease, then developed methods for simultaneous introduction of multiple site-directed mutations and for rapid purification and characterization of the resulting variant proteases. Over 150 modified proteases were evaluated in less than a year, with a success rate for achieving improved stability in excess of 60 percent. Some mutants exhibited an order-of-magnitude improvement in thermal half-life compared to the native protein [155].

Spradlin [156] discusses a range of work directed toward tailoring various enzymes for food processing applications, including engineered enzymes with enhanced thermostability for flavor enhancement during microwave cooking.

9.6.1.4 T4 Lysozyme
A substantial portion of current research in protein engineering is geared toward improving the stability of proteins in harsh environments, for example, extremes of temperature or pH. Perry and Wetzel [149,157] investigated T4 lysozyme, a protein that contains no disulfide bonds and exhibits only limited heat stability at neutral pH. These investigators used computer graphic analysis and X-ray crystallography to select a site for insertion of a disulfide bond between two unpaired cysteine residues. The resulting engineered protein was considerably more stable than the native one. Other temperature-sensitive and temperature-stable forms of T4 lysozyme have also been developed [158,159].

9.6.1.5 Engineered Proteins at the Institute of Food Research
Protein engineering is being studied at the Institute of Food Research (IFR; Shinfield, Reading, UK), whose focus is to understand the role played by proteins and polysaccharides in providing structure and functionality to foods [73]. The Institute's scientists are using advances in molecular biology and analytical methods to generate modified molecules better suited to processing requirements. For example, IFR (in collaboration with Durham University) is using site-specific mutagenesis to modify pea proteins, which are then produced using genetically engineered yeasts and compared to the native protein.

The group at IFR has also designed α-helical polypeptides that have greater stability than naturally occurring molecules. These improved molecules may find widespread use as food emulsifiers; at present they can only be produced in small quantities, but in the next phase of the work the DNA coding for these modified molecules will be introduced into a microorganism amenable to inexpensive large-scale production.

9.6.1.6 Improved Yogurt
Lactobacillus bulgaricus is a prominent member of yogurt starter cultures. The problem addressed by Mainzer and coworkers [160] was the souring of yogurt with time, even at refrigeration temperature. The souring is caused by the conversion of lactose to lactic acid, a reaction catalyzed by *lac* permease (the lactose transport protein) and β-galactosidase. Mainzer and his group were able to modify the native β-galactosidase found in *L. bulgaricus*, producing a modified enzyme that loses activity at refriger-

ation temperature. The gene coding for the native enzyme was deleted and replaced with one coding for the modified enzyme, eventually leading to a yogurt with a substantially longer shelf life.

Another approach to the same problem takes advantage of the drop in pH that occurs during yogurt production. Mainzer et al. [160] are developing a pH-sensitive form of β-galactosidase, which will shut down at or near the final pH of the yogurt, thereby preventing souring.

9.6.2 Designing Enzymes from Scratch

In spite of the successes described in the previous section, selecting changes in the amino acid sequence of a protein that will lead to specific improvements in performance is very difficult, and there is much to learn before the full potential of the technique can be realized. Specifically, existing databases (for both sequence and three-dimensional structure) must be expanded, software for molecular dynamics and energy minimization must be improved [147], and the problem of the relationship between protein structure and function must be solved.

However, these advances will all be in hand in the not-too-distant future, and the potential is mind-boggling. Theoretically, one could write down an amino acid sequence which would define a protein having any set of desired properties! For example, if one were interested in an enzyme with a certain substrate specificity, level of activity, pH optimum, temperature optimum, pH stability, and temperature stability, one would enter these parameters into a computer program, and the output would be an amino acid sequence. One would then synthesize the DNA coding for that sequence, insert the DNA into a suitable host organism, grow the organism, and thereby create a source of the ideal protein for the application of interest. While this type of ability might be considered ''blue sky'' today, it is theoretically possible considering the accomplishments that have been made to date. As stated so eloquently by Neidleman [161], ''Eventually, commercial enzymes will be machine-made with specific properties engineered into molecules by programs based on a detailed understanding of structure-activity relationships.''

9.7 FLUX MAPPING OF BIOCHEMICAL PATHWAYS

A pathway is a series of biochemical reactions defining the conversion of some substrate to one or more products. Pathways can be anabolic, such as synthesis of starch or other complex carbohydrates from simple sugars, or catabolic, such as breakdown of glucose to provide energy for carrying out metabolic activities, a pathway that consists of 20 reactions. Often pathways contain multiple branches and are highly regulated, so that the organism is able to maintain the necessary concentrations of various metabolites. It is important to realize that each reaction of a pathway is catalyzed by a different enzyme.

Flux mapping, as the name implies, is the mapping of the flux of carbon through various branches of a biochemical pathway. To apply the technique, first the biochemical reaction rate of each step of the pathway is measured; once this information is in hand, the rate-limiting steps leading to some desired product can be determined and the appropriate genetic manipulations can be performed to increase the rates of these steps, thereby increasing the overall production rate of the desired product. Flux mapping is also used to determine the relative amounts of various metabolites (including cell mass) formed per unit of substrate consumed. One can then genetically modify the organism to redirect the flux of carbon, enhancing the formation of desired metabolites at the expense of undesired ones. However, redirection of flux distribution in this manner is often difficult because pathways have evolved to resist such changes [162]. Examples of flux mapping follow.

Simon et al. [163] presented a method of flux mapping that gives a dynamic representation of the competition between different branches of a metabolic pathway, using as an example the fermentation of glucose to lysine by *Brevibacterium flavum*. Using measured rates of consumption of glucose and formation of lysine and other metabolites, these workers were able to characterize completely the carbon flow at various points of the fermentation.

Jeong et al. [164] presented a detailed mathematical model for the growth process of *Bacillus subtilis*, which is well suited for use in pathway engineering studies such as those described in Section 9.8. The model describes a number of cellular metabolic processes and gives concentration profiles of 35 biochemical intermediates. It can even simulate the transition between exponential and stationary phases of growth in batch culture.

Stephanopoulos and Vallino [162] used flux mapping to enhance production of lysine by *C. glutamicum*.

Ataai and Jeong [165] developed a detailed model of the major metabolic pathways of *B. subtilis*. They proposed that their model can be used to predict the effect of changing one or more enzymes on the flux of a metabolite of interest or the intracellular concentration of that metabolite. They suggested that this approach could be useful for reducing the inhibitory effect of organic by-products, a common problem in fermentation.

9.8   PATHWAY ENGINEERING

Pathway engineering is defined as manipulation of a pathway to achieve some desired result. Although this definition seems simple, the technique is actually complex and very powerful; indeed, it can be considered the culmination of the techniques described in the preceding sections. To apply the technique, the pathway of interest is altered by manipulating the enzymes catalyzing the component biochemical reactions, that is, by modifying the DNA coding for those enzymes. Genetic engineering, protein engineering, and/or other techniques may be used to achieve highly specific additions or deletions of enzymatic activities in the host organism or to change the regulation of enzyme expression and/or activity. For example, one may want to accelerate a particular reaction, favor one branch over another, or add entire new branches [160].

Consider the pathway illustrated in Figure 9.8a for some hypothetical microorganism. Here compound A is converted to D via intermediates B and C; the respective steps are catalyzed by enzymes 1, 2, and 3. Suppose we are interested in some related compound G; first we write down some reasonable pathway leading to compound G, in this case starting with B and proceeding through E and F, with the respective steps catalyzed by enzymes 4, 5, and 6 (see Figure 9.8b). We then select a suitable organism that contains a pathway as close as possible to the desired one. If there is an organism that contains the desired pathway, the work is done. If not, the question is whether the required enzymes exist in nature somewhere. If they do, the task reduces to a genetic engineering problem—inserting the appropriate genes into the selected organism. If the required enzymes do not exist, the problem is one of protein engineering; we either modify existing similar enzymes or design new ones from scratch, then synthesize the DNA, and insert it into the selected organism. The approach is illustrated in Figure 9.9. Although a microorganism was used for this example, in principle this approach could be used for higher organisms as well.

At this time we are a long way from being able to make such complex manipulations routinely. Among the missing pieces are efficiency with genetic manipulations involving higher organisms (plants and animals), an in-depth

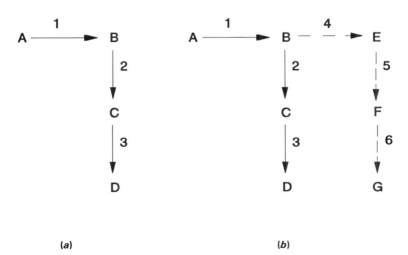

(a)   (b)

FIG. 9.8 Biochemical pathway for some hypothetical microorganism. (a) Unmodified pathway. (b) Engineered pathway.

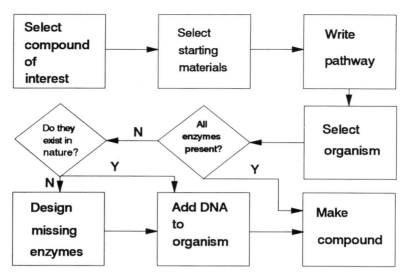

FIG. 9.9 Procedure for pathway engineering.

understanding of the relationship between protein structure and function, and ability to synthesize large pieces of DNA. However, the rudiments are in place, and there are those that believe that in the decades to come, complex engineering of biochemical pathways will become routine.

Furthermore, when its potential is fully realized, pathway engineering will be used to synthesize novel food and flavor ingredients using a powerful proactive approach. That is, given advances in the understanding of the relationship between structure and function of these ingredients, the food technologist will write down the properties desired in a particular ingredient, then use commercially available software to design the molecule. The new ingredient will then be manufactured by an organism modified using the approach outlined in Figure 9.9.

We have already discussed one example of pathway engineering in Section 9.4.6, that is, the work of Backman et al. [79] on enhanced phenylalanine production by *E. coli*. Additional examples from the recent literature are discussed below.

9.8.1 Reduced Acid Production in *Escherichia coli*

Escherichia coli is perhaps one of the most well-studied organisms known and is a popular choice as a host organism in genetic engineering work. The physiology of this organism is such that almost any carbon source can be transformed into acids and other by-products, which in many cases are undesirable, not only because they represent carbon diverted from desired products but because they actually inhibit metabolism [166]. Pathway engineering can be used to redirect carbon flow away from acids in *E. coli*. For example, Ingram et al. [167,168] found that expression of pyruvate decarboxylase and alcohol dehydrogenase from the bacterium *Zymomonas mobilis* in *E. coli* significantly reduced the formation of acids in favor of ethanol.

9.8.2 Acetone–Butanol Fermentation

Perhaps one of the most well known fermentation processes uses *Clostridium acetobutylicum* to produce acetone, butanol, and small amounts of ethanol and isopropanol. At one time, this process was used for commercial production of these solvents, notably to supply the explosives industry during World War I, but the fermentation route has long since been replaced by less expensive petrochemical processes. However, interest in the fermentation route has been revived in recent years for several reasons: (1) these solvents have applications in flavors, and the fermentation route provides natural status; (2) people are anticipating the inevitable increase in the price of oil; and (3) advances in pathway engineering and other techniques discussed in this chapter can potentially be used to increase the efficiency and yield of solvent production by fermentation, allowing the process to again be competitive.

For example, one could conceivably alter the pathway to produce more of one solvent and less of the other, and/or one could use biotechnology to alleviate the inhibition caused by solvent accumulation, thereby allowing higher yields. A recent paper by Mermelstein et al. [169] reported that the pathway had been engineered to enhance the production of acetone and butanol at the expense of acetate and butyrate by inserting an *E. coli* plasmid. Other work has been reported by Lee et al. [170].

9.8.3 Improved Ethanol Production by Baker's Yeast

The conversion of glucose to ethanol by Baker's yeast is well known. Bailey [171] used various experimental techniques to characterize the kinetics of each biochemical step of this pathway, the regulatory characteristics of the process and the sensitivities to certain environmental manipulations like pH. He observed: (1) a major change in pathway flux sensitivities with change in environment; (2) the need for simultaneous manipulation of several pathway steps to achieve large changes in overall metabolic flux; and (3) that the major effect of environment was at the level of gene expression regulation.

9.9 NONAQUEOUS ENZYMATIC CATALYSIS

Because enzymatic reactions are the foundation of the technologies discussed in this chapter so far, it is appropriate to include a section on an important, emerging area of enzymatic catalysis, that is, enzymes in nonaqueous systems.

Most enzymes in nature function in aqueous environments; indeed, water participates in all of the forces that provide structural stability and catalytic activity to enzymes, including van der Waals interactions, hydrogen bonding, salt bridges, and hydrophobic interactions. However, an enzyme in aqueous solution does not "see" the entire 55.5 mol/L of water that surrounds it, which implies that enzyme structure and activity can be maintained with only a thin layer of water around the enzyme. This concept forms the basis of an exciting new area of biotechnology, known as nonaqueous enzymatic catalysis. That is, enzymatic reactions are actually carried out in organic solvents, with only a thin layer of water surrounding each enzyme molecule.

Nonaqueous enzymatic transformations offer a number of advantages over the traditional aqueous reactions, and these advantages (listed in Table 9.13) [172–177] are encouraging a substantial expansion of the use of enzymes in food processing. That is, nonaqueous enzymology is being considered for applications where traditional aqueous enzymology would be unsuitable because, for example, of low yield or poor enzyme stability.

TABLE 9.13   Advantages of Nonaqueous Enzymatic Catalysis[a]

Many compounds of interest in the food and flavor industry are poorly soluble in water. The higher solubility of such compounds in nonaqueous systems makes them more available to the enzyme. For this reason, enzymes often exhibit marked changes in substrate specificity in organic solvents vs. water.

Water often participates in undesirable side reactions.

Thermodynamic equilibria are often unfavorable in aqueous systems, e.g., esterifications catalyzed by lipase [174].

Product recovery from aqueous solution is often more difficult and energy-intensive than recovery from nonaqueous solution, because of the relatively high boiling point of water.

Microbial contamination is not a problem in organic solvents.

Catalytic activity is sometimes higher in organic solvents.

Enzymes are sometimes stable at higher temperatures in organic solvents. This is because most mechanisms of irreversible thermal activation require water. An example is porcine pancreatic lipase, which inactivates almost instantly in water at 100°C, but has been shown to have a half-life of 12 hours at that temperature in systems containing only 0.02 percent water. Furthermore, the enzyme is ten times more active at 100°C than at room temperature.

Enzymes tend to be mobile in water and more rigid in organic solvents. This rigidity allows the formation of an imprint by a ligand, which can alter the active site and change the enzyme's properties, leading to enhanced reactivity with a given substrate. For example, in water, lipase only catalyzes ester hydrolysis, but in nonaqueous systems at least six other reactions can be catalyzed, including transesterification, esterification, amidolysis, and acyl exchange.

Nonaqueous enzymatic catalysis is particularly useful for production of optically active compounds [175–177].

[a]As described in Klibanov [172] and Van Brunt [173], except as noted.

9.9.1 Required Water

As mentioned earlier, in general some water must be present in a nonaqueous system for successful use of enzymes, to preserve structure and catalytic activity [178,179]. However, only a small amount of water is required; Yamane et al. [180] demonstrated lipase activity at initial moisture levels well below the solubility of water in benzene. Interestingly, an optimal water level was observed by these investigators in transesterification reactions, representing a balance between the requirements of hydrating the enzyme and suppressing the reverse reaction.

The requirement for water dictates which solvents are likely to be suitable for enzymatic reactions in nonaqueous systems. Nonpolar solvents usually work well because there is little tendency for the solvent to strip water from the surface of the enzyme. Similarly, more polar solvents don't usually work as well, except for certain enzymes which bind water very tightly, unless the solvent is saturated with water beforehand [173].

Along these lines, Schulze and Klibanov [181] observed that serine proteases (subtilisins) from *Bacillus amyloliquefaciens* and *Bacillus licheniformis* exhibited considerable loss of activity during multihour incubations in *tert*-butyl alcohol and acetonitrile when activities were measured in the solvent, but no loss of activity was observed when measured in water. The loss of activity in the solvent was attributed to stripping of the bound water from the enzyme surface, whereas in water the surface became rehydrated and activity was restored. In support of this view, activity in *tert*-butyl alcohol was restored by the addition of small quantities of water.

Another example is found in the work of Fayolle et al. [182], who studied the esterification reactions of butanol with butyric acid and acetic acid to form butyl butyrate and butyl acetate, respectively. These esterifications were catalyzed by lipase from *Mucor miehei* and were carried out in heptane; octane was also shown to be effective. These investigators found that the reaction rate decreased with increasing substrate concentration, a phenomenon they attributed to either substrate inhibition or increasing hydrophilicity of the system. The latter hypothesis is supported by the observation that the reaction rate decreased by adding 5 g/L of water to the reaction medium. Furthermore, when the kinetics of esterification were followed in detail, it was found that the reaction rate decreased with time, probably due to the water formed. The authors also pointed out that addition of a small amount of water (up to 1 g/L) enhanced activity, consistent with the notion that a small amount of water is necessary in these systems to maintain enzyme structure and function.

The effect of solvent polarity can be quantified using the log P value, defined as the partition coefficient of the solvent in the octanol–water two-phase system. Solvents with high log P values (hydrophobic solvents) are more suitable for enzymatic reactions than solvents with low log P values. Biocatalysis is generally thought to be feasible in solvents with log P values higher than 4, and not feasible in solvents with log P values lower than 2; solvents with log P values between 2 and 4 cannot be generalized [174,183,184].

9.9.2 The Concept of pH

In aqueous systems the pH has a major effect on enzyme activity and function. However, the concept of pH is more abstract in nonaqueous enzymology; indeed, there is no pH in the bulk organic phase, and although a small water layer exists around the enzyme molecule itself, the volume of that water layer is so small that the concept of pH begins to lose meaning. In such a system, the pH is that of the aqueous environment from which the enzyme was taken prior to being transferred to the nonaqueous system. The concept, known as "pH memory," is that protonated groups stay protonated and nonprotonated groups stay nonprotonated. The system is stable as long as no charged species are generated or consumed, otherwise the stability depends on the buffering capacity of the system [185]. Various methods of measuring the pH of the aqueous microenvironment, including NMR and fluorescent molecular probes, are discussed by Mattiasson and Adlercreutz [184] and Carta et al. [174].

9.9.3 Immobilized Enzymes

Use of an enzyme immobilized onto an inert solid support offers several advantages. These are listed in Table 9.14 [186]. The above remarks on the relationship between the polarity of an organic solvent and its suitability for enzymatic reactions can also be applied to solid supports used for immobilized enzyme systems. That is, reaction rates are generally lower when the support has a high affinity for water (known as *aquaphilicity*) [184]. Examples of the use of immobilized enzymes are given in subsequent sections.

9.9.4 Lipases in Organic Solvents

As mentioned in Table 9.13, nonaqueous systems can be useful for shifting the equilibrium of certain reactions in a desirable fashion; an example is lipase-catalyzed reactions. That is, in aqueous media lipases typically catalyze the hydrolysis of esters. However, several investigators have shown that with nonaqueous media such reactions can be run in reverse, (i.e., esters can be syn-

TABLE 9.14 Advantages of Immobilized Enzymes in Nonaqueous Catalysis[a]

Simplified product recovery.
Ability to reuse the enzyme.
Ability to operate in continuous mode, which offers higher productivity than batch operation.
Improved stability.
Avoidance of mass transfer limitations, which occur because of the presence of aggregates. These sometimes form when enzymes are used in organic solvents because of poor solubility of enzymes in such solvents.

[a]As described in Mattiasson and Adlercreutz [184], Carta et al. [174], and Whitaker [186].

thesized). For example, in the reaction of heptanol and butyric acid to form heptanobutyrate, the equilibrium lies only 0.1 percent to the right in water, but 98 percent to the right in hexane [172].

The use of lipases for ester production in organic media has been widely studied. One example, the work of Fayolle et al. [182], was presented in Section 9.9.1. A second example is found in the work of Carta et al. [174], who developed an efficient method of immobilizing lipase which is simple, gives high protein loading (10 mg/g), and gives excellent enzyme stability. These workers used this system to study the formation of various flavor esters from the component acids and alcohols, for example, propionate and butyrate esters of ethanol and isoamyl alcohol. The reactions were carried out in nearly anhydrous hexane, with productivities in excess of 0.02 mol/hour per gram of immobilized protein, and enzyme half-lives in excess of 10 days.

Furthermore, the synthesis of isoamyl propionate in a solvent-free system was demonstrated, with the acid dissolved directly in isoamyl alcohol; maximum productivity was 0.055 mol/hour per gram, obtained at a propionic acid concentration of 3.5 mol/L and a temperature of 25°C. Synthesis of ethyl propionate and isoamyl propionate in a continuous reactor was also demonstrated using a hexane-based system. For the former, a two-stage system consisting of a well-mixed reactor followed by a packed bed reactor was proposed to avoid enzyme inactivation by high concentrations of ethanol.

Welsh and Williams [187] studied ethyl butyrate and butyl butyrate synthesis in hexane using lipases from *Candida cylindracea*, pig pancreas, and two types of *Aspergillus niger*. These investigators considered the effects of pH, temperature, and substrate concentration, and found that the highest yields of both esters were obtained with the *C. cylindricea* lipase. Welsh et al. [188] also studied the lipase-mediated biosynthesis of these and other esters in nonaqueous systems. Interestingly, instead of the more usual organic solvents, various oils were used, including corn, peanut, sunflower, soybean, canola, olive, and butter oil. Butter, soybean, peanut, and olive oil were most effective, with yields of 50 to 60 percent after a 16-hour incubation at 25°C, whereas corn oil was the least effective, with a yield of 38 percent.

Langrand et al. [189] demonstrated the production of 35 short-chain flavor esters using lipases from *M. miehei*, *Aspergillus* sp., *Candida rugosa*, and *Rhizopus arrhizus* in various organic media. Murray and coworkers [190] reported on microbial production of natural acids and alcohols, followed by extraction into organic solvents and then lipase-mediated ester formation, also in organic solvents.

Kanerva et al. [191] demonstrated the utility of nonaqueous systems in the formation of optically active compounds. They used a variety of organic solvents to study the effect of solvent on rate and selectivity of the transesterification of 2-octanol and 1-phenylethanol with 2,2,2-trifluoroethyl butyrate, catalyzed by pig pancreatic lipase, and found that the (R)-ester was obtained at high optical purity in all solvents used. They also found that the solvent affected only the rate of reaction, that is, the enantiomeric purity was not affected by the use of an aliphatic versus an aromatic alcohol. Aliphatic ethers were found to be good solvents for this system.

Lipases can be used in organic media for rearrangement of triglycerides to give different properties, to produce esters and waxes, to form monoglycerides and to modify phospholipids. Advantages of the modified products include improved melting properties (e.g., synthetic cocoa butter), improved nutritional properties (e.g., rearranged positions of ω-3 fatty acids), and improved emulsifying properties. Detailed examples of useful products obtained using various lipases are given by Whitaker [186]. A review of kinetics and mechanisms of reactions catalyzed by immobilized lipases was recently published by Malcata et al. [192].

9.9.5   Other Enzymes in Organic Solvents

If only lipase examples are considered, one might question the generality of the utility of enzymes in nonaqueous systems, since lipases naturally function at lipid–water interfaces. However, a number of studies have been done using other enzymes, including alcohol dehydrogenases, proteases, oxidases, peroxidases, and glycosidases. One example, the work of Schulze and Klibanov [181] on serine proteases, was discussed in Section 9.9.1. Other examples are given below; enzymes discussed in these examples are listed in Table 9.15.

Regarding glycosidases, Laroute and Willemot [193] evaluated the effects of water content, water activity, organic solvent nature, and substrate concentration on α- and β-glucoside synthesis using glucoamylase and β-glucosidase in water–organic mixtures. They demonstrated maximum yields with 10/90 (v/v) water–organic using primary alcohols, and 15/85 using diols.

Miethe et al. [194] studied oxidation of various alcohols to the corresponding aldehydes using yeast alcohol dehydrogenase in a surfactant–hexane system. In this system the surfactant molecules formed aggregates known as liquid crystals, and these aggregates provided pockets of an aqueous environment that housed the enzyme. These workers demonstrasted the oxidation of a number of alcohols, including benzyl alcohol, cinnamyl alcohol, n-butanol, n-hexanol, n-heptanol, n-octanol, n-nonanol, and n-decanol. No activity was observed with methanol, isopropanol, tert-butanol or amyl alcohol. In related work, Legoy et al. [195] reported on the use of immobilized horse liver alcohol

TABLE 9.15   Enzymes with Demonstrated Activity in Organic Solvents

Lipase
Glycosidase
Alcohol dehydrogenase
Protease
Chymotrypsin
Peroxidase
Ligninase

dehydrogenase for the oxidation of geraniol to the flavor aldehyde geranial, in a medium consisting of 91 percent hexane. In this work, an organic solvent made possible the use of an otherwise unstable enzyme and cofactor.

Another exciting application of nonaqueous enzymatic catalysis is in lignin degradation. Lignin, the major structural component of trees and other plants, is the second most abundant organic chemical on Earth, and upon degradation, is the largest source of renewable aromatic chemicals. Lignin degradation also has applications in the pulp and paper industry. Ligninase is produced by certain fungi but collection of commercial quantities is difficult. A more common and less expensive enzyme, horseradish peroxidase, does not degrade lignin in water but it does in 95 percent dioxane [173].

Still another application is peptide synthesis in organic solvents. Moderate concentrations of water-miscible solvents have been used in peptide synthesis for many years, mainly to decrease the acidity of the α-carboxyl group, thereby shifting the equilibrium toward peptide synthesis and away from proteolysis. Clapés et al. [196] studied peptide synthesis reactions between N-protected amino acid esters and leucine amides in organic media, using α-chymotrypsin immobilized onto Celite; their work included an evaluation of the effects of the solvent and the thermodynamic water activity. Substrate specificity was shown to be determined by both the enzyme specificity and the influence of the solvent. High reaction rates were observed in hydrophobic solvents, and the competing hydrolysis of the ester occurred to only a minor extent. Reactions occurred with water activities as low as 0.11, but rate constants were two orders of magnitude higher at the highest water activity tested (0.7).

9.10 CONCLUSION

It is clear from the numerous examples presented in this chapter that genetic engineering and related areas of biotechnology are having a substantial impact on the food and flavor industry. Many products have been touched by these technologies, with results covering the range from laboratory curiosity to commercialization. Indeed, with today's business climate more competitive than ever before, food and flavor companies are turning to technologies that offer a way to increase profits, and biotechnology will continue to capture the attention of the industry. As technical advances are made and the consumer becomes more comfortable with rDNA products, cost-effective production of novel ingredients and lower-cost production of existing ingredients will be achieved using the technologies discussed in this chapter.

REFERENCES

1. Newsletter, Spring. Consulting Resources Corporation, Lexington, MA, 1992.
2. *The Impact of Biotechnology on the Flavors and Fragrances Industries—A Worldwide Study of Market Opportunities in Flavors, Fragrances, Aromatics, and Essential Oils.* Technology Management Group, New Haven, CT, 1988.

3. Moshy, R., in *Biotechnology in Food Processing* (S.K. Harlander and T.P. Labuza, eds.), p. 8. Noyes Publications, Park Ridge, NJ, 1986.

4. Harlander, S.K., *Food Technol.*, pp. 84, 86, 91, 92, 95 (1991).

5. Lewis, R., *Health* **18**, 55–56 (1986).

6. Lehninger, A.L., *Principles of Biochemistry*, pp. 913–944. Worth Publishers, New York, 1982.

7. Walker, J.M., and Gingold, E.B., *Molecular Biology and Biotechnology.* Royal Society of Chemistry, London, 1985.

8. Bu'lock, J., and Kristiansen, B., *Basic Biotechnology.* Academic Press, London, 1987.

9. Emery, A.E.H., *An Introduction to Recombinant DNA.* Wiley, New York, 1984.

10. Skoda, J., and Skodova, H., *Developments in Food Science. 14. Molecular Genetics: An Outline for Food Chemists and Biotechnologists.* Elsevier, New York, 1987.

11. Zabriskie, D.W., and Arcuri, E.J., *Enzyme Microb. Technol.* **8**, 706–717 (1986).

12. Marx, J.L., ed., *A Revolution in Biotechnology.* Cambridge University Press, New York, 1989.

13. Debenham, P.G., *Trends Biotechnol.* **10**, 96–102 (1992).

14. Fox, J.L., *Bio/Technology* **10**, 625 (1992).

15. Lehninger, A.L., *Principles of Biochemistry*, pp. 871–912. Worth Publishers, New York, 1982.

16. Thomas, C.M., *Chem. Ind. (London)*, pp. 674–677 (1991).

16a. Levin, M.A., Kidd, G.H., Zuagg, R.H., and Swarz, J.R., *Applied Genetic Engineering: Future Trends and Problems*, p. 9. Noyes Publications, Park Ridge, NJ, 1983.

17. Harlander, S.K., *Food Biotechnol.* **4**, 515–526 (1990).

18. Froseth, B., Harmon, K., and McKay, L., *Appl. Environ. Microbiol.* **54**, 2136–2139 (1988).

19. Harmon, K., and McKay, L., *Appl. Environ. Microbiol.* **52**, 45–50 (1986).

20. Harmon, K., and McKay, L., *Appl. Environ. Microbiol.* **53**, 1171–1174 (1987).

21. Punzo, C., and Hodgson, J., *Bio/Technology* **10**, 609 (1992).

22. Sautter, C., Waldner, H., Neuhaus-Url, G., Galli, A., Neuhaus, G., and Potrykus, I., *Bio/Technology* **9**, 1080–1085 (1991).

23. Wasserman, B.P., and Montville, T.J., in *Biotechnology Challenges for the Flavor and Food Industry* (R.C. Lindsay and B.J. Willis, eds.), p. 2. Elsevier Applied Science, London, 1989.

24. Buckholz, R.G., and Gleeson, M.A.G., *Bio/Technology* **9**, 1067–1072 (1991).

25. Stewart, G.G., and Russel, I., *Cereal Foods World* **32**, 766–769 (1987).

26. Rank, G.H., Casey, G., and Xiao, W., *Food Biotechnol.* **2**, 1–42 (1988).

27. Deák, T., *Zentralbl. Mikrobiol.* **145**, 327–351 (1990).

28. Esser, K., and Mohr, G., *Food Biotechnol.* **4**, 495–496 (1990).

29. Buchanan, R., in *Biotechnology in Food Processing* (S.K. Harlander and T.P. Labuza, eds.), pp. 209–219. Noyes Publications, Park Ridge, NJ, 1986.

30. Ward, O.P., *Fermentation Biotechnology: Principles, Processes and Products*, pp. 199–200. Prentice-Hall, Englewood Cliffs, NJ, 1989.

31. Wheat, D., and Wheat, B., in *Biotechnology of Food: An Old Industry on a New Vector* (H. Hiebert, ed.), pp. 14–17. Decision Resources, Burlington, MA, 1991.

32. Gocho, S., *Perfum. & Flavor.* **9**, 61–65 (1984).

33. Klausner, A., *Bio/Technology* **3**, 534–536, 538 (1985).

34. Leuchtenberger, A., *Acta Biotechnol.* **12**, 57–65 (1992).

35. Overbeeke, N., in *Biotechnology Challenges for the Flavor and Food Industry* (R.C. Lindsay and B.P. Willis, eds.), pp. 89–96. Elsevier Applied Science, London, 1989.

36. Taylor, R.F., in *Biotechnology of Food: An Old Industry on a New Vector* (H. Hiebert, ed.), p. 76. Decision Resources, Burlington, MA, 1991.

37. Bines, V.E., Young, P., and Law, B.A., *J. Dairy Res.* **56**, 657–664 (1989).

38. Meisel, H., *Milchwissenschaft* **43**, 71–75 (1988).

39. Anonymous, *Food Eng.*, September, vol. 63, p. 14 (1991).

40. Wasserman, B.P., Montville, T.J., and Korwek, E.L., *Food Technol.*, January, pp. 133–146 (1988).

41. McKay, L.L., in *Bacterial Starter Cultures for Foods* (S.E. Gilliland, ed.), p. 160. CRC Press, Boca Raton, FL, 1989.

42. McKay, L.L., *Antonie van Leeuwenhoek* **49**, 259–274 (1983).

43. Gasson, J.J., *Antonie van Leeuwenhoek* **49**, 275–282 (1983).

44. Klaenhammer, T.R., *Abstr. Pap., 192nd Meet. Am. Chem. Soc.*, BTEC47 (1986).

45. Yu, R.S.T., Hung, T.V., Kyle, W.S., and Azad, A.A., *Proc. Int. Congr. Food Sci. Technol., 6th, 1983*, vol. 2, 154–155 (1983).

46. Gasson, M.J., Dodd, H.M., Shearman, C.A., Underwood, H.M., and Anderson, P.H., *Aust. Biotechnol. Conf., 8th Meet.*, pp. 4–13 (1989).

47. Barach, J.T., *Food Technol.* **39**, 73–74, 79, 84 (1985).

48. Shamel, R.E., *Timelines to Commercialization: Biotechnology-Based Food Products.* Decision Resources, Burlington, MA, 1991.

49. Harlander, S.K., *Cereal Foods World* **35**, 1106–1109 (1990).

50. Anonymous, *Genet. Technol. News* **11**, 4 (1991).

51. Kilara, A., *Process Biochem.* **20**, 35–45 (1985).

52. Harlander, S.K., *J. Am. Oil Chem. Soc.* **65**, 1727–1728, 1730 (1988).

53. Jimenez-Flores, R., and Richardson, T., *J. Diary Sci.* **71**, 2640–2654 (1988).

54. Overbeeke, N., Fellinger, A.J., and Hughes, S.G., World Patent 8,707,641 (1987).

55. Fujii, T., Konda, K., Shimizu, F., Sone, H., Tanaka, J.-I., and Inoue, T., *Appl. Environ. Microbiol.* **56**, 997–1003 (1990).

56. Penttila, M., Suihko, M.L., Blomqvist, K., Nikkola, M., Knowles, J.K.C., and Enari, T.M., *Yeast* **4**, 473 (1988).

57. Blomqvist, K., Suihdo, M.-L., Knowles, J., and Penttila, M., *Appl. Environ. Microbiol.* **57**, 2796–2803 (1991).

58. Kielland-Brandt, M.C., Gjermansen, C., Nilsson-Tillgren, T., Holmberg, S., and Petersen, J.G.L., *Yeast* **4**, 470 (1988).

59. Hinchcliffe, E., *Biochem. Soc. Trans.* **16**, 1077–1079 (1988).

60. Kielland-Brandt, M.C., Gjermansen, C., Nilsson-Tillgren, T., Holmberg, S., and Petersen, J.G.L., *Yeast* **4**, Spec. Issue, S470 (1988).

61. MacQueen, H., *New Sci.*, **116**, 66–68 (1987).

62. Strasser, A., Martens, F.B., Dohmen, J., and Hollenberg, C.P., European Patent 257,115 (1988).

63. Tubb, R.S., *Trends Biotechnol.* **4**, 98–104 (1986).

64. Berry, D.R., in *Distill. Beverage Flavour* (J. Raymonds and A. Paterson, eds.) pp. 299–307. Ellis Horwood, Chichester, UK, 1989.

65. Russel, I., M.Sc. Thesis, University of Strathclyde, Glasgow, Scotland (1984).

66. Stewart, G.G., and Russel, I., in *Yeast Biotechnology* (D.R. Berry, I. Russel, and G.G. Stewart, eds.), pp. 277–310. Allen & Unwin, London, 1987.

67. Salmon, J.M., in *Application à l'Oenologie des Progrès Récents en Microbiologie et en Fermentation*, pp. 145–157. La Grande Motte, France (1988).

68. Fukuda, K., Watanabe, M., Asano, K., Ouchi, K., and Takasawa, S., *J. Ferment. Bioeng.* **73**, 366–369 (1992).

69. Trivedi, N.B., *Genet. Eng. News* **6** (2), 1, 11 (1986).

70. Miyagawa, K.-I., Kimura, H., Nakahama, K., Kikuchi, M., Doi, M., Akiyama, S.I., and Nakao, Y., *Bio/Technology* **4**, 225–228 (1986).

71. Fujio, T., Ito, S., Maruyama, A., Nishi, T., and Ozaki, A., World Patent 8,504,187 (1985).

72. Liljestrom-Suominen, P.L., Joutsjoki, V., and Korhola, M., *Appl. Environ. Microbiol.* **54**, 245–249 (1988).

73. Davies, L., *Food Sci. Technol. Today* **2**, 147–150 (1988).

74. Vakeria, D., *Brew. Distill. Int.* **19**, 19–21, 31 (1988).

75. Tubb, R.S., *J. Inst. Brew.* **93**, 91–96 (1987).

76. Hammond, J.R.M., *J. Appl. Bacteriol.* **65**, 169–178 (1988).

77. von Wettstein, D., *J. Microbiol.* **53**, 299–306 (1987).

78. Holmberg, S., *Trends Biotechnol.* **2**, 98–102 (1984).

79. Backman, K., O'Connor, M.J., Maruya, A., Rudd, E., McKay, D., and Balakrishnan, R., *Ann. N.Y. Acad. Sci.* **589**, 15–24 (1990).

80. Ito, H., Sato, K., Matsui, K., Sano, K., Nakamori, S., Tanaka, T., and Enei, H., European Patent 183,175 (1986).

81. Follettie, M.T., and Sinskey, A.J., *Food Technol.* **40**, 88, 90–94 (1986).

82. Venkat, K., Backman, K., and Hatch, R.T., *Food Biotechnol.* **4**, 547–548 (1990).

83. Momose, H., *Dev. Ind. Microbiol.* **23**, 109 (1982).

84. Berger, R.G., Drawert, F., and Haedrich, S., *Bioflavour '87, Proc. Int. Conf., 1987*, pp. 415–434 (1988).

85. Casey, J., and Dobb, R., *Enzyme Microb. Technol.* **14**, 739–747 (1992).

86. Kinzel, B., *Agric. Res.*, January, 18–20, 1992.

87. Javelot, C., Girard, P., Colonna-Ceccaldi, B., and Vladescu, B., *J. Biotechnol.* **21**, 239–252 (1991).

88. Spencer, M.E., Hodge, R., Deakin, E.A., and Ashton, S., World Patent 9,119,801 (1991).

89. Sreekrishna, K., and Dickson, R.C., *Proc. Natl. Acad. Sci. U.S.A.* **82,** 7909 (1985).

90. Walsh, P.M., Haas, M.J., and Somkuti, G.M., *Appl. Environ. Microbiol.* **47,** 253 (1984).

91. Goldberg, I., *Trends Biotechnol.* **6,** 32–34 (1988).

92. Chassy, B.M., *Trends Biotechnol.* **3,** 273–275 (1985).

93. Chassy, B.M., in *Biotechnology in Food Processing* (S.K. Harlander and T.P. Labuza, eds.), pp. 197–207. Noyes Publications, Park Ridge, NJ, 1986.

94. Gasson, M.J., in *Chemical Aspects of Food Enzymes* (A.T. Andrews, ed.), pp. 188–195. Royal Society of Chemistry, London, 1987.

95. Muriana, P.M., and Klaenhammer, T.R., *J. Diary Sci.* **72,** 123 (1989).

96. Kim, S.G., and Batt, C.A., *Food Microbiol.* **5,** 59–74 (1988).

97. Martinez-Arias, A.E., and Casadaban, M.J., *Mol. Cell. Biol.* **3,** 580–586 (1983).

98. Teuber, M., *Food Biotechnol.* **4,** 537–546 (1990).

99. Sandine, W.E., *FEMS Microbiol. Rev.* **46,** 205–220 (1987).

100. Stafford, R.K., *Abstr. Pap., 198th Meet. Am. Chem. Soc.*, MBTD30 (1989).

101. Anderson, S., Marks, C.B., Lazarus, R., Miller, J., Stafford, K., and Seymour, J., *Science* **230,** 144–149 (1985).

102. Robert-Baudouy, J., *Trends Biotechnol.* **9,** 325–329 (1991).

103. Cantarelli, C., *Ital. J. Food Sci.* **2,** 9–24 (1990).

104. Taylor, R.F., and Wheat, D., in *Biotechnology of Food: An Old Industry on a New Vector* (H. Hiebert, ed.), p. 77. Decision Resources, Burlington, MA, 1991.

105. Rosenberg, S., World Patent 8,912,675 (1989).

106. Louffler, A., *Food Technol.* **40,** 63–79 (1986).

107. Lei, S.P., and Lai, J., *Biotechnol. USA 1987*, pp. 324–329 (1987).

108. Heckl, K., Spevak, W., Ostermann, E., Zophel, A., Krystek, E., Maurerfogy, I., Wichecasta, M.J., and Stratow, C., European Patent 282,899 (1988).

109. Haas, M.J., *Bio/Technology* **1,** 575–578 (1983).

110. Pitcher, W.H., *Food Technol.* **40,** 62, 63, 69 (1986).

111. Edens, L., and van der Wel, H., *Trends Biotechnol.* **3,** 61–64 (1985).

112. Illingworth, C., Larson, G., and Hellerkant, G., *J. Ind. Microbiol.* **4,** 37–42 (1989).

113. Matt, J., Edens, L, Ledeboer, A.M., Toonen, M., Visser, C., Bom, I., and Verrips, C.T., Miami Winter Symposia **19,** From *Gene to Protein; Translation into Biotechnology; Proceedings* (F. Ahmad, ed.), p. 540, Academic Press, New York, 1982.

114. Edens, L., Heslinger, L., Klok, R., Ledeboer, A.M., Matt, J., Toonen, M.Y., Visser, C., and Verrips, C.T., *Gene* **18,** 1–12 (1982).

115. Hallborn, J., Walfridsson, M., Airaksinen, U., Ojamo, H., Hahn-Hägerdal, B., Penttitä, M., and Keräanen, S., *Bio/Technology* **9,** 1090–1095 (1991).

116. Hasegawa, S., *ACS Symp. Ser.* **405,** 84–96 (1989).

117. Bar-Peled, M., Lewinsohn, E., Fluhr, R., and Gressel, J., *J. Biol. Chem.* **266** (31), 20953–20959 (1991).

118. Rehm, H.J., Single Cell Protein Production from Petroleum Derivatives and Its Utilization as Food and Feed—Using Recombinant DNA Techniques for SCP Production, in *Perspective in Biotechnology and Applied Microbiology* (DI Alani, and M. Moo-Young, eds.). *Inst. Mikrobiol.* Univ. Muenster, Muenster, FRG, (1986).

119. Cocking, E.C., and Davey, M.R., *Chem. Ind. (London)*, pp. 833–835 (1991).

120. Lee, C.Y., and Kime, R.W., *J. Apic. Res.* **23,** 45–49 (1984).

121. Finn, R.K., *Food Biotechnol.* **4,** 1–13 (1990).

122. Gormer, P., and Kotulak, R., *Food Technol.*, August, pp. 46, 48, 50, 53, 103 (1991).

122a. Chakrabarty, A.M., in *Kirk-Othmer Encyclopedia of Chemical Technology* (M. Grayson and D. Eckroth, eds.), 3rd ed., Vol. 11, pp. 730–745. Wiley, New York, 1980.

123. Hardy, R.W.F., *ACS Symp. Ser.* **362,** 312–319 (1988).

124. DeJong, D.W., and Phillips, M., *ACS Symp. Ser.* **362,** 258–261 (1988).

125. Fraley, R., *Bio/Technology* **10,** 40–43 (1992).

126. Knight, P., *Bio/Technology* **7,** 1233–1237 (1989).

127. Gasser, C.S., and Fraley, R.T., *Science* **244,** 1293–1299 (1989).

128. Anonymous, *Food Eng.*, January, p. 26 (1992).

129. Bird, C.R., Ray, J., Schuich, W., Grierson, D., Smith, C.J.S., Morris, P.C., and Watson, G., *J. Sci. Food Agric.* **54,** 159–160 (1991).

130. Roberts, L., *Science* **241,** 1290 (1988).

131. Kim, J., *USA Today*, April 28 (1992).

132. Hoyle, R., *Bio/Technology* **10,** 629 (1992).

133. Kadlec, D., *USA Today,* August 20 (1992).

134. Marymont, M., *Wilmington News J.*, July 10 (1992).

134a. Darlin, D., *Forbes* **152,** (8), Oct. 11, pp. 88–89, 1993.

135. Field, P., Taylor, R.F., and Wheat, D., in *Biotechnology of Food: An Old Industry on a New Vector* (H. Heibert, ed.), p. 42. Decision Resources, Burlington, MA, 1991.

136. Kridl, J. et al., 3rd Int. Congr. Plant Mol. Biol., Tucson, AZ, *1991*, Abstr. 723 (1991).

137. Friedt, W., *Fette Wiss. Technol.* **90,** 51–55 (1988).

138. Hardin, B., *Agric. Res.* **37,** 10–12 (1989).

139. Scowcroft, W.R., *J. Am. Oil Chem. Soc.* **66,** 455 (1989).

140. Rattray, B.M., *J. Am. Oil Chem. Soc.* **61,** 648 (1984).

141. Hibberd, K., Anderson, P., and Barker, M., U.S. Patent 4,581,847 (1985).

142. Davies, C.S., Nielsen, S.S., and Nielsen, N.C., *J. Am. Oil Chem. Soc.* **64,** 1428–1433 (1987).

143. Stark, D. et al., *3rd Int. Congr. Plant Mol. Biol.*, Tucson, AZ, *1991*, Abstr. (1991).

144. Worrell, A., Bruneau J.M., Summerfelt, K., Boersig M., and Volker, T.A., *Plant Cell* **3,** 1121–1130 (1991).

145. Bushuk, W., *Food Technol. Aust.* **36,** 533–534 (1984).

146. Lehninger, A.L., *Principles of Biochemistry*, pp. 121*ff*. Worth Publishers, New York, 1982.

147. Anonymous, *Trends Biotechnol.* **10**, 141–144 (1992).

148. Lee, R.H., *Nature (London)* **356**, 543–544 (1992).

149. Wetzel, R., in *Biotechnology in Food Processing* (S.K. Harlander and T.P. Labuza, eds.), pp. 57*ff*. Noyes Publications, Park Ridge, NJ, 1986.

150. Richardson, T., *J. Dairy Sci.* **68**, 2753–2762 (1985).

151. Branner-Jörgensen, S., Schneider, P., and Eigtved, P., U.S. Patent 4,357,357 (1982).

152. Richardson, T., and Jimenez-Flores, J.A., *J. Am. Oil. Chem. Soc.* **65**, 483 (1988).

153. Estell, D.A., Graycar, T.P., and Wells, J.A., *J. Biol. Chem.* **260**, 6518–6521 (1985).

154. Chen, F.H., and Arnold, K.Q., *Bio/Technology* **9**, 1073–1077 (1991).

155. Ladin, B.F., Cognis, Inc., Santa Rosa, CA, personal communication (1992).

156. Spradlin, J.E., *ACS Symp. Ser.* **389**, 24–43 (1989).

157. Perry, L.J., and Wetzel, R., *Science* **226**, 555–557 (1984).

158. Hawkes, R., Gruetter, M.G., and Schellman, J., *J. Mol. Biol.* **175**, 195–211 (1984).

159. Albert, S., and Wozniak, J.A., *Proc. Natl. Acad. Sci. U.S.A.* **82**, 747–750 (1985).

160. Mainzer, S.E., Yoast, S., Palombella, A., Silva, R.A., Poolman, B., Chassy, A.B.M., Boizet, B., and Schmidt, B.F., in *Yogurt: Nutritional and Health Properties* (R.C. Chandan, ed.), 41–45. Conference, New York, 1989.

161. Neidleman, S., in *Biotechnology in Food Processing* (S.K. Harlander and T.P. Labuza, eds.), p. 46. Noyes Publications, Park Ridge, NJ, 1986.

162. Stephanopoulos, G., and Vallino, J.J., *Science* **252**, 1675–1681 (1991).

163. Simon, J.L., Engasser, J.M., and Germain, P., *3rd Meet. Eur. Congr. Biotechnol.* **2**, 49–54 (1984).

164. Jeong, J.W., Snay, J., and Ataai, M.M., *Biotechnol. Bioeng.* **35**, 160–184 (1990).

165. Ataai, M.M., and Jeong, J.W., *199th Nat. Meet. Am. Chem. Soc.*, Boston, *1990*.

166. Diaz-Ricci, J.C., Regan, L., and Bailey, J.E., *Biotechnol. Bioeng.* **38**, 1318–1324 (1991).

167. Ingram, L.O., Conway, T., Clark, D.P., Swell, G.W., and Preston, J.F., *Appl. Environ. Microbiol.* **53**, 2420–2425 (1987).

168. Ingram, L.O., and Conway, T., *Appl. Environ. Microbiol.* **54**, 397–404 (1988).

169. Mermelstein, L.D., Petersen, D.J., Bennett, G.N., and Papoutsakis, E.T., *Annu. Meet. Amer. Inst. Chem. Eng.*, Miami Beach, FL, *1992*, Pap. 1492 (1992).

170. Lee, S.Y., Mermelstein, L., Bennett, G.N., and Papoutsakis, E.T., *Abstr. Pap.*, *198th Meet. Am. Chem. Soc.*, MBTD176 (1989).

171. Bailey, J.E., *200th Nat. Meet. Am. Chem. Soc.*, Washington, DC, *1990*.

172. Klibanov, A.M., in *Biotechnology Challenges for the Flavor and Food Industry* (R.C. Lindsay and B.J. Willis, eds.), pp. 35–40. Elsevier Applied Science, London, 1989.

173. Van Brunt, J., *Bio/Technology* **4**, 611 (1986).

174. Carta, G., Gainer, J.L., and Benton, A.H., *Biotechnol. Bioeng.* **37**, 1004–1009 (1991).

175. Chen, C.-S., and Sih, C.J., *Angew. Chem., Int. Ed. Engl.* **28**, 695–707 (1989).

176. Dordick, J.S., *Enzyme Microb. Technol.* **11**, 194–211 (1989).

177. Klibanov, A.M., *Acc. Chem. Res.* **23**, 114–120 (1990).

178. Abramowicz, D.A., and Keese, C.R., *Biotechnol. Bioeng.* **33**, 149 (1989).

179. Klibanov, A.M., CHEMTECH, June, p. 354 (1986).

180. Yamane, T., Kojima, Y., Ichiryu, T., Nagata, M., and Shimizu, S., *Biotechnol. Bioeng.* **34**, 838–843 (1989).

181. Schulze, B., and Klibanov, A.M., *Biotechnol. Bioeng.* **38**, 1001–1006 (1991).

182. Fayolle, F., Marchal, R., Frederic, M., Blanchet, D., and Ballerini, D., *Enzyme Microb. Technol.* **13**, 215–219 (1991).

183. Laane, C., Boeren, S., Vos, K., and Veeger, C., *Biotechnol. Bioeng.* **30**, 81 (1987).

184. Mattiasson, B., and Adlercreutz, P., *Trends Biotechnol.* **9**, 394–398 (1991).

185. Zaks, A., and Klibanov, A.M., *Proc. Natl. Acad. Sci. U.S.A.* **82**, 3192–3196 (1985).

186. Whitaker, J.R., *Food Biotechnol.* **4**, 669–697 (1990).

187. Welsh, F.W., and Williams, R.E., *Enzyme Microb. Technol.* **12**, 743–748 (1990).

188. Welsh, F.W., Williams, R.E., Chang, S.C., and Dicaire, C.J., *J. Chem. Technol. Biotechnol.* **52**, 201–209 (1991).

189. Langrand, G., Rondot, N., Triantaphylides, C., and Baratti, J., *Biotechnol. Lett.* **12**, 581–586 (1990).

190. Murray, W.D., Duff, S.J.B., Lanthier, P.H., Armstrong, D.W., Welsh, F.W., and Williams, R.E., *Dev. Food Sci.* **17**, 1–18 (1988).

191. Kanerva, L.T., Vihanto, J., Halme, M.H., Loponen, J.M., and Euranto, E.K., *Acta Chem. Scand.* **44**, 1032–1035 (1990).

192. Malcata, F.X., Reyes, H.R., Garcia, H.S., Hill, C.G., Jr., and Amundson, C.H., *Enzyme Microb. Technol.* **14**, 426–446 (1992).

193. Laroute, V., and Willemot, R.-M., *Biotechnol. Lett.* **14**, 169–174 (1992).

194. Miethe, P., Gruber, R., and Voss, H., *Biotechnol. Lett.* **11**, 449–454 (1989).

195. Legoy, M.D., Kim, H.S., and Thomas, D., *Process Biochem.* **20**, 145–148 (1985).

196. Clapés, P., Valencia, G., and Adlercreutz, P., *Enzyme Microb. Technol.* **14**, 575 (1992).

RETURN TO: CHEMISTRY LIBRARY
100 Hildebrand Hall • 510-642-3753

LOAN PERIOD 1	2	1-MONTH USE
4	5	6

ALL BOOKS MAY BE RECALLED AFTER 7 DAYS.

Renewals may be requested by phone ~~or, using GLADIS,~~ type Inv ~~followed by your patron ID number.~~

DUE AS STAMPED BELOW.

DEC 5 – '94

FORM NO. DD 10
3M 7-08

UNIVERSITY OF CALIFORNIA, BERKELEY
Berkeley, California 94720–6000